Elements of Mathematical Ecology

Elements of Mathematical Ecology provides an introduction to classical and modern mathematical models, methods, and issues in population ecology. The first part of the book is devoted to simple, unstructured population models that, for the sake of tractability, ignore much of the variability found in natural populations. Topics covered include density dependence, bifurcations, demographic stochasticity, time delays, population interactions (predation, competition, and mutualism), and the application of optimal control theory to the management of renewable resources. The second part of this book is devoted to structured population models, covering spatially structured population models (with a focus on reaction-diffusion models), age-structured models, and two-sex models. Suitable for upper level students and beginning researchers in ecology, mathematical biology and applied mathematics, the volume includes numerous line diagrams that clarify the mathematics, relevant problems throughout the text that aid understanding, and supplementary mathematical and historical material that enrich the main text.

MARK KOT is Associate Professor in the Department of Applied Mathematics at the University of Washington, Seattle, USA.

Elements of Mathematical Ecology

MARK KOT

Department of Applied Mathematics,
University of Washington

CAMBRIDGE UNIVERSITY PRESS
Cambridge, New York, Melbourne, Madrid, Cape Town, Singapore, São Paulo

Cambridge University Press
The Edinburgh Building, Cambridge CB2 8RU, UK

Published in the United States of America by Cambridge University Press, New York

www.cambridge.org
Information on this title: www.cambridge.org/9780521802130

First published 2001
Reprinted 2003 with corrections

A catalogue record for this publication is available from the British Library

Library of Congress Cataloguing in Publication data
Kot, Mark, 1956–
Elements of mathematical ecology / Mark Kot.
    p.   cm.
Includes bibliographical references (p.).
ISBN 0 521 80213 X – ISBN 0 521 00150 1 (pb.)
1. Population biology – Mathematical models. 2. Ecology – Mathematical models.
3. Population biology – Mathematics. 4. Ecology – Mathematics. I. Title.
QH352.K66 2001
577'.01'51 – dc21   00-065165

ISBN 978-0-521-80213-0 hardback
ISBN 978-0-521-00150-2 paperback

Transferred to digital printing 2008

Contents

Preface

Ecology is an old discipline. The discipline was christened in 1866 by Ernst Haeckel, a well-known German evolutionary biologist. Haeckel was a neologist – he loved to invent new scientific terms. His most famous gems are *phylogeny* and *oecologie*. *Oecologie* and *ecology* take their derivation from the Greek *oikos*, house or dwelling place. Ecology, as envisioned by Haeckel, is the study of the houses and the housekeeping functions of plants and animals. It is the scientific study of the interrelationships of organisms, with each other, and with their physical environment. The *idea* of ecology is even older (Worster, 1994). It is closely related to 18th century notions of the balance or economy of nature reflected, most clearly, in Linnaeus's 1749 essay *Oeconomia Naturae* (Stauffer, 1960).

Ecology is also a diverse discipline. After all, it has all of life to account for. In the old days, it was common to divide ecology into two subdisciplines: *autecology*, the ecology of individual organisms and of populations, and *synecology*, the study of plant and animal communities. Ecology is now divided into many subdisciplines (see Table 1).

Several subdisciplines use mathematics. For example, behavioral ecology makes extensive use of game theory and of other brands of optimization. It is impossible to cover all of these subdisciplines in one short book. Instead, I focus on population ecology and engage in occasional forays into community ecology and evolutionary ecology. This book could, and perhaps should, have been entitled *The Dynamics of Biological Populations*.

The material in this book has been used to teach a two-semester course. There is, therefore, a dichotomy in these notes. The first semester of the course is devoted to unstructured population models, models that, in effect, treat organisms as 'homogeneous green gunk'. Unstructured population models have the advantage, at first, of simplicity. As one adds extra bits

Table 1. *Branches of ecology*

Synecology	Landscape ecology
	Systems ecology
	Community ecology
Autecology	Population ecology
	Evolutionary ecology
	Behavioral ecology
	Physiological ecology
	Chemical ecology

of biology, these models become more realistic and more challenging. The topics in the first half of the book include density dependence, bifurcations, demographic stochasticity, time delays, population interactions (predation, competition, and mutualism), and the application of optimal control theory to the management of renewable resources.

Variety, and variability, are the spice of life. We frequently ascribe differences in the success of individuals to differences in age, space (spatial location), or sex. The second half of this book is devoted to structured population models that take these variables into account. I begin with spatially-structured population models and focus on reaction-diffusion models. There is also tremendous interest in metapopulation models, coupled lattice maps, integrodifference equations, and interacting particle systems (Turchin, 1998; Hanski, 1999). However, my colleagues and I tend to leave this material for our advanced course. I follow with an overview of age-structured population models in which I compare integral equations, discrete renewal equations, matrix population models, and partial differential equations. I conclude with a brief introduction to two-sex models.

The emphasis in these notes is on strategic, not tactical, models (Pielou, 1981). I am interested in simple mechanistic models that generate interesting hypotheses or explanations rather than in detailed and complex models that provide detailed forecasts. You will also find many equations, but few formal theorems and proofs. Applied scientists and pure mathematicians both have reason to be offended ! Because of the interdisciplinary nature of my class and because of my own preference for solving problems over proving theorems, I have tried to hold to a middle course that should appear natural to applied mathematicians and to theoretical biologists. I hope that this middle course will appeal to a broad range of (present and future) scientists. Failing that, I hope that you, gentle reader, can use this book as a springboard for more detailed applied and theoretic investigations.

Acknowledgments

I have been blessed with excellent teachers and students. I wish to thank all my teachers, but especially William K. Smith, W. Tyler Estler, Richard H. Rand, Simon A. Levin, William M. Schaffer, Paul Fife, Jim Cushing, Stephen B. Russell, and Hanno Rund.

Stéphane Rey coauthored Chapter 9. Other former students, Michael G. Neubert, Emily D. Silverman, and Eric T. Funasaki, will recognize work that we published together. Michael Neubert used this material in a class and provided a number of useful comments and criticisms.

Several cohorts of students studied this material as Mathematics or Ecology 581 and 582 at the University of Tennessee or as Applied Mathematics 521 at the University of Washington. I thank these students for their enthusiasm and hard work. I am grateful to the University of Tennessee and the University of Washington and to my colleagues at these institutions for the chance to teach these courses.

The early drafts of this book could not have been written without several valuable pieces of software. I thank Joseph Osanna for troff, Brian W. Kernighan and Lorinda L. Cherry for eqn, Jon L. Bentley and Brian W. Kernighan for grap, Michael Lesk for ms, tbl, and refer, James J. Clark for groff, Bruce R. Musicus for numeqn, Nicholas B. Tufillaro for ode, and Ralph E. Griswold, Madge T. Griswold, and the Icon Project for the Icon programming language.

It has been a pleasure working with Cambridge University Press. I wish to thank Alan Crowden, Maria Murphy, Jayne Aldhouse, Zoe Naylor, and especially Sandi Irvine for all their efforts.

Finally, I want to thank my parents for their encouragement and interest and my wife, Celeste, for her encouragement, support, and desire to purchase the movie rights.

Knoxville, Tennessee   Mark Kot

Part I Unstructured population models

Section A
SINGLE-SPECIES MODELS

1 Exponential, logistic, and Gompertz growth

Tradition dictates that we begin with a simple homogeneous population. This population is that 'homogeneous green gunk' that I referred to in the preface. I will represent the number (or sometimes the density) of individuals in this population by $N(t)$. I will also make frequent reference to the rate of change, dN/dt, and to the per capita rate of change, $(1/N)\,dN/dt$, of this population.

Let us assume that all changes in this population result from births and deaths and that the per capita birth rate b and per capita death rate d are constant:

$$\frac{1}{N}\frac{dN}{dt} = b - d. \tag{1.1}$$

The difference between the per capita birth and death rates, $r \equiv b - d$, plays a particularly important role and is known as the *intrinsic rate of growth*. Equation (1.1) is commonly rewritten, in terms of r, as

$$\frac{dN}{dt} = rN. \tag{1.2}$$

One must also add an initial condition, such as

$$N(0) = N_0, \tag{1.3}$$

that specifies the number of individuals at the start of the process.

Equation (1.2) is a linear, first-order differential equation. It is easily integrated, either as a separable equation or with an integrating factor, and it possesses the solution

$$N(t) = N_0 e^{rt}. \tag{1.4}$$

This solution grows exponentially for positive intrinsic rates of growth and

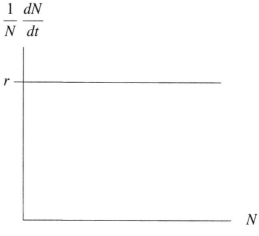

Fig. 1.1. Per capita growth rate.

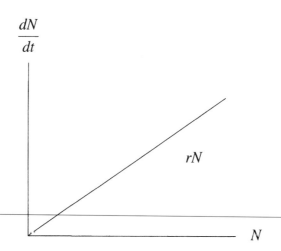

Fig. 1.2. Population growth rate.

decays exponentially for negative intrinsic rates of growth. It remains constant when births balance deaths.

Three different graphs capture the behavior of this system. In Figure 1.1, I have plotted the per capita growth rate as a function of the population size. The per capita growth rate remains constant for all population sizes: crowding has no effect on individuals. However, the growth rate for the entire population (Figure 1.2) increases with number as each new individual adds its own undiminished contribution to the total growth rate. The result (Figure 1.3) is a population that grows at ever-increasing rates.

The population size $N^* = 0$ is an *equilibrium point*. Since there is no

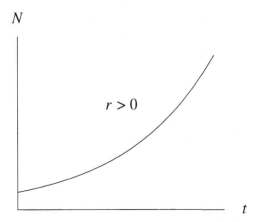

Fig. 1.3. Population trajectory.

immigration or emigration in this model, populations that start at zero stay at zero. For positive r, this equilibrium is *unstable*. After small perturbations, the population moves away from zero. For negative r, this equilibrium is *asymptotically stable*. Small perturbations now decay back to zero. I will say more about equilibria and stability later.

Problem 1.1 *Monod's† nightmare*
Escherichia coli is a bacterium that has been used extensively in microbiological studies. *Escherichia coli* cells are rod shaped; they are 0.75 μm wide and 2 μm long. Under ideal conditions, a population of *E. coli* doubles in just over 20 minutes.

(1)   What is r for *E. coli?*
(2)   If $N_0 = 1$, how long would it take for an exponentially growing population of *E. coli* experiencing ideal conditions to fill your classroom?

There are several defects with this simple exponential model:

(1)   The model has constant per capita birth and death rates and generates limitless growth. This is patently unrealistic.
(2)   The model is deterministic; we have ignored chance or stochastic effects. Stochastic effects are particularly important at small population sizes.

---

† Jacques Monod (1910–1976) was the recipient of a 1965 Nobel Prize for Medicine for his work on gene regulation. He also conducted innovative experimental studies on the kinetics and stoichiometry of microbial growth (Panikov, 1995).

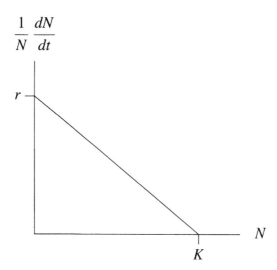

Fig. 1.4. Decreasing per capita growth rate.

(3)   The model ignores lags. The growth rate does not depend on the past. More-over, the population responds *instantaneously* to changes in the current popu-lation size.

(4)   We have ignored temporal and spatial variability.

Let us start with the first defect.

What are the factors that regulate the growth of populations? There have been two schools of thought. In 1933, A. J. Nicholson, an Australian entomologist, published a seminal paper in which he stressed the importance of density-dependent population regulation. Nicholson (1933), the British ornithologist David Lack (1954), and others argued that populations are regulated by biotic factors such as competition and disease that have a disproportionately large effect on high-density populations. The opposing view, promulgated by the Australian entomologists H. G. Andrewartha and L. C. Birch (1954), is that populations are kept in check by abiotic, density-independent factors, such as vagaries in the weather, that have as adverse an effect on low-density populations as they do on high-density populations.

The dispute between these two schools occupied ecology for most of the 1950s (Tamarin, 1978; Kingsland, 1985; Sinclair, 1989). Density-dependent and density-independent factors may both be important in regulating pop-ulations. From a modeling perspective, however, it is easier to start with density-dependent regulation.

Consider a per capita growth rate that decreases linearly with population

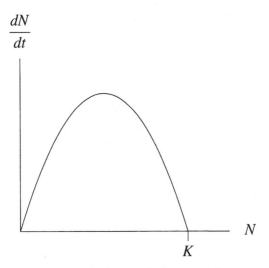

$\dfrac{dN}{dt}$

Fig. 1.5. Parabolic population growth rate.

size,

$$\frac{1}{N}\frac{dN}{dt} = r\left(1 - \frac{N}{K}\right) \qquad (1.5)$$

(see Figure 1.4). This decrease in the per capita growth rate may be thought of as an extremely simple form of density-dependent regulation. Note that the per capita growth rate falls to zero at the *carrying capacity K*.

The population's growth rate,

$$\frac{dN}{dt} = rN\left(1 - \frac{N}{K}\right), \qquad (1.6)$$

is now a quadratic function of population size (see Figure 1.5). Equation (1.6) is known as the *logistic equation* or, more rarely, as the *Pearl–Verhulst equation*. It has an exact analytical solution. Figure 1.6 illustrates this solution for two different initial conditions. You are asked to find this closed-form solution in Problem 1.2. Since few nonlinear differential equations can be solved so easily, I will concentrate on a general method of analysis that emphasizes the qualitative features of the solution.

Equation (1.6) has two equilibria, $N^* = 0$ and $N^* = K$; at each of these two values, the growth rate for the population is equal to zero. Near $N^* = 0$, N^2/K is small compared to N so that

$$\frac{dN}{dt} \approx rN. \qquad (1.7)$$

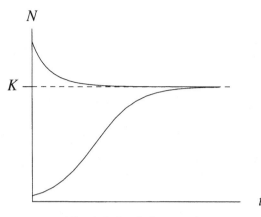

Fig. 1.6. Logistic growth.

For $r > 0$, small perturbations about $N^* = 0$ grow exponentially; the equilibrium $N^* = 0$ is unstable.

Problem 1.2 *Exact solution of the logistic equation*
Show that the logistic equation has the solution

$$N(t) = \frac{K}{1 + \left(\frac{K}{N_0} - 1\right) e^{-rt}} \tag{1.8}$$

(1) treating the logistic equation as a separable equation, and
(2) treating the logistic equation as a Bernoulli equation.

Close to $N^* = K$, we instead introduce a new variable that measures the deviation of N from K:

$$x \equiv N - K. \tag{1.9}$$

Substituting $N = K + x$ into equation (1.6) gives us

$$\frac{dx}{dt} = -rx - \frac{r}{K}x^2, \tag{1.10}$$

and since x is small for N close to K, we have that

$$\frac{dx}{dt} \approx -rx. \tag{1.11}$$

For $r > 0$, small perturbations about $N^* = K$ *decay* exponentially; the equilibrium $N^* = K$ is asymptotically stable. For positive r, solutions to the logistic equation (see Figure 1.6) are essentially a combination of exponential growth, close to zero, and of exponential decay, close to the carrying capacity.

Equations (1.7) and (1.11) imply that the solution of the logistic equation is concave up just above the origin and concave down just below the carrying capacity. It stands to reason that an inflection point lies between the origin and the carrying capacity. This inflection point can be found by setting the derivative of both sides of logistic equation (1.6) equal to zero:

$$\frac{d^2N}{dt^2} = r\left(1 - \frac{2N}{K}\right)\frac{dN}{dt} = 0. \tag{1.12}$$

It follows that the inflection point is at $N = K/2$.

Mathematical meanderings

Consider the differential equation

$$\frac{dN}{dt} = f(N). \tag{1.13}$$

This equation is *autonomous* in that f does not contain any explicit dependence on t. I have introduced several concepts that are useful, not only for the logistic differential equation, but for many autonomous, first-order, ordinary differential equations. Let's formalize these concepts.

Definition We say that $N = N^*$ is an *equilibrium point* (also known as a *fixed point, critical point, rest point*) if

$$f(N^*) = 0. \tag{1.14}$$

Definition An equilibrium point N^* is *Lyapunov stable* if, for any (arbitrarily small) $\epsilon > 0$, there exists a $\delta > 0$ (depending on ϵ) such that, for all initial conditions $N(t_0) = N_0$ satisfying $|N_0 - N^*| < \delta$, we have $|N(t) - N^*| < \epsilon$ for all $t > t_0$. In other words, an equilibrium is stable if starting close (enough) guarantees that you stay close.

Definition An equilibrium point N^* is *asymptotically stable (in the sense of Lyapunov)* if it is stable and if there exists a $\rho > 0$ such that

$$\lim_{t \to \infty} |N(t) - N^*| = 0 \tag{1.15}$$

for all N_0 satisfying

$$|N_0 - N^*| < \rho. \tag{1.16}$$

Thus an equilibrium is asymptotically stable if all sufficiently small initial deviations produce small excursions that eventually return to the equilibrium.

At this point, the only interesting question is a practical one: is a given equilibrium point stable or unstable?

Theorem Suppose that N^* is an equilibrium point and that $f(N)$ is a continuously differentiable function. Suppose also that $f'(N^*) \neq 0$. Then the equilibrium point N^* is asymptotically stable if $f'(N^*) < 0$, and unstable if $f'(N^*) > 0$.

Proof Consider an equilibrium for which $f'(N^*) < 0$ and let $x(t) \equiv N(t) - N^*$. If we expand $f(N)$ about N^*, we obtain

$$\frac{dx}{dt} = f(N^*) + f'(N^*)x + g(x). \tag{1.17}$$

Since N^* is an equilibrium, equation (1.17) reduces to

$$\frac{dx}{dt} = f'(N^*)x + g(x), \tag{1.18}$$

which may be viewed as a perturbation of a linear, constant-coefficient differential equation. Note that $g(x)$ consists of higher-order terms. In particular, $g(x)$ satisfies $g(0) = 0$ and also $g'(0) = 0$. This, along with the continuity of $g'(x)$ (which follows from the continuity of $f'(N)$), guarantees us that for each $\epsilon > 0$ there is a small δ neighborhood about zero wherein $|g'(x)| < \epsilon$. As a result,

$$g(x) = \int_0^x g'(s)\,ds \leq \epsilon |x|. \tag{1.19}$$

It follows that

$$\frac{dx}{dt} \leq f'(N^*)x + \epsilon |x|. \tag{1.20}$$

For small enough δ and ϵ and $f'(N^*) \neq 0$, the higher-order terms cannot change the sign of dx/dt. As a result, small enough perturbations will decay; the equilibrium is asymptotically stable. A similar argument can be made to show that $f'(N^*) > 0$ implies instability. $\square$

Take another look at Figure 1.5. You should be able to ascertain the stability of the two equilibria *by inspection* with this theorem. What do you think happens when $f'(N^*) = 0$?

Historical hiatus

The concepts of exponential and logistic growth arose gradually. A few people played especially important roles in the development of these concepts.

John Graunt (1662) was a 'collector and classifier of facts' (Hutchinson, 1980). He was also the inventor of modern scientific demography. Graunt tabulated the Weekly Bills of Mortality for London. These bills listed births and deaths; they were used as an early warning system for the plague. Using these bills, Graunt estimated a doubling time for London of 64 years. This is an extremely short period of time. Graunt posited that if the descendants of Adam and Eve (created in 3948 BC, according to Scaliger's chronology) doubled in number every 64 years, the world should be filled with 'far more People, than are now in it.' By my calculation, this would amount to

$$2^{(1662 + 3948)/64} \approx 2^{87.7} \approx 2.5 \times 10^{26} \approx 200 \text{ million people/cm}^2. \quad (1.21)$$

Graunt was clearly aware of the power of exponential growth.

Sir William Petty (1683) faulted Graunt for ignoring the biblical flood. He started the clock with Noah ($t_0 = 2700$ BC with $N_0 = 8$). Petty also felt that Graunt's estimate for a doubling time was misleading, since much of London's increase was due to immigration. Petty estimated the doubling time for England to be between 360 and 1200 years. However, a doubling time of 360 years left Petty with a population projection,

$$8 \times 2^{12.175} \approx 36\,994, \quad (1.22)$$

that was far too small. Petty therefore proposed that the human growth rate had fallen steadily in postdiluvian times. He produced a table 'shewing how the people might have doubled in the several ages of the world'; the table exhibited a steady increase in the doubling time, much as one would expect for logistic growth.

The **Reverend Thomas Robert Malthus** (1798) is famous for having written *An Essay on the Principle of Populations*. The essence of this book may be represented with a simple quasi-equation:

$$\begin{array}{ccccc} \text{a geometrically} & & \text{an arithmetically} & & \text{much} \\ \text{growing} & + & \text{growing} & = & \text{human .} \\ \text{population} & & \text{food supply} & & \text{misery} \end{array} \quad (1.23)$$

Many of Malthus's conclusions had already been anticipated by Graunt, Petty, and others. However, Malthus is justly famous for stating the case so clearly. Malthus's book had tremendous influence on Charles Darwin and Alfred Russel Wallace and, in effect, provided them with the material basis for natural selection.

Pierre-François Verhulst (1845) was a Belgian who presented an entirely modern derivation of the logistic equation. His work went unappreciated during his own lifetime and he died in relative obscurity.

Raymond Pearl and **Lowell Reed** (1920) rediscovered the logistic equation and launched a crusade to make the logistic equation a 'law of nature' (Kingsland, 1985). They published over a dozen papers between 1920 and 1927 promulgating this law. Some of their extrapolations and *ad hoc* pastings of logistic curves were questionable, but they did make the logistic equation famous.

The logistic equation allows us to handle limited growth in a natural way. This equation also has a long history. However, there is nothing sacred about this equation. Other models, derived differently, exhibit many of the same properties.

Problem 1.3 *The Gompertz (1825) equation*
Consider a population that grows with an intrinsic rate of growth that decays exponentially:

$$\frac{dN}{dt} = r_0 e^{-\alpha t} N. \tag{1.24}$$

Solve this *nonautonomous* ordinary differential equation. Sketch typical solution curves. What is the 'carrying capacity' and how does it differ from the carrying capacity of the logistic equation? Where is the inflection point for your solution? How does this differ from the logistic equation? Show that the Gompertz equation can be rewritten as

$$\frac{dN}{dt} = \alpha N \ln\left(\frac{K}{N}\right), \tag{1.25}$$

where K is the carrying capacity.

The Gompertz equation was originally formulated as a law of decreasing survivorship (Gompertz, 1825; see also Easton, 1995). It was popularized as a growth curve by Winsor (1932). It has been used to model the growth of plants (Causton and Venus, 1981) and of tumors (Wheldon, 1988). It also appears in fisheries ecology in the Fox (1970) surplus yield model.

Recommended readings

Banks (1994) is encyclopedic in his treatment of logistic-like growth models. Arrowsmith and Place (1992), Drazin (1992), and Hale and Kocak (1991) provide useful introductions to the qualitative analysis of ordinary differential equations. Cole (1957), Hutchinson (1980), and Kingsland (1985) recount the fascinating histories of demography and of the logistic differential equation.

2 Harvest models: bifurcations and breakpoints

In Chapter 1, I emphasized the equilibrial properties of some simple population models. First-order autonomous differential equations are dominated by their equilibria. Solutions tend to stay at equilibria, move monotonically towards or away from equilibria, or go off monotonically to infinity. This emphasis on equilibria may seem boring. It isn't. If a model has a parameter, and you change that parameter, equilibria may collide; they may be destroyed or they may survive and exchange stability. The resulting qualitative changes in the behavior of a dynamical system are called *bifurcations*.

Consider a population of fish that is growing logistically and that is being harvested. Let us assume that the catch of fish per unit effort is proportional to the stock level N,

$$\frac{dN}{dt} = rN\left(1 - \frac{N}{K}\right) - qEN \tag{2.1}$$

(Schaefer, 1954; Clark, 1990). The harvest rate is the product of three terms: the fishing effort E, a proportionality constant q that measures catchability, and the stock level N. The product of the catchability and of the effort, qE, is the *fishing mortality*; it has the same dimensions as r and will play an important role in what follows.

We have equilibria, N^*, whenever the growth rate of the fish population equals the harvest rate:

$$rN^*\left(1 - \frac{N^*}{K}\right) = qEN^* \tag{2.2}$$

(see Figure 2.1). There are two equilibria,

$$N^* = K\left(1 - \frac{qE}{r}\right) \tag{2.3}$$

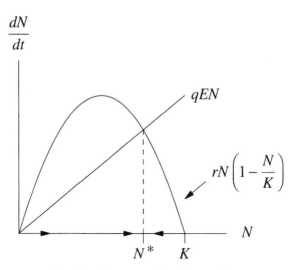

$$\frac{dN}{dt}$$

qEN

$rN\left(1 - \frac{N}{K}\right)$

N

N^* K

Fig. 2.1. Low-mortality harvesting.

and

$$N^* = 0. \tag{2.4}$$

The second equilibrium corresponds to extirpation of the stock.

For small values of the fishing mortality, equilibrium (2.3) is asymptotically stable. This is clear from Figure 2.1. For perturbations to the right of this equilibrium, the harvest rate is greater than the growth rate and the stock level decreases. For small perturbations to the left of this equilibrium, the growth rate exceeds the harvest rate and the stock increases. In both instances, perturbations decay and the stock level returns to N^*, as suggested by the arrows along the abscissa. Equilibrium (2.4), in turn, is unstable.

As we begin to increase the fishing mortality (Figure 2.2), equilibrium (2.3) shifts to the left but maintains its stability. However, as we continue to increase the fishing mortality, we eventually reach a point (Figure 2.3) where the harvest exceeds the growth rate for all positive stock levels. Suddenly, the equilibrium at the origin, corresponding to extinction, is stable.

I have drawn the location of the equilibria as a function of the fishing mortality in Figure 2.4. In doing so, I have continued the first equilibrium into the fourth quadrant. This does not make any biological sense, but it does make eminent mathematical sense. I have represented stable branches of equilibria with a thin line, and unstable branches with a thick line. Branching diagrams that illustrate how the location and stability of solutions depend on a parameter are called *bifurcation diagrams*. Figure 2.4 has a *transcritical bifurcation* at $qE = r$: two branches 'collide' and exchange

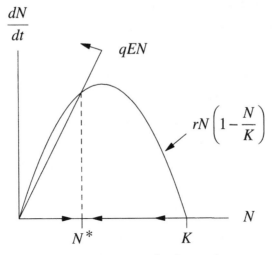

Fig. 2.2. High-mortality harvesting.

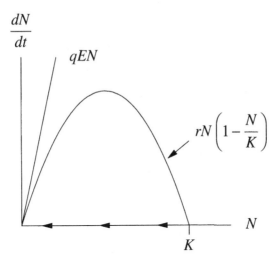

Fig. 2.3. Severe overfishing.

stability. At this bifurcation there is a qualitative change in the behavior of the system.

Figure 2.4 depicts changes in the equilibrial stock level with changes in the fishing mortality. The equilibrial harvest rate or sustainable yield is equally important. The sustainable yield at equilibrium (2.3) is

$$Y = qEN^* = qEK\left(1 - \frac{qE}{r}\right). \qquad (2.5)$$

The graph of this function is a parabola (see Figure 2.5). Increasing fishing

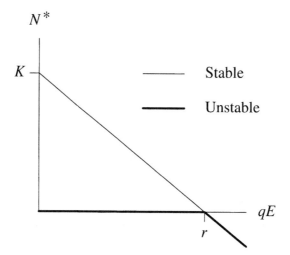

Fig. 2.4. Transcritical bifurcation.

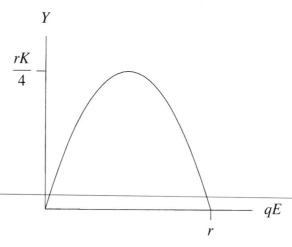

Fig. 2.5. Yield–effort curve.

effort increases sustainable yield – up to a point. Further increases in effort lower the yield as the stock becomes increasingly overexploited and depleted. For fixed catchability, the maximum sustainable yield (MSY) occurs when

$$\frac{dY}{dE} = qK\left(1 - \frac{2qE}{r}\right) = 0. \tag{2.6}$$

The corresponding optimal level of effort,

$$E_{\text{MSY}} = \frac{r}{2q}, \tag{2.7}$$

$\dfrac{dN}{dt}$

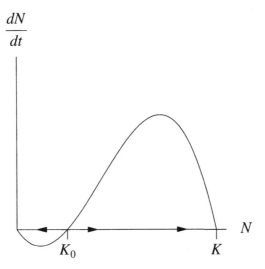

Fig. 2.6. Critical depensation.

produces the maximum sustainable yield

$$\text{MSY} = \frac{rK}{4}.$$
(2.8)

Problem 2.1 *Fox surplus yield model*
Find the equilibria, the yield curve, and the maximum sustainable yield for
a fish population that is growing according to the Gompertz equation and
that is being harvested so that the catch per unit effort is proportional to
the stock size:

$$\frac{dN}{dt} = \alpha N \ln\left(\frac{K}{N}\right) - qEN.$$
(2.9)

Our entire discussion has been premised on the fact that fish populations
grow logistically. However, some populations possess a threshold to growth.
Consider the simple differential equation

$$\frac{dN}{dt} = rN\left(\frac{N}{K_0} - 1\right)\left(1 - \frac{N}{K}\right).$$
(2.10)

By plotting the right-hand side of equation (2.10) as a function of N (Figure 2.6), we see that there is a threshold at K_0. Solutions with initial conditions above K_0 approach the carrying capacity K, while those with initial
conditions below this threshold decay to zero. Since the net growth rate is

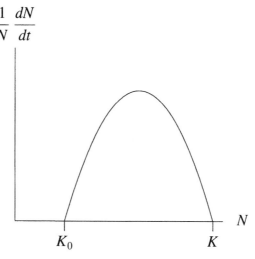

Fig. 2.7. Allee effect.

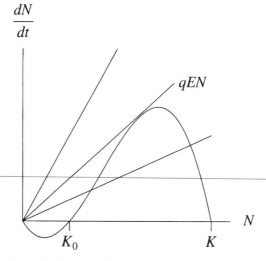

Fig. 2.8. Critical depensation with harvesting.

negative at low population levels, this model exhibits *critical depensation*. In addition, the per capita growth rate is no longer a monotonically decreasing function of density but instead shows an *Allee effect*, or an increase in the per capita growth rate, over certain ranges of density (see Figure 2.7).

If we add harvesting to model (2.10), we obtain

$$\frac{dN}{dt} = rN \left(\frac{N}{K_0} - 1\right) \left(1 - \frac{N}{K}\right) - qEN \qquad (2.11)$$

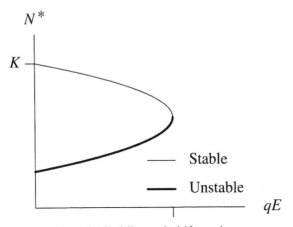

Fig. 2.9. Saddle–node bifurcation.

(see Figure 2.8). Equilibria occur when

$$rN^* \left(\frac{N^*}{K_0} - 1 \right) \left(1 - \frac{N^*}{K} \right) = qEN^*. \tag{2.12}$$

There is a trivial equilibrium at $N^* = 0$. The remaining equilibria satisfy the quadratic equation

$$r \left(\frac{N^*}{K_0} - 1 \right) \left(1 - \frac{N^*}{K} \right) = q E. \tag{2.13}$$

We can solve for $N^*(qE)$ with the quadratic formula. Alternatively, we may plot the fishing mortality qE as a function of N^* and reflect our graph about the 45° line. Either way, we quickly generate an interesting bifurcation diagram (see Figure 2.9).

There are three branches of equilibria (including $N^* = 0$). The middle branch consists of unstable equilibria that act as *breakpoints* (May, 1977) between the equilibria on the two stable branches. As we increase the fishing mortality, equilibria on the middle and upper branches approach each other. Eventually they collide and disappear in a *fold, tangent,* or *saddle-node bifurcation*. A fishery poised close to this bifurcation point may undergo a catastrophic collapse after a small increase in fishing mortality. This precipitous state of affairs is also reflected in the yield curve for this model (see Figure 2.10). This yield curve shows a catastrophic collapse at high enough mortality levels. Bifurcations and breakpoints pervade mathematical ecology: small changes often have large effects.

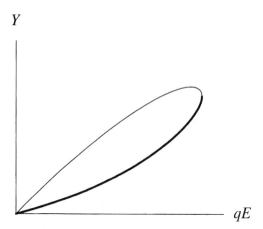

Fig. 2.10. Catastrophic yield curve.

Problem 2.2 *Constant-rate harvesting*
Consider the logistic differential equation with constant-rate harvesting:

$$\frac{dN}{dt} = r N \left(1 - \frac{N}{K} \right) - h. \qquad (2.14)$$

Treat the harvest rate h as a bifurcation parameter. Find the critical value of h for a bifurcation to occur. Which bifurcation is this? Sketch the bifurcation diagram. What are the effects of overharvesting this system?

Problem 2.3 *Some basic bifurcations*
Let x be the state variable and let μ be a bifurcation parameter. Sketch bifurcations diagrams for the following differential equations:

(1)   saddle-node bifurcation

$$\frac{dx}{dt} = \mu - x^2, \qquad (2.15)$$

(2)   transcritical bifurcation

$$\frac{dx}{dt} = \mu x - x^2, \qquad (2.16)$$

(3)   supercritical pitchfork bifurcation

$$\frac{dx}{dt} = \mu x - x^3, \qquad (2.17)$$

(4) subcritical pitchfork bifurcation

$$\frac{dx}{dt} = \mu x + x^3. \tag{2.18}$$

Be sure to show the stability of each branch.

Problem 2.4 *More than one bifurcation*

Sketch the bifurcation diagram and characterize the bifurcations for the following differential equation in which x is the state variable and μ is the bifurcation parameter:

$$\frac{dx}{dt} = \mu x + x^3 - x^5. \tag{2.19}$$

Problem 2.5 *A bunch of bifurcations*

Sketch the bifurcation diagram for the following differential equation in which x is the state variable and μ is the bifurcation parameter:

$$\frac{dx}{dt} = x(9 - \mu x)(\mu + 2x - x^2)\left[(\mu - 10)^2 + (x - 3)^2 - 1\right]. \tag{2.20}$$

Be sure to show the stability of each branch.

All the bifurcations that we have considered are codimension 1. (The *codimension* is the minimum number of control parameters that are needed to characterize the bifurcation.) However, not all codimension-1 bifurcations are equally robust. Small perturbations to the structure of the underlying differential equations may affect different bifurcations in different ways.

To analyze the *structural stability* of a bifurcation that arises in a differential equation of the form

$$\frac{dN}{dt} = f(N, \mu), \tag{2.21}$$

we consider the perturbed differential equation

$$\frac{dN}{dt} = f(N, \mu) + \epsilon g(N), \tag{2.22}$$

where ϵ is the small amplitude of the structural perturbation and $g(N)$ is some arbitrary function. We expand $g(N)$ in a Taylor series about the bifurcation point and keep only the lower-order terms (since we are only interested in changes close to the bifurcation point). The function $g(N)$

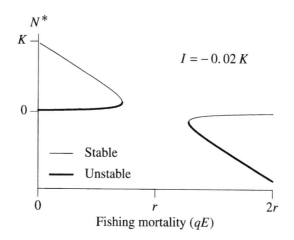

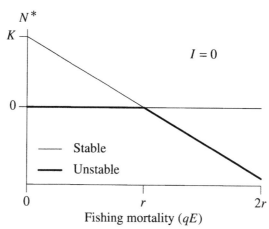

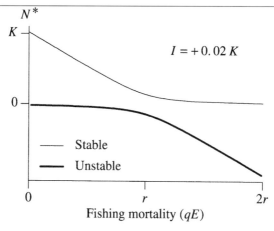

Fig. 2.11. Migration as an imperfection.

may be thought of as an *imperfection*. We would like to know whether this imperfection causes qualitative changes in the bifurcation diagram.

The structural instability of transcritical bifurcations can be highlighted by adding a small constant to harvest model (2.1),

$$\frac{dN}{dt} = rN\left(1 - \frac{N}{K}\right) + I - qEN. \tag{2.23}$$

For I positive, we have immigration. For I negative, we have emigration. Figure 2.11 shows the bifurcation diagram for this equation for small emigration ($I = -0.02\,K$), no migration ($I = 0.0$), and small immigration ($I = +0.02\,K$). For small emigration, the transcritical bifurcation is replaced by two fold bifurcations. The first fold bifurcation leads to a catastrophic collapse of the fishery. For small immigration, in contrast, there is no bifurcation. Since the transcritical bifurcation undergoes such a profound qualitative change in its behavior when we subject the underlying model to a small perturbation, we consider this bifurcation to be structurally unstable.

We have looked only at small, constant perturbations. One might also consider small linear perturbations or, for the pitchfork bifurcation, small quadratic perturbations. Also, a structurally unstable bifurcation may often be made structurally stable by embedding it in a higher-dimensional extended control parameter and phase space. However, this takes us to topics such as the *versal unfolding* of bifurcations and *catastrophe theory* that are beyond the scope of this book.

Problem 2.6 *Imperfection theory*

Consider the three differential equations

(1) $\quad \dot{x} = f(x, \mu, \epsilon) = \epsilon + \mu x - x^2,$

(2) $\quad \dot{x} = f(x, \mu, \epsilon) = \epsilon + \mu x - x^3,$

(3) $\quad \dot{x} = f(x, \mu, \epsilon) = \epsilon + \mu - x^2.$

For each equation, plot the bifurcation diagram for $\epsilon = -0.25$, $\epsilon = 0.0$, and $\epsilon = 0.25$. Show that transcritical and pitchfork bifurcations are structurally unstable to small constant perturbations.

Recommended readings

This section borrows heavily from chapter 1 of Clark's (1990) classic book on the optimal management of renewable resources. May (1977) wrote a review article on breakpoints that preceded Ludwig *et al.*'s (1978) extremely important qualitative analysis of bifurcations and outbreaks in the spruce budworm system. Wiggins (1990) may be consulted for a more thorough introduction to bifurcation theory, codimension, and structural stability.

3 Stochastic birth and death processes

The deterministic equations

$$\frac{dN}{dt} = r N \tag{3.1}$$

and

$$\frac{dN}{dt} = r N \left(1 - \frac{N}{K}\right) \tag{3.2}$$

ignore elements of chance that may be important in determining the growth of populations, especially at small population sizes. Chance may enter *environmentally*, due to storms, freezes, etc., or *demographically*, due to natural variation in the birth and death rates about their averages. We will consider some simple models that include demographic stochasticity. Stochastic models are more informative, but harder to analyze, than deterministic models.

Linear stochastic processes pose fewer problems than their nonlinear counterparts. I will begin with a simple linear birth process that was introduced by Yule (1924) to model the evolution of new species and by Furry (1937) to model particle creation in cosmic ray showers. The Yule–Furry process will act as a stepping-stone to an even more realistic (Feller–Arley) process in which deaths are also possible. We will see that extinction may occur even if the average birth rate exceeds the average death rate ($\beta > \mu$). We will also derive the odds of extinction for the simple birth and death process in terms of the average birth rate, β, the average death rate, μ, and the initial population size n_0.

I will concentrate on linear stochastic models. However, nonlinear stochastic models are also important. In analyzing nonlinear deterministic models, I focused on simple equilibria. The stochastic analog of an equilibrium is a *statistically stationary state*, a limiting equilibrium distribution of population sizes. Unfortunately, most closed (no immigration or emigration) density-dependent birth and death processes only have the trivial stationary state. If you cannot continue to grow, it is just a matter of time before a string of bad luck knocks you to extinction. Even so, many nonlinear processes possess two time scales. Over the short term, populations approach a *statistically quasistationary state*. This is a stationary state for a conditional probability distribution – conditional on no extinction. Over the long term, this quasi-stationary state 'leaks' to extinction. In certain cases, one can determine the mean time to extinction. I will briefly touch on one or two topics from the theory of nonlinear stochastic processes towards the end of this chapter.

Linear birth process

The Yule–Furry process is a *Markov process* – the future is independent of the past, given the present – with a continuous parameter (time). The state space for this process is the potential number of individuals in the population at any instant of time and, since this space is countable, it is customary to refer to this Markov process as a continuous-time *Markov chain*.

As usual, I will let $N(t)$ be the number of individuals at time t. $N(t)$ is now a random variable. Accordingly, I will let

$$p_n(t) = P[N(t) = n], \quad n = 0, 1, 2,\ldots, \tag{3.3}$$

represent the probability that the population size, $N(t)$, takes the value n.

I will begin by allowing births, but not deaths. I will assume that each individual can give birth to new individuals and that each individual acts independently of all others. For a *single* individual, I will also assume that

$$P \{1 \text{ birth in } (t, t + \Delta t] \mid N(t) = 1\} = \beta \, \Delta t + o(\Delta t), \tag{3.4a}$$

$$P \{>1 \text{ birth in } (t, t + \Delta t] \mid N(t) = 1\} = o(\Delta t), \tag{3.4b}$$

$$P \{0 \text{ births in } (t, t + \Delta t] \mid N(t) = 1\} = 1 - \beta \, \Delta t + o(\Delta t). \tag{3.4c}$$

Thus the probability of a birth to an individual in some small time is (to a first approximation) proportional to that time, and the probability of more than one birth is higher order in that time. The order symbol $o(\Delta t)$ denotes quantities for which

$$\lim_{\Delta t \to 0} \frac{o(\Delta t)}{\Delta t} = 0. \tag{3.5}$$

For exactly one birth amongst n individuals, we must have one individual that gives birth and $n-1$ individuals who do not give birth. This can happen in n ways. Thus

$$P\{1\,\text{birth in}\,(t, t + \Delta t)\,|N(t) = n\} = n\,[\beta\,\Delta t + o(\Delta t)]\,[1 - \beta\,\Delta t + o(\Delta t)]^{n-1}$$
$$= n\beta\,\Delta t + o(\Delta t). \qquad (3.6)$$

Similarly,

$$P\{>1\,\text{birth in}\,(t, t + \Delta t)\,|N(t) = n\} = o(\Delta t) \qquad (3.7)$$

and

$$P\{0\,\text{births in}\,(t, t + \Delta t)\,|N(t) = n\} = 1 - n\beta\,\Delta t + o(\Delta t). \qquad (3.8)$$

We are now ready to write down an equation for the $p_n(t)$ of equation (3.3). There can be n individuals at time $t + \Delta t$ if there were $n - 1$ individuals at time t and one birth occurred, or if there were already n individuals at time t and no births occurred,

$$p_n(t + \Delta t) = p_{n-1}(t)\,P\{1\,\text{birth in}\,(t, t + \Delta t)\,|N(t) = n - 1\}$$
$$+ p_n(t)\,P\{0\,\text{births in}\,(t, t + \Delta t)\,|N(t) = n\}. \qquad (3.9)$$

We thus have that

$$p_n(t + \Delta t) = (n - 1)\beta\,\Delta t\,p_{n-1}(t) + (1 - n\beta\,\Delta t)\,p_n(t) + o(\Delta t). \qquad (3.10)$$

After some simple algebra, we obtain

$$\frac{p_n(t + \Delta t) - p_n(t)}{\Delta t} = -n\beta\,p_n(t) + (n - 1)\beta\,p_{n-1}(t) + \frac{o(\Delta t)}{\Delta t}. \qquad (3.11)$$

In the limit as Δt goes to zero this reduces to

$$\frac{dp_n}{dt} = -n\beta\,p_n + (n - 1)\beta\,p_{n-1}. \qquad (3.12)$$

This chain of ordinary differential equations must be augmented with the initial condition

$$p_n(0) = \begin{cases} 1, & n = n_0, \\ 0, & n \neq n_0. \end{cases} \qquad (3.13)$$

Also, since this is a pure birth process,

$$p_n(t) = 0, \; n < n_0. \qquad (3.14)$$

Since the rate of change of $p_n(t)$ depends only upon $p_n(t)$ and on the preceding $p_{n-1}(t)$, we can work our way up this chain. Say that we wish to

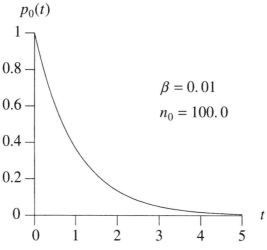

Fig. 3.1. Probability of no births.

find the probability of staying at n_0. By equations (3.12), (3.13), and (3.14), we have that

$$\frac{dp_{n_0}}{dt} = -\beta\, n_0\, p_{n_0}, \tag{3.15a}$$

$$p_{n_0}(0) = 1. \tag{3.15b}$$

This has the straightforward solution

$$p_{n_0}(t) = e^{-\beta\, n_0\, t} \tag{3.16}$$

(see Figure 3.1). The probability of staying at the initial condition thus decreases exponentially with time. This probability decreases more rapidly for large initial populations and for large birth rates.

Okay, how about the probability of being at $n_0 + 1$? In light of equation (3.16), equations (3.12) and (3.13) reduce to

$$\frac{dp_{n_0+1}}{dt} = -\beta\,(n_0 + 1)\,p_{n_0+1} + \beta\, n_0\, e^{-\beta\, n_0\, t}, \tag{3.17a}$$

$$p_{n_0+1}(0) = 0. \tag{3.17b}$$

This linear, nonhomogeneous, first-order differential equation can be solved using an integrating factor, leading to

$$p_{n_0+1}(t) = n_0\, e^{-\beta\, n_0\, t}\left(1 - e^{-\beta t}\right) \tag{3.17c}$$

(see Figure 3.2). The probability of $n_0 + 1$ individuals first increases, but

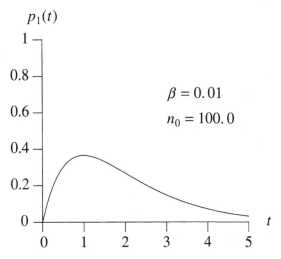

Fig. 3.2. Probability of one birth.

then decreases (as added births occur). It peaks at

$$t = \frac{1}{\beta} \ln \left(1 + \frac{1}{n_0} \right). \tag{3.18}$$

This process can be continued by induction:

$$p_{n_0+m} = \binom{n_0 + m - 1}{n_0 - 1} e^{-\beta n_0 t} \left(1 - e^{-\beta t} \right)^m. \tag{3.19}$$

Equivalently,

$$p_n(t) = \binom{n - 1}{n_0 - 1} e^{-\beta n_0 t} \left(1 - e^{-\beta t} \right)^{n - n_0} \tag{3.20}$$

for each $n \geq n_0$ (see Figure 3.3). This is a *negative binomial distribution* in which the chance of success in a single trial, $\exp(-\beta t)$, decreases exponentially with time.

With the probabilities $p_n(t)$ in hand, it is easy to verify that the expected value, variance, and coefficient of variation are given by

$$E[N(t)] \equiv \sum_{n=0}^{\infty} n \, p_n(t) = n_0 \, e^{\beta t}, \tag{3.21}$$

$$\text{Var}[N(t)] \equiv \sum_{n=0}^{\infty} n^2 \, p_n(t) - E^2[N(t)]$$

$$= n_0 \left(1 - e^{-\beta t} \right) e^{2\beta t}, \tag{3.22}$$

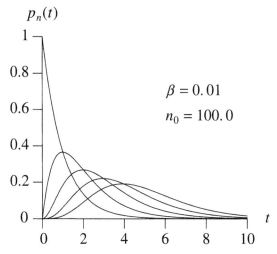

Fig. 3.3. Probabilities of 0, 1, 2, 3, and 4 births.

$$V[N(t)] \equiv \frac{\sqrt{\text{Var}[N(t)]}}{E[N(t)]} = \sqrt{\frac{1 - e^{-\beta t}}{n_0}}, \tag{3.23}$$

so that

$$\lim_{t \to \infty} V[N(t)] = \sqrt{\frac{1}{n_0}}. \tag{3.24}$$

(The expected value and the variance may be derived directly, or by deriving differential equations for each of these quantities. I will soon present an alternative method for computing the mean and variance that uses the probability generating function.)

Problem 3.1 *Mean and variance*
Derive the formulae for the expected value and the variance.

For positive β, the mean population size, $E[N(t)]$, increases exponentially. The variance also increases. However, the coefficient of variation tends towards a constant that depends only on the initial population size. This coefficient is small if the initial population is large. If the initial population size is instead small, any single run of the stochastic process is likely to differ greatly from the corresponding deterministic process.

Figures 3.4 and 3.5 show two sets of 10 simulations of the birth process

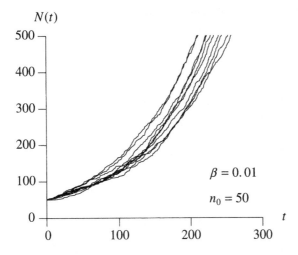

Fig. 3.4. Ten realizations of a birth process with large n_0.

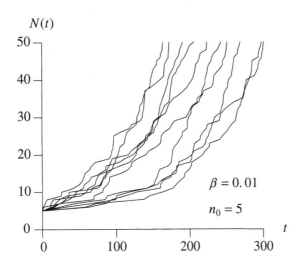

Fig. 3.5. Ten realizations of a birth process with small n_0.

for the same parameter ($\beta = 0.01$) but for two different initial conditions ($n_0 = 50$ and $n_0 = 5$). The second set clearly has more variability.

Problem 3.2 *The simple death process*

Construct and analyze a continuous-time Markov chain in which individuals persist until they die. Do not replace dead individuals. Assume that an individual alive at time t dies with probability $\mu \Delta t + o(\Delta t)$, in $(t, t + \Delta t]$ and that each individual acts independently of all others.

Problem 3.3 *Extinction time*
Assume that there are n_0 individuals present at the start of a simple death process. Let T be the extinction time for this process. What is the probability density function for T? What is the mean time to extinction?

Linear birth and death process

The obvious drawback to the simple birth process is that individuals never die. Life, in contrast, is risky: some individuals die before siring young. Accordingly, we will assume that for each individual

$$P\{1 \text{ birth in } (t, t + \Delta t] \,|\, N(t) = 1\} = \beta \, \Delta t + o(\Delta t), \tag{3.25a}$$

$$P\{1 \text{ death in } (t, t + \Delta t] \,|\, N(t) = 1\} = \mu \, \Delta t + o(\Delta t), \tag{3.25b}$$

$$P\{\text{no change in } (t, t + \Delta t] | N(t) = 1\} = 1 - (\beta + \mu) \, \Delta t + o(\Delta t). \tag{3.25c}$$

There now is some probability of dying as well as of giving birth. The probability of several events (births and/or deaths) is taken to be $o(\Delta t)$. Individuals are still assumed to be independent and so, amongst n individuals,

$$P\{1 \text{ birth in } (t, t + \Delta t] \,|\, N(t) = n\} = n \, [\beta \, \Delta t + o(\Delta t)]$$
$$\times [1 - (\beta + \mu) \Delta t + o(\Delta t)]^{n-1}. \tag{3.26}$$

As a result,

$$P\{1 \text{ birth in } (t, t + \Delta t] \,|\, N(t) = n\} = n \beta \, \Delta t + o(\Delta t).$$

Similarly,

$$P\{1 \text{ death in } (t, t + \Delta t] \,|\, N(t) = n\} = n \mu \, \Delta t + o(\Delta t) \tag{3.27}$$

and

$$P\{\text{no change in } (t, t + \Delta t] \,|\, N(t) = n\} = 1 - n (\beta + \mu) \, \Delta t + o(\Delta t). \tag{3.28}$$

With these probabilities in hand, we are now ready to derive an equation for the probability, $p_n(t)$, that the population is of size n. There can be n individuals at time $t + \Delta t$ if there were $n - 1$ individuals at time t and one birth occurred, if there were $n + 1$ individuals at time t and one death occurred, or if there were already n individuals at time t and no births or deaths occurred,

$$p_n(t + \Delta t) = (n - 1)\beta \Delta t p_{n-1}(t) + (n + 1)\mu \Delta t \, p_{n+1}(t)$$
$$+ [1 - n(\beta + \mu)\Delta t] \, p_n(t) + o(\Delta t). \qquad (3.29)$$

After some simple algebra and after taking the limit as Δt goes to zero, we obtain

$$\frac{dp_n}{dt} = \beta (n - 1) p_{n-1} + \mu (n + 1) p_{n+1} - (\beta + \mu) n \, p_n. \qquad (3.30)$$

We must augment this differential equation with the initial condition

$$p_n(0) = \begin{cases} 1, & n = n_0, \\ 0, & n \neq n_0, \end{cases} \qquad (3.31)$$

and with the stipulation that $p_n(t) = 0$ for $n < 0$.

You may wonder why we bothered with the simple birth process when the (substantially more realistic) birth and death process is not much harder to derive. It is not much harder to derive, but it is much harder to solve. For the simple birth process, the rate of change of $p_n(t)$ depends only on $p_n(t)$ and on the preceding $p_{n-1}(t)$. For the birth and death process, this same $p_n(t)$ depends not only on $p_n(t)$ and $p_{n-1}(t)$, but also on the, as yet, unsolved-for $p_{n+1}(t)$.

The cure for this problem is to solve for all the $p_n(t)$ as one. The easiest way do so is to introduce the *probability generating function*

$$F(t, x) = \sum_{n=0}^{\infty} p_n(t) x^n. \qquad (3.32)$$

(Alternatively, one can introduce a moment generating function.) Since the probabilities $p_n(t)$ are functions of time, $F(t, x)$ is a bivariate function. However, we will often think of F as a univariate function of x, with the probabilities as the coefficients in the power series expansion of this function. Since every $p_n(t)$ is less than or equal to unity, the generating function converges for all $|x| < 1$ (by direct comparison with the geometric series). Since the $p_n(t)$ sum to 1, $F(t, x)$ also converges for $x = 1$. There are a number of other properties of the probability generating function $F(t, x)$ that are worth highlighting:

(1)   The probability of being extinct at time t, $p_0(t)$, is given by

$$p_0(t) = F(t, 0). \qquad (3.33)$$

(2) Taylor's theorem permits us to expand a function in terms of its derivatives at zero. Thus, by calculating derivatives, we can find the sequence of probabilities associated with a given probability generating function,

$$p_n(t) = \frac{1}{n!} \frac{\partial^n F}{\partial x^n}\bigg|_{x=0}. \tag{3.34}$$

(3) The probability generating function allows us to compute the average or expected value of $N(t)$ without tedious calculations involving discrete sums:

$$E[N(t)] \equiv \sum_{n=0}^{\infty} n\, p_n(t) = \frac{\partial F}{\partial x}\bigg|_{x=1}. \tag{3.35}$$

(4) The probability generating function also allows us to compute the variance of $N(t)$ with only a little more effort. In particular,

$$\frac{\partial^2 F}{\partial x^2}\bigg|_{x=1} = \sum_{n=0}^{\infty} (n^2 - n)\, p_n(t)$$

$$= E[N^2(t)] - E[N(t)]. \tag{3.36}$$

However,

$$\mathrm{Var}\,[N(t)] = E[N^2(t)] - E^2[N(t)], \tag{3.37}$$

so that

$$\mathrm{Var}\,[N(t)] = \left[\frac{\partial^2 F}{\partial x^2} + \frac{\partial F}{\partial x} - \left(\frac{\partial F}{\partial x}\right)^2\right]_{x=1}. \tag{3.38}$$

We can, in other words, compute all the probabilities and statistics that we need in a straightforward way with the probability generating function.

So, how do we find this wonderful function? Well, the probability generating function $F(t, x)$ arises as the solution to a partial differential equation. If we differentiate the generating function, equation (3.32), with respect to time,

$$\frac{\partial F}{\partial t} = \sum_{n=0}^{\infty} \dot{p}_n\, x^n, \tag{3.39}$$

and make use of differential equations (3.30) for the derivatives of the probabilities,

$$\frac{\partial F}{\partial t} = \sum_{n=0}^{\infty} \beta\,(n-1)\, p_{n-1}\, x^n + \sum_{n=0}^{\infty} \mu\,(n+1)\, p_{n+1}\, x^n - \sum_{n=0}^{\infty} (\beta + \mu)\, n\, p_n\, x^n, \tag{3.40}$$

$$\frac{\partial F}{\partial t} = \beta x^2 \sum_{n=0}^{\infty} (n-1)\, p_{n-1}\, x^{n-2} + \mu \sum_{n=0}^{\infty} (n+1)\, p_{n+1}\, x^n$$

$$- (\beta + \mu)\, x \sum_{n=0}^{\infty} n\, p_n\, x^{n-1}, \tag{3.41}$$

we arrive at the partial differential equation

$$\frac{\partial F}{\partial t} = [\beta x^2 - (\beta + \mu) x + \mu] \frac{\partial F}{\partial x}. \tag{3.42}$$

Equivalently,

$$\frac{\partial F}{\partial t} + (\beta x - \mu)(1 - x) \frac{\partial F}{\partial x} = 0. \tag{3.43}$$

This partial differential equation has the initial condition

$$F(0, x) = x^{n_0} \tag{3.44}$$

since the population starts at n_0 with probability 1.

Problem 3.4 *Moment generating function*

The moment generating function, $M(t, \theta)$, is defined as

$$M(t, \theta) = \sum_{n=0}^{\infty} p_n(t) e^{n\theta}. \tag{3.45}$$

Show that

$$E[N(t)] = \left. \frac{\partial M}{\partial \theta} \right|_{\theta=0}, \tag{3.46}$$

$$\text{Var}[N(t)] = \left[\frac{\partial^2 M}{\partial \theta^2} - \left(\frac{\partial M}{\partial \theta} \right)^2 \right]_{\theta=0}, \tag{3.47}$$

and that the moment generating function for the linear birth and death process satisfies

$$\frac{\partial M}{\partial t} = [\beta(e^\theta - 1) + \mu(e^{-\theta} - 1)] \frac{\partial M}{\partial \theta}. \tag{3.48}$$

Mathematical meanderings

To solve partial differential equation (3.43) we must use the *method of characteristics*. Consider the change of variables $(t, x) \rightarrow (u, v)$. We will use u as an independent variable that measures distance along a characteristic curve and v to parametrize the initial conditions. The value of v will specify which characteristic curve we are on. Along each characteristic curve (for fixed v), F will depend solely on u.

Assume $\beta \neq \mu$. By the chain rule,

$$\frac{dF}{du} = \frac{\partial F}{\partial t}\frac{dt}{du} + \frac{\partial F}{\partial x}\frac{dx}{du}. \tag{3.49}$$

A comparison of equations (3.43) and (3.49) suggests that equations (3.43) and (3.44) are equivalent to three ordinary differential equations,

$$\frac{dt}{du} = 1, \tag{3.50a}$$

$$\frac{dx}{du} = (\beta x - \mu)(1 - x), \tag{3.50b}$$

$$\frac{dF}{du} = 0, \tag{3.50c}$$

with the initial conditions

$$t(0, v) = 0, \tag{3.51a}$$

$$x(0, v) = v, \tag{3.51b}$$

$$F(0, v) = v^{n_0}. \tag{3.51c}$$

Equations (3.51a), (3.51b) and (3.51c) are a parametric representation of initial condition (3.44).

Equations (3.50a) and (3.50c) are easy to solve. They give

$$t = u, \tag{3.52a}$$

$$F = v^{n_0}. \tag{3.52b}$$

All we have to do now is to solve equation (3.50b), invert the solution so that we have $v(t, x)$, and plug the result into equation (3.52b).

Equation (3.50b) is separable. We separate variables to obtain

$$\left(\frac{1}{1 - x} + \frac{\beta}{\beta x - \mu}\right) dx = (\beta - \mu)\, du. \tag{3.53}$$

Integrating both sides and exponentiating produces

$$\left|\frac{\beta x - \mu}{1 - x}\right| = c\, e^{ru}, \tag{3.54}$$

with

$$r \equiv \beta - \mu \tag{3.55}$$

and c a constant. Initial condition (3.51b), in turn, implies that

$$\left|\frac{\beta x - \mu}{1 - x}\right| = \left|\frac{\beta v - \mu}{1 - v}\right| e^{rt}. \tag{3.56}$$

(We have also replaced u with t.) Solving for v gives us

$$v = \frac{\mu(1 - x)e^{rt} - (\mu - \beta x)}{\beta(1 - x)e^{rt} - (\mu - \beta x)}. \tag{3.57}$$

Plugging v back into equation (3.52b) produces the solution

$$F(t, x) = \left[\frac{\mu(1 - x)e^{rt} - (\mu - \beta x)}{\beta(1 - x)e^{rt} - (\mu - \beta x)} \right]^{n_0}. \tag{3.58}$$

For $\beta = \mu$, everything is the same except that equation (3.50b) is replaced by

$$\frac{dx}{du} = -\beta(x - 1)^2. \tag{3.59}$$

Equation (3.59) is separable. If we separate variables, we obtain

$$\frac{dx}{(x - 1)^2} = -\beta\,du. \tag{3.60}$$

Integrating both sides yields

$$\frac{1}{x - 1} = \beta u + c \tag{3.61}$$

and from equation (3.52a) and initial condition (3.51b) we have that

$$\frac{1}{x - 1} = \beta t + \frac{1}{v - 1}. \tag{3.62}$$

Solving for v, gives us

$$v = \frac{\beta t + (1 - \beta t)x}{(1 + \beta t) - \beta t x}. \tag{3.63}$$

Plugging v back into equation (3.52b) now produces the solution

$$F(t, x) = \left[\frac{\beta t + (1 - \beta t)x}{(1 + \beta t) - \beta t x} \right]^{n_0}. \tag{3.64}$$

Partial differential equation (3.43) with initial condition (3.44) has the solution

$$F(t, x) = \begin{cases} \left[\dfrac{\mu(1 - x)e^{rt} - (\mu - \beta x)}{\beta(1 - x)e^{rt} - (\mu - \beta x)} \right]^{n_0}, & \beta \neq \mu, \\[4mm] \left[\dfrac{\beta t + (1 - \beta t)x}{(1 + \beta t) - \beta t x} \right]^{n_0}, & \beta = \mu. \end{cases} \tag{3.65}$$

It follows that the probability of being extinct at time t is

$$p_0(t) = F(t, 0) = \begin{cases} \left[\dfrac{\mu(e^{rt} - 1)}{\beta e^{rt} - \mu} \right]^{n_0}, & \beta \neq \mu, \\[4mm] \left(\dfrac{\beta t}{1 + \beta t} \right)^{n_0}, & \beta = \mu, \end{cases} \tag{3.66}$$

and that the asymptotic probability of extinction is

$$\lim_{t \to \infty} p_0(t) = \begin{cases} \left(\dfrac{\mu}{\beta}\right)^{n_0}, & \beta > \mu, \\ 1, & \beta \leq \mu. \end{cases} \tag{3.67}$$

(In taking this limit, be sure to pay attention to the sign of r.) The probability of extinction, equation (3.66) is a monotonically increasing function that rises up from zero to the asymptotic limit of equation (3.67). For death rates that equal or exceed the birth rate, extinction is certain. However, even for positive growth rates, there is some finite probability of extinction. Populations with small initial numbers are especially susceptible to extinction.

The probability generating function, equation (3.65), can also be used to derive the expected value, equation (3.35), and variance, equation (3.38), of the population. These are simply

$$E[N(t)] = \left. \frac{\partial F}{\partial x} \right|_{x=1} = n_0 e^{rt}, \tag{3.68}$$

$$\text{Var}[N(t)] = \begin{cases} n_0 \dfrac{(\beta + \mu)}{(\beta - \mu)} e^{rt}(e^{rt} - 1), & \beta \neq \mu, \\ 2n_0\beta t, & \beta = \mu. \end{cases} \tag{3.69}$$

The expected population size and the variance both grow exponentially for $\beta > \mu$. The coefficient of variation,

$$V[N(t)] = \sqrt{\frac{1}{n_0} \frac{(\beta + \mu)}{(\beta - \mu)}(1 - e^{-rt})}, \quad \beta > \mu, \tag{3.70}$$

increases monotonically towards a constant. For $\beta = \mu$, the variance increases linearly. For $\beta < \mu$, the variance first increases, but then decreases exponentially.

Problem 3.5 *Waterfowl dynamics*

A population of B ducks lives on two ponds, one large, one small. Let $N(t)$ be the number of birds on the small pond. You may assume that there are $B - N(t)$ birds on the large pond. Let the probability of a departure from the small pond be proportional to the departure rate r_d, to the interval, and to the number of birds on the small pond,

$$Pr[N(t + \Delta t) = n - 1 \,|\, N(t) = n] = r_d n \Delta t + o(\Delta t). \tag{3.71}$$

Similarly, assume that the probability of an arrival onto the small pond is proportional to the arrival rate r_a, to the interval, and to the number of birds off the small pond,

$$Pr[N(t + \Delta t) = n + 1 \,|\, N(t) = n] = r_a (B - n) \Delta t + o(\Delta t). \qquad (3.72)$$

(1) Derive a system of differential equations for p_n, the probability of n birds on the small pond (Silverman and Kot, 2000).
(2) Derive a partial differential equation for the probability generating function.

Problem 3.6 *Birth, death, and immigration*

Consider a birth and death process with constant immigration,

$$P[\Delta N = +1 \,|\, N(t) = n] = \beta n \Delta t + I \Delta t + o(\Delta t^2), \qquad (3.73a)$$

$$P[\Delta N = -1 \,|\, N(t) = n] = \mu n \Delta t + o(\Delta t^2), \qquad (3.73b)$$

$$P[\Delta N = 0 \,|\, N(t) = n] = 1 - [(\beta + \mu) n + I] \Delta t + o(\Delta t^2). \qquad (3.73c)$$

Assume that the mortality rate is greater than the birth rate, $\mu > \beta$.

(1) Derive differential equations for the probabilities.
(2) Derive a partial differential equation for the probability generating function.
(3) Find the probability generating function for the equilibrium probability distribution. To wit, set the time derivative in your partial differential equation equal to zero. Solve the corresponding ordinary differential equation. Choose the generating function so that the corresponding probabilities sum to 1.
(4) What is the expected population size at equilibrium?
(5) What is the variance of the equilibrium population sizes?
(6) What are the odds that the population is of size zero at equilibrium?

Problem 3.7 *A statistician's revenge*

Fifty ungulates (hoofed, grazing mammals) are introduced into an uninhabited island. These ungulates have the following characteristics:

$$\beta = 0.15 \text{ ungulates per individual per year}, \qquad (3.74a)$$

$$\mu = 0.05 \text{ ungulates per individual per year}. \qquad (3.74b)$$

How long should you wait to revisit the island to be 95% sure that the population has doubled?

Hint Think of each of the 50 introduced ungulates as a separate birth and death process. Feel free to use the central limit theorem.

Nonlinear birth and death processes

It is easy enough to extend system (3.30) so that it models density-dependent growth. We need merely let the per capita birth and death rates, β and μ, depend on the population size,

$$\frac{dp_0}{dt} = \mu_1 p_1, \tag{3.75a}$$

$$\frac{dp_1}{dt} = 2\mu_2 p_2 - (\beta_1 + \mu_1) p_1, \tag{3.75b}$$

$$\frac{dp_n}{dt} = [(n-1)\beta_{n-1}] p_{n-1} + [(n+1)\mu_{n+1}] p_{n+1}$$
$$-n(\beta_n + \mu_n) p_n, \quad n > 1. \tag{3.75c}$$

We will assume that the density-dependent per capita birth and death rates, β_n and μ_n, prevent unlimited growth.

I stressed the importance of equilibria for the nonlinear deterministic models in the previous two chapters. It is natural to ask whether these equations possess the stochastic analog of an equilibrium, a *statistically stationary state*. It is clear from equation (3.75a), that if $p_0(t)$ is constant, $p_1(t)$ must be zero. If $p_1(t)$ is zero, it follows, from equation (3.75b), that the only stationary state of $p_2(t)$ is also zero. Indeed, it soon follows that the only way for $p_n(t)$ to be constant for any $n > 0$ is for it to be zero. Thus, for a closed population (without migration), the only finite stationary state is one that corresponds to extinction. If the population is closed and bounded, all probability 'leaks' to the absorbing $N = 0$ state; the ultimate probability of extinction is 1.

If a population leaks to extinction, the expected time to extinction may still be much longer than the time for 'typical' population changes. There may well be a *statistically quasistationary state*. To make this concept more rigorous, we introduce the conditional probabilities

$$q_n(t) \equiv \frac{p_n(t)}{1 - p_0(t)}, \tag{3.76}$$

conditional on no extinction. If we differentiate equation (3.76), we obtain

$$\frac{dq_n}{dt} = \frac{1}{1 - p_0(t)} \frac{dp_n}{dt} + \frac{p_n(t)}{[1 - p_0(t)]^2} \frac{dp_0}{dt}. \tag{3.77}$$

We may evaluate the right-hand side of equation (3.77), using equations (3.75a), (3.75b), (3.75c), and definition (3.76) to obtain a system of equations,

$$\frac{dq_1}{dt} = 2\mu_2 q_2 - (\beta_1 + \mu_1) q_1 + \mu_1 q_1^2, \tag{3.78a}$$

$$\frac{dq_n}{dt} = [(n-1)\beta_{n-1}]q_{n-1} + [(n+1)\mu_{n+1}]q_{n+1}$$
$$- [n(\beta_n + \mu_n)]q_n + \mu_1 q_1 q_n, \quad n > 1, \tag{3.78b}$$

that may now possess a set of equilibria q_n^*.

Nisbet and Gurney (1982) recommend solving for the quasistationary states iteratively:

(1) Guess q_1^*.
(2) Calculate $q_2^*, q_3^*, \ldots$ by repeated application of equations (3.78a) and (3.78b) with dq_n/dt set to zero. Stop at a value of n high enough that q_n^* is negligible.
(3) Calculate $q_1^* / \sum q_n^*$. If the result differs greatly from the old value of q_1^*, repeat (2) with the revised value of q_1^*. Continue iterating until you get the accuracy you want.

Nisbet and Gurney (1982), Ricciardi (1986), and Renshaw (1991) discuss a number of other methods for analyzing and simulating density-dependent birth and death processes. Most of these methods are beyond the scope of this book. However, one special case merits attention. If an ensemble of populations starts at or quickly reaches its quasistationary distribution, then

$$\frac{p_1(t)}{1 - p_0(t)} \approx q_1^* \tag{3.79}$$

is the probability that $N(t) = 1$, given that no extinction has occurred. Equation (3.75a) then simplifies to

$$\frac{dp_0}{dt} \approx \mu_1 q_1^* (1 - p_0), \tag{3.80}$$

so that

$$p_0(t) \approx 1 - \exp(-\mu_1 q_1^* t). \tag{3.81}$$

Since $p_0(t)$ is the probability of extinction through time t, dp_0/dt is the probability density of extinction at time t. Thus the mean time to extinction is

$$T_{\text{extinction}} = \int_0^\infty t \, \frac{dp_0}{dt} \, dt \approx \int_0^\infty t \, \mu_1 q_1^* \exp(-\mu_1 q_1^* t) \, dt \tag{3.82}$$

or

$$T_{\text{extinction}} \approx \frac{1}{\mu_1 q_1^*}. \tag{3.83}$$

If μ_1 and q_1^* are both small, this time to extinction will be large.

Recommended readings

Tuckwell (1988) introduces the reader to birth and death processes. Bharucha-Reid (1997) considers this topic in greater detail. Bailey (1964) and Ricciardi (1986) cover the same ground using moment generating functions rather than probability generating functions. Many combinatorics and probability books (Riordan, 1958; Chiang, 1980) discuss generating functions. There is even a book dedicated to generatingfunctionology (Wilf, 1994). Nisbet and Gurney (1982) and Renshaw (1991) provide numerous examples of nonlinear birth and death processes in ecology.

4 Discrete-time models

For many organisms, births occur in regular, well-defined 'breeding seasons'. This contradicts our earlier assumption that births occur continuously. I begin this section by describing the life histories of some organisms with discrete reproduction. We will then consider some simple discrete-time models that are well-suited to these organisms.

Plants

Herbs often flower in their first year and then die, roots and all, after setting seed. Plants that flower once and then die are *monocarpic*. Many monocarps are annuals, but a few species have long lives. Bamboos are grasses, but they grow to unusually large size. Many species of bamboo grow vegetatively for 20 years before flowering and dying; other species have flowering times of 1, 3, 11, 15, 30, 48, and 60 years (McClure, 1967). One Japanese species, *Phyllostachys bambusoides*, waits 120 years to flower (Janzen, 1976). A few species of bamboo also synchronize reproduction within cohorts. This may have disastrous consequences for organisms that subsist on bamboo. In the spring of 1983, the simultaneous mass flowering and death of *Fargasia spathacea* and *Sinarundinaria fangiana* resulted in the starvation of many pandas within their main reserve in China. Another long-lived, monocarpic plant is the desert agave, *Agave deserti* (one of about 10 plants referred to as century plants), which routinely lives 20 to 25 years before flowering.

Most trees flower repeatedly. However, Foster (1977) has characterized *Tachigalia versicolor* as a 'suicidal neotropical tree'. After reaching heights of 30–40 m, it flowers once and then dies.

Insects

Semelparity† is for animals what monocarpy is for plants. There are semel-parous insects that are *univoltine*, with one generation per year, *bivoltine*, with two generations per year, and *multivoltine*, with more than two generations per year. Mayflies, or day-flies, are famous for their semelparity. Close to 2000 species of mayflies are found in the order Ephemeroptera. Most species are univoltine. Either eggs hatch quickly and larvae develop slowly over a one-year period, or, more commonly, the eggs undergo diapause and the larvae develop quickly over two to four months. Some species are bivoltine or multivoltine, especially in warmer climates (Thorp and Covich, 1991). The larvae undergo many (>12) molts. After development, the larvae rise to the surface and molt into winged subadults (subimagos). Within the day, the subadults molt, one last time, into adults (imagos). The adults do not feed and they generally live a short time (a few hours to a few days). In cooler climates, the emergence of winged mayflies is often highly synchronous over a wide area, leading to enormous swarms for short periods of the year. Trends in water temperature frequently provide the cues that determine the timing of emergence (Peters *et al.*, 1987).

Some semelparous insects are long lived. Three species of periodic cicadas of the eastern United States, *Magicicada septendecim*, *M. cassini*, and *M. septendecula*, have broods that take 13 or 17 years to develop. The slow growth is usually attributed to the low caloric and nutritional content of the xylem fluids (99.9% water) that cicadas suck from the roots of trees and grasses (White and Strehl, 1978; Lloyd and White, 1987). The 17-year broods grow extremely slowly during their first four years, but then grow at the same rate as the 13-year broods (White and Lloyd, 1975); each brood has five instars. The slow development is terminated by the synchronous emergence of adults within the same one- to two-day period. The adults live four to six weeks (or for less than 1% of the lifespan). The 13-year broods are found in the Mississippi Valley and in southern states; 17-year broods are found north and west of the 13-year range. Historical evidence suggests that 13-year cicadas have been spreading northward and that the range of 17-year cicadas has been contracting. The 17- and 13-year broods evidently hybridized in 1868 (Lloyd *et al.*, 1983). For geographically and temporally overlapping broods, hybridization can recur every 221 years ($221 = 13 \times 17$).

---

† Semele, in Greek mythology, was a mortal paramour of Zeus. At the urging of Hera (Zeus's wife), Semele begged Zeus to appear before her in all his celestial splendor. Sadly, the spectacle was too much for Semele and she was consumed by flames. Zeus rescued her child, Bacchus, from her womb and bore it in his own thigh until Bacchus's birth.

Fish

A small fraction (<1%) of the 22 000 species of teleost fish are semelparous and die soon after spawning (Finch, 1990). Nearly all semelparous species are also *diadromous* (with migrations between freshwater and marine habitats) or of diadromous ancestry. Pacific salmon of the genus *Oncorhyncus* fall into the *anadromous* camp. They either spawn in fresh water and feed in the ocean, or are recently land locked. Most of the species in this genus also show spectacular senescence at first spawning. The freshwater European eel, *Anguilla anguilla*, falls into the *catadromous* camp. Individuals of this species spend most of their first 10–15 years in freshwater European lakes and rivers. They then migrate out to the Sargasso Sea near Bermuda where they spawn and die.

Birds

I know of no examples of semelparity in birds; all birds are *iteroparous*. At the same time, breeding seasons can be quite short. Many species of shorebirds and waterfowl breed north of the Arctic circle where the climate forces surprising synchrony. For example, the largest and best-known breeding area for Greater Snow Geese (variously *Anser caerulescens atlantica* or *Chen caerulescens atlantica*) is on Bylot Island off the northeast coast of Baffin Island in the Northwest Territories of Canada. In 1957, 15 000 Greater Snow Geese nested on the island (Lemieux, 1959). Egg laying started on 8 June, stopped on 20 June, and peaked between 12 June and 17 June. Virtually all hatchings occurred between 8 and 13 July.

Mammals

Nine small marsupials in the genera *Antechinus* and *Phascogale* are semelparous. These species ovulate once a year and produce a single litter. A remarkable feature of these creatures' biology is the abrupt and total mortality amongst males following mating (Braithwaite and Lee, 1979). In *A. stuartii* and *A. swainsonii*, males become sexually active after 11 months and die three to four weeks later. Births are highly synchronized, often to within a day or two within a population. These species live in environments that have predictable insect 'blooms' in a given season of the year.

The examples that I have discussed include species, such as cicadas, that have discrete, nonoverlapping generations, and other species, such as Snow Geese, that have synchronized reproduction but overlapping generations. Species with overlapping generations are best described by age-structured models. For now, I will concentrate on models with discrete, nonoverlapping generations.

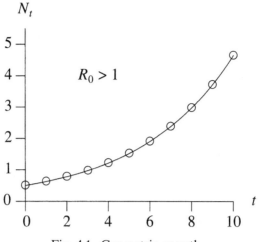

Fig. 4.1. Geometric growth.

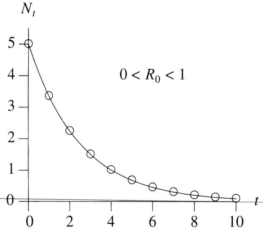

Fig. 4.2. Geometric decay.

Density-independent growth

Let N_t be the size of a population in year (or generation) t. We will census the population each year at the same stage of the life cycle. Imagine that each individual leaves R_0 offspring before dying. I will call R_0 the *net reproductive rate*. It follows that

$$N_{t+1} = R_0 N_t. \tag{4.1}$$

Equation (4.1) is a linear, first-order, constant-coefficient difference equation. Since we have assumed that individuals always leave the same number

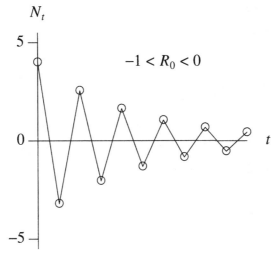

Fig. 4.3. Decaying oscillations.

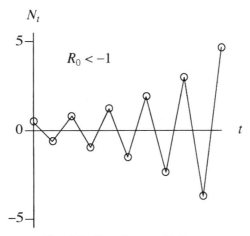

Fig. 4.4. Growing oscillations.

of offspring, it follows that

$$N_1 = R_0 N_0, \tag{4.2a}$$

$$N_2 = R_0 N_1 = R_0 (R_0 N_0) = R_0^2 N_0, \tag{4.2b}$$

$$\vdots$$

$$N_t = R_0^t N_0. \tag{4.2c}$$

The solution of equation (4.1) is thus one of geometric growth or decay. If $R_0 > 1$, each individual leaves more than one descendant, and the population grows geometrically (see Figure 4.1). If $0 < R_0 < 1$, individuals leave, on

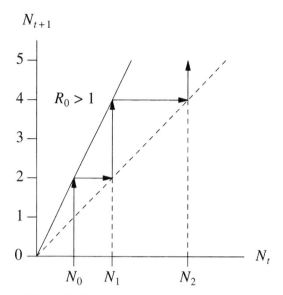

Fig. 4.5. Geometric growth à la cobweb.

average, fewer than one descendant, and the population declines geometrically (see Figure 4.2). These figures resemble those for exponential growth
and decay. And indeed, for positive R_0, we may identify $R_0 \equiv e^r$ and rewrite
the solution as an equation of exponential growth. Remember, however, that
t is now integer valued.

Individuals cannot leave a negative number of offspring. However, nothing
prevents us from pondering this possibility mathematically. For $-1 < R_0 < 0$,
we get decaying oscillations (see Figure 4.3). For $R_0 < -1$, we get growing
oscillations (see Figure 4.4). Figures 4.1, 4.2, 4.3, and 4.4 suggest that solutions
approach the origin if R_0 is less than 1 *in magnitude* and that these solutions
diverge if this coefficient is greater than 1 in magnitude.

There is a second method for plotting the solutions to equation (4.1).
This second method is *cobweb analysis*. Figure 4.5 shows N_{t+1} plotted as a
function of N_t for $R_0 > 1$. This curve is clearly a straight line. Figure 4.5
also contains the 45° dashed line, $N_{t+1} = N_t$. We may iterate our difference
equation by repeatedly (a) moving up (or down) to the curve and then (b)
bouncing off the 45° line (so that we reset $N_{t+1} = N_t$). This approach will
be extremely helpful later, in our analyses of nonlinear difference equations.

Zero is an equilibrium point for equation (4.1). Equilibria are simply the
fixed points of a difference equation. At an equilibrium, we expect that

$$N_{t+1} = N_t = N^*. \tag{4.3}$$

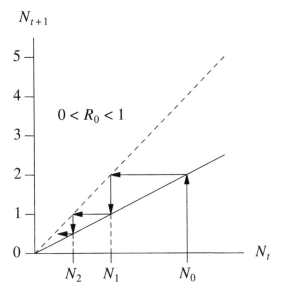

Fig. 4.6. Geometric decay *à la* cobweb.

The trivial equilibrium in Figure 4.5 is unstable; the trivial equilibrium in Figure 4.6 is stable.

A linear, density-independent difference equation may have a nonzero equilibrium if we allow for immigration or emigration. For example, the difference equation

$$N_{t+1} = \frac{3}{4} N_t + 10 \tag{4.4}$$

has an equilibrium that satisfies

$$N^* = \frac{3}{4} N^* + 10. \tag{4.5}$$

In this instance, immigration balances low survivorship at

$$N^* = 40. \tag{4.6}$$

If we introduce a new variable that measures our deviation from this equilibrium,

$$x_t \equiv N_t - 40, \tag{4.7}$$

equation (4.4) reduces to

$$x_{t+1} = \frac{3}{4} x_t. \tag{4.8}$$

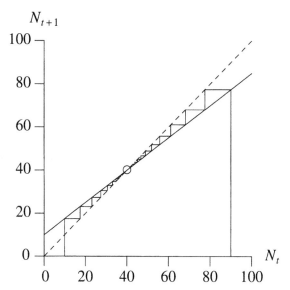

Fig. 4.7. Cobwebbing to a stable equilibrium.

This has the obvious solution

$$x_t = x_0 \left(\frac{3}{4}\right)^t, \tag{4.9}$$

so that

$$N_t = 40 + (N_0 - 40)\left(\frac{3}{4}\right)^t. \tag{4.10}$$

Small perturbations about the equilibrium decay; the equilibrium is asymptotically stable. This stability also comes out in cobweb analyses (see Figure 4.7).

Density-dependent growth

Density dependence occurs if the number of offspring per adult varies with density. We have density dependence whenever our underlying difference equation is nonlinear. In Chapter 1, we analyzed the nonlinear logistic differential equation. Can we find a discrete-time analog of the logistic differential equation? There are two such analogs. These two models possess different forms of density dependence. They also exhibit radically different behaviors. This warrants investigation.

The first model assumes that the per capita number of offspring is inversely

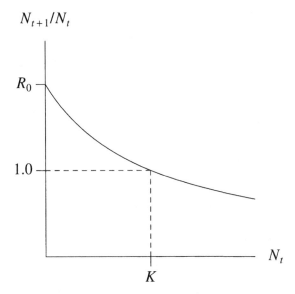

N_{t+1}/N_t

R_0

1.0

K

N_t

Fig. 4.8. Compensation.

proportional to a linearly increasing function of the number of adults,

$$\frac{N_{t+1}}{N_t} = \frac{R_0}{1 + [(R_0 - 1)/K] N_t} \tag{4.11}$$

(see Figure 4.8). The resulting difference equation,

$$N_{t+1} = \frac{R_0 N_t}{1 + [(R_0 - 1)/K] N_t}, \tag{4.12}$$

is known as the *Beverton–Holt stock-recruitment curve* (Beverton and Holt, 1957) and is a monotonically increasing hyperbolic mapping (see Figure 4.9).

The second model is more clearly similar to the logistic differential equation in form, if less so in solution. It is derived as a direct approximation to the logistic differential equation. Starting with

$$\frac{dN}{dt} = r N \left(1 - \frac{N}{K} \right), \tag{4.13}$$

we approximate the derivative on the left-hand side with a finite-difference quotient,

$$\frac{\Delta N}{\Delta t} = r N \left(1 - \frac{N}{K} \right). \tag{4.14}$$

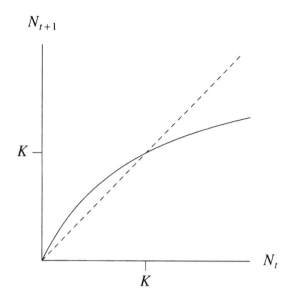

Fig. 4.9. Beverton–Holt stock-recruitment curve.

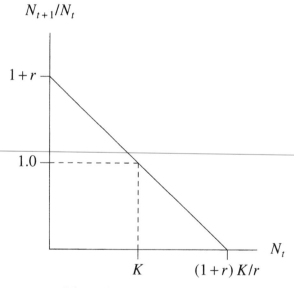

Fig. 4.10. Overcompensation.

For the organisms in question, the time step is one generation, $\Delta t = 1$, so that equation (4.14) reduces to

$$N_{t+1} - N_t = r N_t \left(1 - \frac{N_t}{K}\right). \tag{4.15}$$

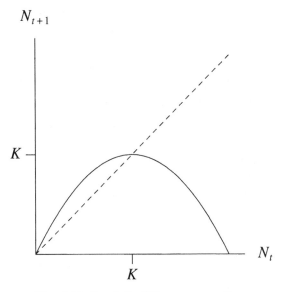

Fig. 4.11. Logistic difference equation.

We are thus left with the *logistic difference equation,*

$$N_{t+1} = (1 + r) N_t - \frac{r}{K} N_t^2. \tag{4.16}$$

For this logistic difference equation, the number of offspring per adult,

$$\frac{N_{t+1}}{N_t} = (1 + r) - \frac{r}{K} N_t, \tag{4.17}$$

is a linearly decreasing function (see Figure 4.10). The mapping itself is parabolic (see Figure 4.11).

Mathematicians frequently rescale equation (4.16) by letting

$$x_t \equiv \frac{r}{(1 + r)} \frac{N_t}{K}, \tag{4.18a}$$

$$\mu \equiv 1 + r. \tag{4.18b}$$

Equation (4.16) is then replaced by

$$x_{t+1} = \mu x_t (1 - x_t). \tag{4.19}$$

Equations (4.16) and (4.19) each have their advantages. Equation (4.16) has the advantage that it always possesses a nontrivial equilibrium at $N_t = K$. Equation (4.19), in turn, has the advantage that the interval $0 \le x \le 1$ is invariant for $0 \le \mu \le 4$.

The Beverton–Holt stock-recruitment curve and the logistic difference equations have two different forms of density dependence. The Beverton–Holt curve is *compensatory* (Neave, 1953): it increases monotonically, but with ever-decreasing slope. An increase in density leads to a decrease in per capita reproduction but does not reduce the recruitment of the entire population. In contrast, the density dependence in the logistic difference equation can become so intense that recruitment for the entire population does decrease with increasing density. This is *overcompensation* (Clark, 1990). Normal compensation and overcompensation are also related to the two extreme forms of intraspecific competition that Nicholson (1954) recognized as *contest competition* and *scramble competition*. In a contest, some individuals win, while others lose. The contest may allow the winners to sequester enough resources to reproduce, irrespective of density. The contest could, for example, be over exclusive use of a feeding territory. Recruitment will then level off, without actually decreasing, as adult numbers increase. In a scramble, all individuals divide the resources equally. If the density is high enough, it may happen that no one sequesters enough resources to reproduce. This can easily lead to overcompensation.

The Beverton–Holt stock-recruitment curve is one of those rare nonlinear difference equations for which there is an exact, closed-form solution (see Problem 4.1). For most nonlinear difference equations, we must resort to qualitative methods of analysis. In just a moment, then, I will outline the discrete-time analogs of the definitions and stability criteria of Chapter 1.

Problem 4.1 *Beverton–Holt stock-recruitment curve*

Derive an exact, closed-form solution for the Beverton–Holt difference equation,

$$N_{t+1} = \frac{R_0 N_t}{1 + [(R_0 - 1)/K] N_t}. \tag{4.20}$$

Show that you can define R_0 so that your solution matches solution (1.8) for the logistic differential equation.

Hint Show that the substitution

$$u_t \equiv \frac{1}{N_t} \tag{4.21}$$

gives rise to a linear difference equation.

Consider the autonomous difference equation

$$N_{t+1} = f(N_t). \tag{4.22}$$

Definition We say that $N = N^*$ is an *equilibrium point* (also known as a *fixed point* or *rest point*) if

$$N^* = f(N^*). \tag{4.23}$$

Definition An equilibrium point N^* is said to be *Lyapunov stable* if, for any (arbitrarily small) $\varepsilon > 0$, there exists a $\delta > 0$ (depending on ε) such that

$$|N_0 - N^*| < \delta \Rightarrow |N_t - N^*| < \varepsilon, \text{ for all } t > t_0. \tag{4.24}$$

Definition An equilibrium point N^* is said to be *asymptotically stable* if it is stable and if, in addition, there exists some $\rho > 0$, such that

$$\lim_{t \to \infty} |N_t - N^*| = 0 \tag{4.25}$$

for all N_0 satisfying

$$|N_0 - N^*| < \rho. \tag{4.26}$$

As with simple differential equations, we may study the stability of an equilibrium by considering the linearization of the difference equation at the equilibrium. However, the criterion for stability for difference equations differs from that of differential equations.

Theorem Suppose that N^* is an equilibrium point and that $f(N)$ is a continuously differentiable function. Suppose also that $|f'(N^*)| \neq 1$. Then the equilibrium point N^* is asymptotically stable if $|f'(N^*)| < 1$, and unstable if $|f'(N^*)| > 1$.

Proof The proof is more or less identical with that in Chapter 1, except that in the end we are left with a linear constant-coefficient difference equation and with the conditions for growth or decay that arise from such an equation.

With these results in hand, we are now set to compare the Beverton–Holt stock-recruitment curve and the logistic difference equation. $\qquad\square$

EXAMPLE Beverton–Holt stock-recruitment curve
The Beverton–Holt stock-recruitment curve,

$$N_{t+1} = \frac{R_0 N_t}{1 + [(R_0 - 1)/K] N_t}, \tag{4.27}$$

has equilibria that satisfy

$$N^* = \frac{R_0 N^*}{1 + [(R_0 - 1)/K] N^*}.$$ (4.28)

There is thus the trivial equilibrium at

$$N^* = 0$$ (4.29)

and a nontrivial equilibrium at

$$N^* = K.$$ (4.30)

We may determine the stability of each equilibrium by investigating

$$f'(N^*) = \frac{R_0}{\left[1 + \left(\frac{R_0 - 1}{K}\right) N^*\right]^2}.$$ (4.31)

At $N^* = 0$,

$$f'(0) = R_0.$$ (4.32)

The trivial equilibrium is thus unstable if $R_0 > 1$. (I am restricting our attention to biologically realistic, nonnegative values of R_0.)
At $N^* = K$,

$$f'(K) = \frac{1}{R_0}.$$ (4.33)

The carrying capacity K is thus asymptotically stable for all $R_0 > 1$.

For $R_0 > 1$, small perturbations from the trivial equilibrium grow, at first, geometrically. Small enough perturbations from the carrying capacity, in turn, decay geometrically. If you have tried Problem 4.1, you also know that the closed-form solution for the Beverton–Holt model is virtually indistinguishable from that of the logistic differential equation. Finally, cobweb analysis (Figure 4.12) confirms that the Beverton–Holt model has the same steady monotonic approach to a carrying capacity (Figure 4.13) that we associate with the logistic differential equation. ◇

EXAMPLE   Logistic difference equation
The logistic difference equation,

$$N_{t+1} = (1 + r) N_t - \frac{r}{K} N_t^2,$$ (4.34)

has equilibria that satisfy

$$N^* = (1 + r) N^* - \frac{r}{K} N^{*2}.$$ (4.35)

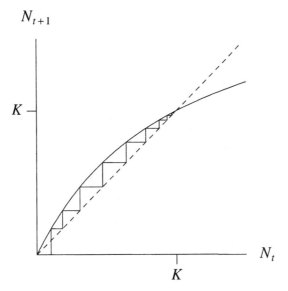

Fig. 4.12. Beverton–Holt cobweb.

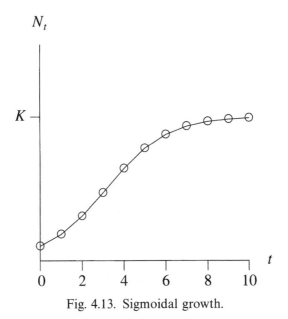

Fig. 4.13. Sigmoidal growth.

There are, once again, two equilibria: the trivial equilibrium,

$$N^* = 0, \tag{4.36}$$

and the nontrivial equilibrium at the carrying capacity,

$$N^* = K. \tag{4.37}$$

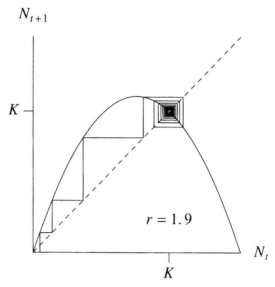

Fig. 4.14. Logistic cobweb.

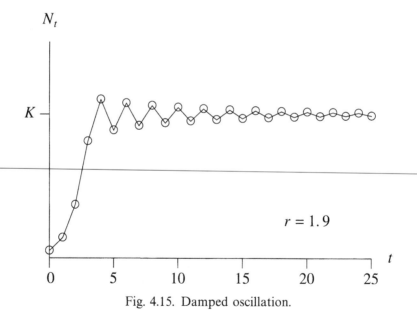

Fig. 4.15. Damped oscillation.

Stability is determined by the coefficient

$$f'(N^*) = (1 + r) - 2\frac{r}{K}N^*. \tag{4.38}$$

For $N^* = 0$,

$$f'(0) = 1 + r. \tag{4.39}$$

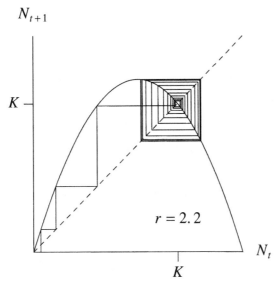

Fig. 4.16. Cobwebbing to a 2-cycle.

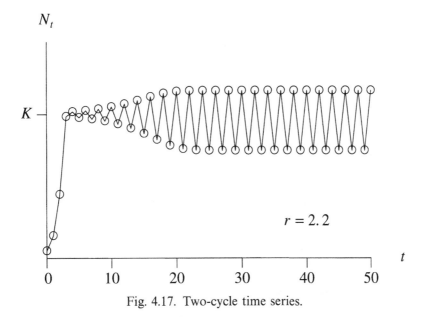

Fig. 4.17. Two-cycle time series.

The trivial equilibrium is thus unstable for $r > 0$. For $N^* = K$,

$$f'(K) = 1 - r. \tag{4.40}$$

The behavior of the logistic difference equation now departs from that of the Beverton–Holt stock-recruitment curve. The carrying capacity is asymptoti-

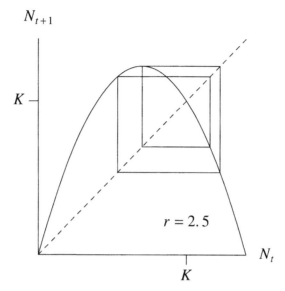

Fig. 4.18. Cobwebbing a 4-cycle.

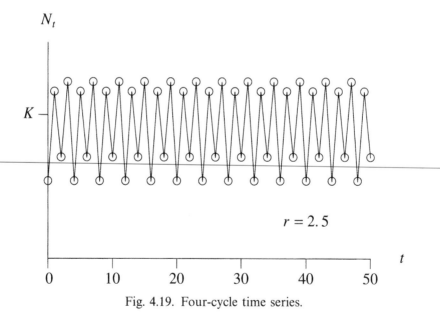

Fig. 4.19. Four-cycle time series.

cally stable only for $0 < r < 2$. For $0 < r < 1$, solutions tend monotonically towards the carrying capacity. For larger r, $1 < r < 2$, small perturbations about the carrying capacity still return to the carrying capacity, but in an oscillatory manner (see Figures 4.14 and 4.15). Finally, when $r > 2$, small perturbations from the carrying capacity diverge, with oscillations.

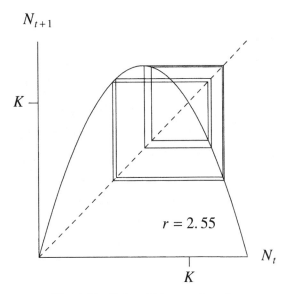

Fig. 4.20. Cobwebbing an 8-cycle.

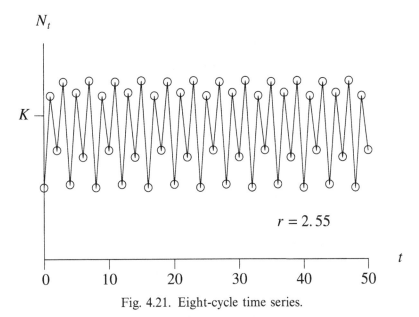

Fig. 4.21. Eight-cycle time series.

What happens to the trajectories that move away from the carrying capacity for $r > 2$? A cobweb analysis shows that for small $r > 2$, trajectories tend towards a stable *2-cycle*, i.e., towards a stable periodic solution of period 2 (see Figures 4.16 and 4.17). As r increases, this 2-cycle increases in amplitude until it ultimately loses stability, at $r = \sqrt{6}$, to a 4-cycle (Figures 4.18 and

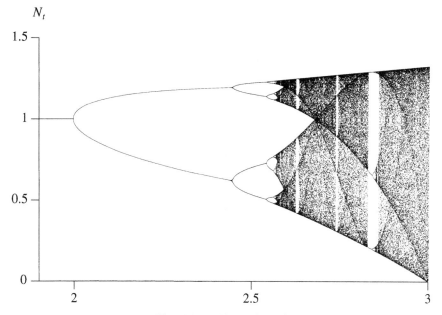

Fig. 4.22. Bifurcation diagram.

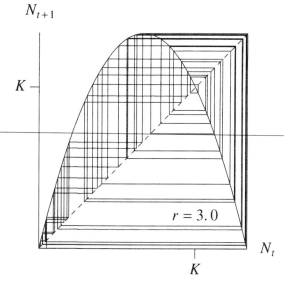

Fig. 4.23. Chaotic cobweb.

4.19). As we continue to increase r, the 4-cycle grows in amplitude until it too loses stability, to an 8-cycle (see Figures 4.20 and 4.21).

Figure 4.22 displays all of these period doublings on a single bifurcation diagram; for each r we plot the stable attractor, be it a single point (cor-

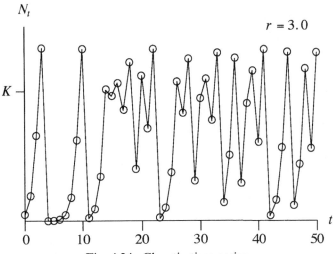

N_t

$r = 3.0$

K

t

Fig. 4.24. Chaotic time series.

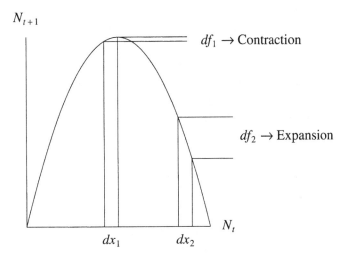

N_{t+1}

$df_1 \rightarrow$ Contraction

$df_2 \rightarrow$ Expansion

N_t

dx_1 dx_2

Fig. 4.25. Constriction or dilation of small differences in initial conditions.

responding to a stable equilibrium), two points (corresponding to a stable 2-cycle), four points (corresponding to a stable 4-cycle), or whatever. The bifurcation diagram clearly shows the sequence of period-doubling *flip* bifurcations that takes us from an equilibrium, to a 2-cycle, to a 4-cycle, and so on. There are an infinite number of period doublings, corresponding to all cycles of period 2^n. However, the range of r values over which each of these cycles is stable decreases sufficiently quickly that this infinite sequence actually converges, at $r_\infty = 2.5699456\ldots$.

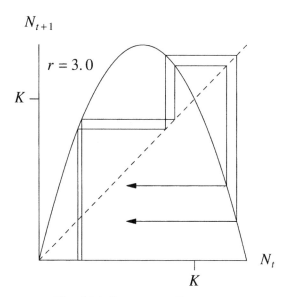

Fig. 4.26. Long-term divergence.

For $r > r_\infty$, we find a densely interwoven region with both periodic and chaotic orbits. Figures 4.23 and 4.24 show part of the chaotic trajectory that arises for $r = 3.0$. I refer to this trajectory as *chaotic* for two reasons. First, it is aperiodic. However long I run it, it will not repeat itself (once allowances are made for the finite precision of my computer). In addition, it exhibits *sensitive dependence on initial conditions*. This feature is highlighted in Figures 4.25, 4.26, and 4.27.

Figure 4.25 emphasizes the fact that a nonlinear mapping such as the logistic difference equations will either stretch or shrink small differences in initial conditions (e.g., measurement errors) depending on the magnitude of the local derivative. For a chaotic trajectory, expansion, on average, outweighs contraction so that nearby trajectories begin to diverge (see Figure 4.26). This sensitivity on initial conditions can be dramatic. Figure 4.27 shows two time series that result from simulating the logistic difference equation (with $r = 3.0$) with two initial conditions (0.05 and 0.050001) that differ by one in the sixth decimal place. The two time series track each other for about 15 iterates. After that, they diverge. The existence of sensitive dependence on initial conditions in simple (but chaotic) models implies that we have lost our ability to make long-term predictions for these models. We can still make short-term predictions, but small measurement error or round-off error will eventually become magnified to the point that long-term predictions are worthless.

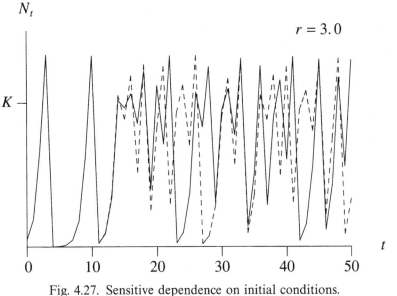

Fig. 4.27. Sensitive dependence on initial conditions.   ◇

Problem 4.2 *Ricker (1954) spawner–recruit curve*

Consider the density-dependent difference equation

$$N_{t+1} = N_t \, e^{r[1-(N_t/K)]}. \tag{4.41}$$

Find the fixed points for this mapping. Determine their stability. Draw this mapping for one or two different values of r and iterate by cobwebbing.

Problem 4.3 *Generalized Beverton–Holt curve*

Consider a generalized Beverton–Holt curve of the form

$$N_{t+1} = \frac{R_0^{\beta} \, N_t}{\{1 + [(R_0 - 1)/K] \, N_t\}^{\beta}}. \tag{4.42}$$

Find the fixed points of this mapping. Determine their stability. Determine and sketch the portions of the (R_0, β) plane that correspond to monotonic damping to the carrying capacity, oscillatory damping to the carrying capacity, and instability of the carrying capacity.

Historical hiatus

Ecologists' perceptions about mathematical models were long colored by the population regulation debate of the 1950s and 1960s. Earlier, in Chapter 1, I described this as a debate between density-independent and density-dependent mechanisms of population regulation. However, this was also a debate about methods of modeling. The density-independent school often viewed nature as best described by stochastic models: density-independent factors such as vagaries in the weather were best thought of as unpredictable or random events that led to noisy or erratic population time series. The density-dependent school, in turn, favored simple deterministic models, such as the logistic differential equation, that exhibited equilibrial behavior. Increasing r in the logistic difference equation is tantamount to increasing density dependence. So, it came as something of a shock for ecologists to realize that extremes in Nicholsonian density dependence could lead to chaotic dynamics that looked, at least superficially, like the stochastic and noisy time series of density-independent regulation. Most ecologists were exposed to chaos through the pioneering papers of Robert May (May, 1974, 1975, 1976; May and Oster, 1976). Ecologists' belief in the importance of chaos has waxed and waned through time (Hassell *et al.*, 1976; Schaffer and Kot, 1986a; Berryman and Millstein, 1989; Hastings *et al.*, 1993).

Mathematical meanderings

Our discussion of periodic orbits and sensitive dependence on initial conditions can be made more formal.

Definition N_0 is called a *point of period n* or an *n-cycle point* if it is a fixed point of the n-fold composed mapping,

$$f^n(N_0) \equiv f \circ f \circ f \circ \cdots \circ f(N_0) = N_0. \tag{4.43}$$

Definition   If N_0 is an *n*-cycle point, then the orbit based at N_0,

$$\{N_0, f(N_0), \ldots, f^{n-1}(N_0)\} = \{N_0, N_1, \ldots, N_{n-1}\}, \tag{4.44}$$

is called a *periodic orbit of period n* or an *n-cycle*.

Theorem   Let f be a C^1 mapping. An *n*-cycle point N_0 is *asymptotically stable* if

$$\left| \frac{df^n}{dN}(N_0) \right| < 1 \tag{4.45}$$

and is *unstable* if

$$\left| \frac{df^n}{dN}(N_0) \right| > 1. \tag{4.46}$$

Proof Since an *n*-cycle point is simply a fixed point of an *n*-fold composed map, we may use our earlier criterion, for the stability of equilibria, directly on the *n*-fold composed mapping. ☐

If a periodic point is stable, the whole orbit is stable. For the 2-cycle point N_0, the chain rule implies that

$$\left|\frac{df^2}{dN}(N_0)\right| = \left|\frac{df}{dN}(f(N_0)) \cdot \frac{df}{dN}(N_0)\right|$$

$$= \left|\frac{df}{dN}(N_1)\right| \cdot \left|\frac{df}{dN}(N_0)\right|. \tag{4.47}$$

The chain rule also gives the same product for N_1. For an *n*-cycle,

$$\left|\frac{df^n}{dN}(N_0)\right| = \prod_{i=0}^{n-1} \left|\frac{df}{dN}(N_i)\right|. \tag{4.48}$$

The *n*th root of this quantity,

$$\Lambda = \left(\prod_{i=0}^{n-1} \left|\frac{df}{dN}(N_i)\right|\right)^{1/n}, \tag{4.49}$$

is the geometric mean rate of stretching or shrinking along the entire orbit and is called the *characteristic (or Floquet) multiplier*. The associated logarithm,

$$\lambda = \frac{1}{n} \sum_{i=0}^{n-1} \ln\left(\left|\frac{df}{dN}(N_i)\right|\right), \tag{4.50}$$

is the *characteristic (or Floquet) exponent*.

We are also interested in chaotic orbits. These may be thought of as periodic orbits in the limit as the period goes to infinity. It is also natural to define a generalization of the characteristic exponent. This new exponent is called a *Lyapunov exponent*.

Definition   If

$$\lambda_{N_0} = \lim_{n \to \infty} \frac{1}{n} \sum_{i=0}^{n-1} \ln\left(\left|\frac{df}{dN}(N_i)\right|\right) \tag{4.51}$$

exists, we refer to it as the *Lyapunov exponent* of the orbit based at N_0. If a group of solutions (e.g., with different initial conditions) are attracted to the same attractor, then their Lyapunov exponents will be the same and we can drop the subscript.

The Lyapunov exponent is both a measure of sensitive dependence and an indicator of chaos. If it is positive, small differences are, on average, magnified and the resulting orbit is chaotic. Figure 4.28 shows a plot of the Lyapunov

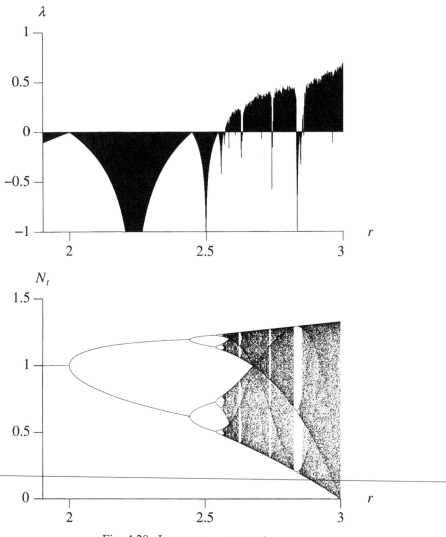

Fig. 4.28. Lyapunov exponents.

exponent, for different values of r, atop the corresponding bifurcation diagram. The chaotic regions are easy to identify.

Problem 4.4 *Two-cycles*

Determine the 2-cycle points of the logistic difference equation, as a function of r, analytically. Determine the range of r where the 2-cycle is asymptotically stable.

Hint The two-fold composition of the logistic difference equation yields a quartic equation and we are interested in the fixed points of this quartic. Fortunately, the trivial equilibrium and carrying capacity are 'trivial' 2-cycles (why?) and may be factored out of the quartic.

Recommended readings

May and Oster (1976) provide a wonderful introduction to the complex dynamics that can arise in simple ecological models. Chapter 8 of Berge *et al.* (1984), chapter 3 of Schuster (1988), and chapter 4 of Jackson (1989) are well-written summaries of the dynamic properties of simple difference equations. Elaydi (1996) is a recent textbook devoted to difference equations.

5 Delay models

Why were the nonlinear difference equations in the last chapter so wild? One answer is that they were wild because of an implicit time lag. Systems governed by difference equations cannot respond instantaneously. They must wait until the next discrete time to make 'course corrections'. Solutions may thus overshoot equilibria, causing the dramatic sequence of damped oscillations, periodic orbits, and chaotic orbits that we observed.

If implicit delays cause instability, what are the effects of explicit delays? Do they aggravate the problem? To answer these questions, we will consider simple models with both explicit and implicit lags. In particular, we will consider models of the form

$$N_{t+1} = f(N_t, N_{t-T}) = N_t F(N_{t-T}), \tag{5.1}$$

where T is an explicit lag in density dependence. Examples include the *lagged Beverton–Holt model*,

$$N_{t+1} = \frac{R_0 N_t}{1 + [(R_0 - 1)/K] N_{t-T}}, \tag{5.2}$$

the *lagged logistic difference equation*,

$$N_{t+1} = \left[1 + r \left(1 - \frac{N_{t-T}}{K} \right) \right] N_t, \tag{5.3}$$

and the *lagged Ricker curve*,

$$N_{t+1} = N_t \exp \left[r \left(1 - \frac{N_{t-T}}{K} \right) \right]. \tag{5.4}$$

These are really higher-order difference equations, but I will often refer to them as difference-delay equations (Levin and May, 1976).

I have written the examples so that

$$F(K) = 1 \tag{5.5}$$

and $N_t = N_{t-T} = K$. The carrying capacity is thus an equilibrium. To determine the stability of this equilibrium, let us proceed in the usual manner. Let

$$x_t \equiv N_t - K. \tag{5.6}$$

The linearization of equation (5.1) about the carrying capacity may now be written

$$x_{t+1} = \left.\frac{\partial f}{\partial N_t}\right|_K x_t + \left.\frac{\partial f}{\partial N_{t-T}}\right|_K x_{t-T}. \tag{5.7}$$

For the lagged logistic difference equation and the lagged Ricker curve,

$$x_{t+1} = x_t - r\,x_{t-T}. \tag{5.8}$$

Equation (5.8) is linear. If I try a solution of the form

$$x_t = x_0\,\lambda^t, \tag{5.9}$$

I am led to the characteristic equation

$$r = \lambda^T\,(1 - \lambda). \tag{5.10}$$

This algebraic equation has $T + 1$ roots λ_i. If the roots all satisfy

$$|\lambda_i| < 1, \tag{5.11}$$

the carrying capacity is asymptotically stable.

I will refer to the largest-modulus root as the *dominant* root. If it is real, positive, and smaller than 1 in modulus, we expect small perturbations about the carrying capacity to decay monotonically to zero. If it is complex or negative and smaller than 1 in modulus, we expect damped oscillations. If the dominant root has a modulus greater than 1, the equilibrium is unstable.

When do damped oscillations first appear? Let us look at the real roots of the characteristic equation. If we plot each side of equation (5.10) as a function of λ, the real roots occur at the intersection of the two curves (see Figure 5.1). For small positive r, there are two positive roots. The larger of the two positive roots is the dominant eigenvalue (Levin and May, 1976). As we increase r, the two roots converge. Eventually, they coalesce and disappear. Oscillations set in at the critical value, $r = r_0$, where the two positive roots disappear.

This critical value is the maximum of the right-hand side of characteristic equation (5.10) over the interval $0 < \lambda < 1$. To determine this maximum, we set

$$\frac{dr}{d\lambda} = T\lambda^{T-1} - (T + 1)\lambda^T = 0. \tag{5.12}$$

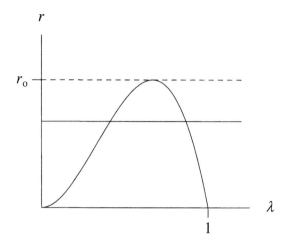

Fig. 5.1. Positive real roots.

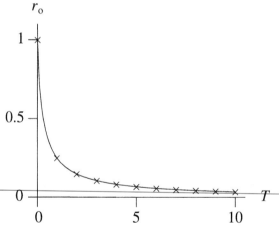

Fig. 5.2. Critical values for oscillations.

The maximum occurs at

$$\lambda = \frac{T}{T+1} \tag{5.13}$$

and is equal to

$$r_o = \left(\frac{T}{T+1}\right)^T \left(1 - \frac{T}{T+1}\right) = \frac{T^T}{(T+1)^{T+1}} \tag{5.14}$$

(see Figure 5.2). The critical value is a decreasing function of T that tends to zero in the limit of large delay,

$$\lim_{T \to \infty} r_o(T) = 0. \tag{5.15}$$

It thus becomes easier to set off oscillations as we increase the delay.

For $r > r_o$ and T odd, all of the roots of equation (5.10) are complex. For $r > r_o$ and T even, one real root lies between -1 and zero. In both cases, stability is lost as complex roots pass through the unit circle,

$$\lambda = e^{i\theta}, \tag{5.16}$$

in the complex plane. If we imagine that this occurs at $r = r_u$, we may rewrite characteristic equation (5.10) at this value as

$$\lambda = 1 - r_u \lambda^{-T}, \tag{5.17}$$

By equation (5.16), we have that

$$e^{+i\theta} = 1 - r_u e^{-i\theta T}. \tag{5.18}$$

If we multiply this equation by its complex conjugate,

$$e^{-i\theta} = 1 - r_u e^{+i\theta T}, \tag{5.19}$$

we quickly conclude that

$$r_u = 2 \cos \theta T. \tag{5.20}$$

If we plug the complex form of equation (5.20) back into equation (5.18) we obtain

$$e^{i\theta} = -e^{-2i\theta T} \tag{5.21}$$

or

$$e^{i\theta} = e^{[i(2n+1)\pi - 2i\theta T]}. \tag{5.22}$$

Therefore,

$$\theta = (2n + 1)\pi - 2\theta T. \tag{5.23}$$

Solving for θ yields

$$\theta = \frac{(2n + 1)}{(2T + 1)}\pi, \quad n = 0, 1, 2, \dots. \tag{5.24}$$

Substituting θ back into equation (5.20) for r_u produces

$$r_u = 2 \cos \frac{(2n + 1)\pi T}{(2T + 1)}. \tag{5.25}$$

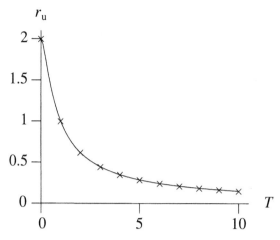

Fig. 5.3. Critical values for instability.

Finally, since we are interested in the first roots that cross the unit circle, we set $n = 0$ to obtain

$$r_u = 2 \cos \frac{\pi T}{2T + 1}. \tag{5.26}$$

A closer examination of equation (5.26) shows that r_u is a decreasing function of the lag T (see Figure 5.3). Thus, for both the lagged logistic equation and the lagged Ricker curve, it becomes easier to destabilize the nontrivial equilibrium as the lag increases.

Analysis of the lagged Beverton–Holt model shows that it too has the same increased susceptibility to damped oscillations and instability with increases in the lag. Explicit time lags can most certainly cause instability.

Problem 5.1 *The lagged Beverton–Holt model*
Analyze the lagged Beverton–Holt model, equation (5.2), and determine the values of R_0 that lead to oscillations and instability.

Mathematical meanderings

The lagged logistic difference equation (with a lag of 1) was introduced by Maynard Smith (1968) as a growth model for herbivorous populations. Aronson *et al.* (1980, 1982) examined this model in greater detail and showed that it can exhibit some truly fantastic dynamics. We have only scratched the surface with our equilibrium stability analysis!

Let

$$u_t \equiv \frac{r}{(1+r)K} N_t, \tag{5.27a}$$

$$a \equiv 1 + r. \tag{5.27b}$$

The lagged logistic difference equation, with $T = 1$, may now be rewritten as

$$u_{t+1} = a u_t (1 - u_{t-1}). \tag{5.28}$$

If we let

$$x_t = u_{t-1}, \tag{5.29a}$$

$$y_t = u_t, \tag{5.29b}$$

our second-order difference equation may be rewritten as a system of two first-order difference equations,

$$x_{t+1} = y_t, \tag{5.30a}$$

$$y_{t+1} = a y_t (1 - x_t). \tag{5.30b}$$

This and similar systems are often referred to as mappings of the plane: each point in the plane, (x_t, y_t), has a forward image, (x_{t+1}, y_{t+1}), under the mapping. The recursive iteration of this mapping, starting with some initial (x_0, y_0), leads to a sequence of points (or a *positive semi-orbit*) in the plane. What do these orbits look like?

We know, from our previous analysis, that equations (5.30a) and (5.30b) have an equilibrium (a fixed point of the mapping) at $[(a - 1)/a, (a - 1)/a]$, that small perturbations about this equilibrium exhibit oscillations starting with $a = 1.25$ ($r_o = 0.25$), and that this equilibrium loses stability at $a = 2$ ($r_u = 1$). Figure 5.4 shows a typical orbit for $a = 1.95$. Figure 5.5 shows the corresponding time series. Since this is a discrete-time model, both the orbit and the time series consist of a set of distinct points (represented by the open circles); I have included lines connecting these points to make the patterns clearer. Both figures show damped oscillations that approach the stable equilibrium.

For $T = 1$ and $a = 1 + r$, the roots of characteristic equation (5.10) are

$$\lambda(a) = \frac{1 \pm \sqrt{5 - 4a}}{2}. \tag{5.31}$$

For $1 < a < 2$, $|\lambda(a)| < 1$. However, at $a = 2$, two roots pass through the unit circle in the complex plane at two of the sixth roots of unity,

$$\lambda(2) = \frac{1 \pm i\sqrt{3}}{2}. \tag{5.32}$$

Think of the local dynamics as a six-fold rotation of the plane. For $a > 2$, the nontrivial equilibrium is repelling. For sufficiently small $a > 2$ and for large enough distances from the equilibrium, this repulsion is balanced by nonlinear terms that we ignored in our linearization. The result is a small attracting

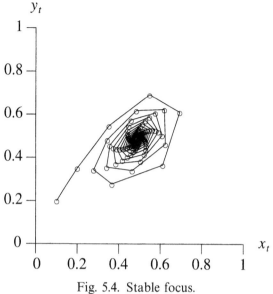

Fig. 5.4. Stable focus.

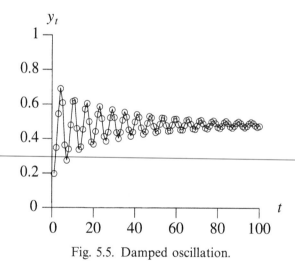

Fig. 5.5. Damped oscillation.

invariant circle in the (x_t, y_t) plane. (Circle is used loosely here for anything topologically similar to a circle.) This circle arises via a *Hopf bifurcation*. I will discuss this bifurcation more fully in later chapters.

The invariant circle grows as we increase *a*. It also changes shape. Figure 5.6 shows invariant circles for several increasing values of *a*. (I had to be selective in how I chose *a* to get this nice sequence of circles – more on this later.) Figure 5.7, in turn, shows the time series corresponding to the attracting circle at *a* = 2.04. The time series is clearly oscillatory.

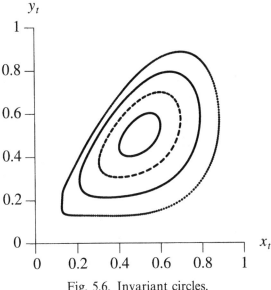

Fig. 5.6. Invariant circles.

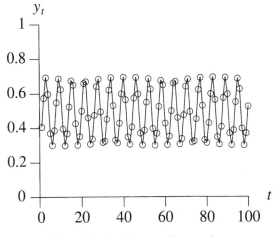

Fig. 5.7. Oscillatory time series.

For large *a*, the attracting invariant circles vanish. Indeed, for *a* larger than $a^* \approx 2.271$ there is no attractor in the first quadrant. This dramatic change is linked to a *homoclinic tangency*. In effect, an unstable manifold emanating from the origin intersects the stable manifold for the origin at a point of tangency on the positive *x*-axis. The forward image of the point of tangency is the origin; that portion of the attractor between the point of tangency and the origin is mapped into a loop, with both ends tethered at the origin. This loop is stretched out along the unstable manifold by subsequent iterates of the mapping.

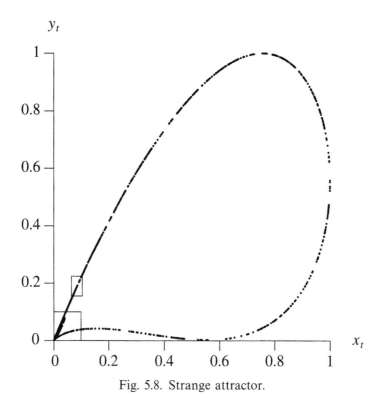

Fig. 5.8. Strange attractor.

Figure 5.8 shows the attractor for $a = 2.27$, just prior to the homoclinic tangency. Superficially, this attractor looks like an invariant circle. However, it is far more complicated. The incipient tangency, and the extra loop in the lower left-hand corner (enlarged in Figure 5.9) are clearly evident. Enlargements of the second small box in Figure 5.8 (see Figures 5.10, 5.11, and 5.12) are even more revealing. They show that this attractor has a fractal 'filo dough' structure – corresponding to the stretching and folding of the loop out along the unstable manifold.

The breakdown and loss of quasiperiodic invariant circles and the resulting increase in complexity begins prior to $a = 2.27$. Figure 5.13 shows a bifurcation diagram for equations (5.30a) and (5.30b). I iterated this system for a thousand steps for each of 370 values of the parameter a between 1.9 and 2.27. I plotted the last 250 iterated values of y_t for each value of a. This gives us a one-dimensional projection of the attractor for each a. An equilibrium appears as a single point, a quasiperiodic invariant circle gives rise to a dense line interval, and a simple periodic orbit gives rise to a small number of distinct points.

Looking at the bifurcation diagram, we see a Hopf bifurcation at $a = 2.0$. Below $a = 2.0$, there is a simple stable equilibrium. From $a = 2.0$ to about $a \approx 2.18$, there is a growing invariant circle. At $a \approx 2.18$, solutions lock on to a simple 7-cycle. From then on (see Figures 5.14 and 5.15), the pattern is quite

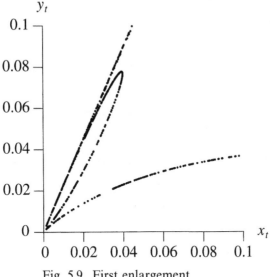

Fig. 5.9. First enlargement.

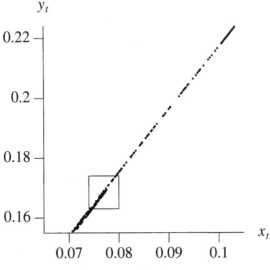

Fig. 5.10. Second enlargement.

complicated: quasiperiodic invariant circles give way to periodic solutions that period double or that 'period bubble' and revert back to quasiperiodic orbits. Beyond $a \approx 2.235$, there is no obvious sign of quasiperiodicity.

The attractor in Figure 5.8 is chaotic. Bifurcation diagrams 5.13, 5.14, and 5.15 also show a great deal of chaotic dynamics. In many ways, though, this chaos

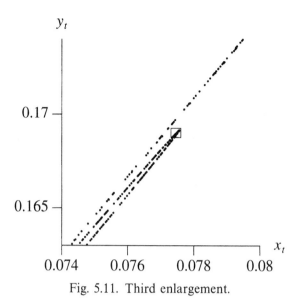

Fig. 5.11. Third enlargement.

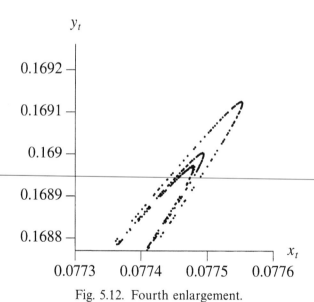

Fig. 5.12. Fourth enlargement.

is quite innocuous. A typical stretch of Figure 5.8's time series (Figure 5.16) could easily be misidentified as periodic.

I have argued that explicit lags in discrete-time models can lead to instability and to complex or chaotic dynamics. Crone (1997) sheds additional light

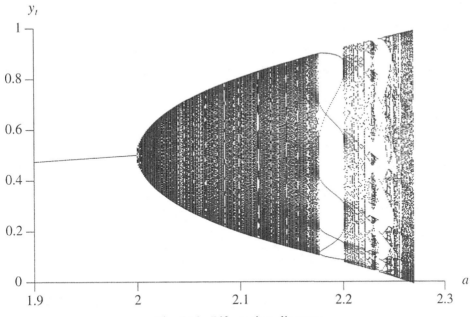

Fig. 5.13. Bifurcation diagram.

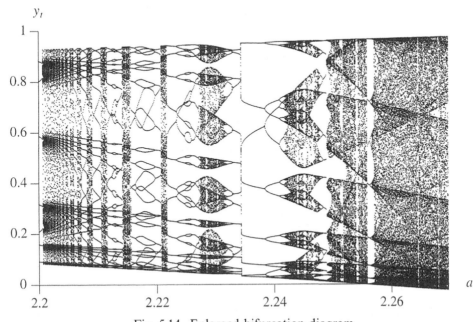

Fig. 5.14. Enlarged bifurcation diagram.

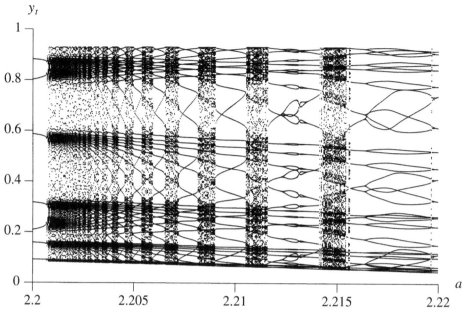

Fig. 5.15. Greatly enlarged bifurcation diagram.

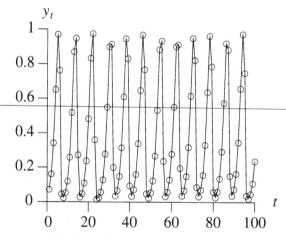

Fig. 5.16. Subtly chaotic time series.

on this topic. Turchin (1990; Turchin and Taylor, 1992) has argued that these lags also play an important role in the analysis of empirical data sets. In 1990, he examined the population dynamics of 14 forest insects; he showed that eight insects exhibited clear evidence of delayed density dependence and lag-induced oscillations. In contrast, he detected direct (unlagged) density dependence in only one population.

We have concentrated on discrete-time models. However, it is appropriate to ask whether explicit time lags can also have a destabilizing effect on continuous-time models? To answer this question, I will start with our old friend the logistic differential equation,

$$\frac{dN}{dt} = r N \left(1 - \frac{N}{K}\right). \tag{5.33}$$

There are several ways of reformulating this model so as to incorporate delays. The first is to rewrite equation (5.33) as a *delay-differential equation*,

$$\frac{dN}{dt} = r N(t) \left[1 - \frac{N(t - \tau)}{K}\right], \tag{5.34}$$

with a single explicit lag of exactly τ time units. (Delay-differential equation is, arguably, the common English designation. However, these equations are also referred to as *differential-difference equations* or as *functional differential equations*. The common Russian name for these equations translates roughly to *differential equations with deviating arguments of the retarded type*.) Equation (5.34) is frequently referred to as the *Hutchinson–Wright equation* after G. Evelyn Hutchinson (1948), a well-known limnologist and ecologist, and E. M. Wright (1946), a number-theorist.

An alternative approach to incorporating delays is to rewrite equation (5.34) as a (*Volterra*) *integrodifferential equation*,

$$\frac{dN}{dt} = r N(t) \int_0^\infty k(\tau) \left[1 - \frac{N(t - \tau)}{K}\right] d\tau, \tag{5.35}$$

where

$$\int_0^\infty k(\tau)\, d\tau = 1. \tag{5.36}$$

Here we weigh history with the kernel $k(\tau)$. This kernel can take many different forms.

Let us start with the Hutchinson–Wright equation,

$$\frac{dN}{dt} = r N(t) \left[1 - \frac{N(t - \tau)}{K}\right], \tag{5.37a}$$

$$N(t) = N_0(t), \quad -\tau \le t \le 0. \tag{5.37b}$$

Note that we need to specify an initial condition on an entire interval of length τ. This is our first clue that this equation is, in some sense, analogous to an infinite-order ordinary differential equation.

Equation (5.37a) possesses two equilibria, $N = 0$ and $N = K$. I will

Table 5.1. *Dynamic regimes*

	Behavior
$0 < r\tau < \frac{1}{e}$	Monotonically damped stable equilibrium
$\frac{1}{e} < r\tau < \frac{\pi}{2}$	Oscillatorily damped stable equilibrium
$r\tau > \frac{\pi}{2}$	Unstable equilibrium

ascertain the stability of these equilibria in the usual manner. Near the trivial equilibrium,

$$\frac{dN}{dt} \approx r N. \tag{5.38}$$

Thus, for positive r and at low densities, the population grows exponentially fast; the trivial equilibrium is unstable.

For the second equilibrium, I introduce a variable that measures the deviation from the carrying capacity,

$$x(t) \equiv N(t) - K, \tag{5.39}$$

and linearize,

$$\frac{dx}{dt} = \left.\frac{\partial f}{\partial N(t)}\right|_{N=K} x(t) + \left.\frac{\partial f}{\partial N(t-\tau)}\right|_{N=K} x(t-\tau), \tag{5.40}$$

to obtain

$$\frac{dx}{dt} = -r\,x(t-\tau). \tag{5.41}$$

Since this is a linear equation (albeit a linear delay-differential equation), it seems reasonable to try an exponential solution,

$$x(t) = x_0\,e^{\lambda t}. \tag{5.42}$$

This produces a transcendental characteristic equation,

$$\lambda = -r\,e^{-\lambda\tau}, \tag{5.43}$$

with an infinite number of complex roots λ.

What does transcendental equation (5.43) tell us about the stability of the carrying capacity? One can show that there are, in fact, three different regimes (see Table 5.1), depending on the magnitude of the product $r\tau$. It is rather remarkable that the thresholds in Table 5.1 involve constants such as e and π. Let us see where these thresholds come from.

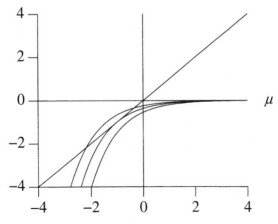

Fig. 5.17. Loss of real roots.

Let us first make the change of variables $\mu \equiv \lambda\tau$ so that the characteristic equation takes the form

$$\mu = -r\tau e^{-\mu}. \tag{5.44}$$

Since we are dealing with positive delay τ, the real part of μ will always have the same sign as the real part of λ; they are equally good indicators of stability. By plotting the left- and the right-hand sides of equation (5.44) as real functions of μ, we obtain Figure 5.17. There may be zero, one, or two intersections between the diagonal line and the exponential on the right-hand side of equation (5.44), depending upon the magnitude of $r\tau$. A double root separates a regime with two distinct real roots from one with no real roots; it also marks the transition from exponential to oscillatory decay. Since the two curves are tangent at this double root, we have that

$$1 = r\tau e^{-\mu}. \tag{5.45}$$

Equations (5.44) and (5.45) together imply that

$$\mu = -1 \tag{5.46}$$

and hence that

$$r\tau = \frac{1}{e}. \tag{5.47}$$

Damped oscillations about the carrying capacity give way to growing oscillations away from the carrying capacity once a pair of complex conjugate roots cross the imaginary axis. If we place a root on the imaginary axis,

$$\mu = i\omega, \tag{5.48}$$

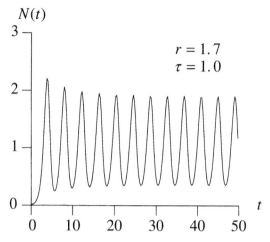

Fig. 5.18. Approach to a periodic limit cycle.

equation (5.44) reduces to

$$i\omega = -r\tau(\cos\omega - i\sin\omega). \qquad (5.49)$$

By equating the real and the imaginary parts of equation (5.49), we obtain

$$0 = -r\tau\cos\omega, \qquad (5.50a)$$

$$\omega = r\tau\sin\omega. \qquad (5.50b)$$

We can now solve for ω and for $r\tau$:

$$\omega = \left(n + \frac{1}{2}\right)\pi, \quad n = 0, 1, 2, \ldots, \qquad (5.51a)$$

$$r\tau = \frac{\omega}{\sin\omega}. \qquad (5.51b)$$

Since I am interested in the earliest traversal of the imaginary axis, I will go ahead and set $n = 0$. The carrying capacity loses stability at $r\tau = \pi/2$.

To determine the behavior of the Hutchinson–Wright equation for $r\tau > \pi/2$, we must resort to numeric simulations. Hoppensteadt (1993) has described an extremely useful method for integrating delay-differential equations. I obtained a solution to equation (5.37a) for $r = 1.7$, $\tau = 1$, and $N_0(t) = 0.01$ by means of this method (see Figure 5.18). The solution rapidly converges to a periodic limit cycle. Other parameter values also give limit cycles. Indeed, there is no sign of anything wilder than a stable limit cycle for any $r\tau > \pi/2$. We do not pick up any of the dynamic complexity and chaos that occurred for the lagged logistic difference equation. However, an explicit lag in this continuous-time system has caused instability and oscillations.

Many delay-differential equations have appeared in the ecological literature. For example, Gurney *et al.* (1980) used a model with delayed recruitment and instantaneous death,

$$\frac{dN}{dt} = \alpha N(t - \tau) e^{-\beta N(t-\tau)} - \delta N(t), \tag{5.52}$$

to describe the periodic oscillations in Nicholson's (1954) classic laboratory experiments with the Australian sheep blowfly, *Lucilia cuprina*. May (1980), in turn, studied the somewhat similar

$$\frac{dN}{dt} = \mu N(t - \tau) \left(1 + q \left\{ 1 - \left[\frac{N(t - \tau)}{K} \right]^z \right\} \right) - \mu N(t) \tag{5.53}$$

as a model for the population dynamics of baleen whales. Interestingly enough, equation (5.53) can exhibit chaotic dynamics for sufficiently high z.

Problem 5.2 *Nicholson's blowflies revisited*
Consider the following model for a laboratory population of blowflies:

$$\frac{dN}{dt} = \alpha N(t - \tau) e^{-\beta N(t-\tau)} - \delta N(t). \tag{5.54}$$

(1) Determine the nontrivial equilibrium. Linearize about the nontrivial equilibrium.
(2) From your linearization, draw boundaries in the $(\delta\tau, \alpha\tau)$ plane that distinguish between regions where small perturbations:
   (a) grow monotonically,
   (b) decay monotonically,
   (c) decay in an oscillatory manner,
   (d) grow in an oscillatory manner.

We may also consider the Volterra integrodifferential equation

$$\frac{dN}{dt} = r N(t) \int_0^\infty k(\tau) \left[1 - \frac{N(t - \tau)}{K} \right] d\tau, \tag{5.55a}$$

$$N(t) = N_0(t), \quad -\infty < t \le 0, \tag{5.55b}$$

with

$$\int_0^\infty k(\tau) \, d\tau = 1. \tag{5.56}$$

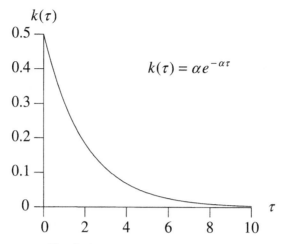

Fig. 5.19. Exponential distribution.

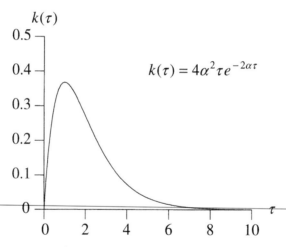

Fig. 5.20. Humped distribution.

The kernel $k(\tau)$ specifies how we weigh the past. A simple exponential distribution (see Figure 5.19) assigns great weight to recent events. Figure 5.20, in turn, shows a humped kernel that assigns greater weight to events in the past. It is tempting to think of the first distribution as indicating a weak delay and the second as indicating a strong delay. In reality, both distributions have been chosen so that the mean delay is $\tau_{\text{average}} = 1/\alpha$.

The kernels in Figures 5.19 and 5.20 are special cases of a family of kernels of the form

$$k(\tau) = \frac{\alpha n (\alpha n \tau)^{n-1}}{(n-1)!} e^{-\alpha n \tau}. \tag{5.57}$$

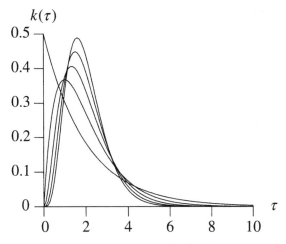

Fig. 5.21. Gamma distributions.

This family of gamma distributions (see Figure 5.21) has been written in such a way that the mean delay for each n is $\tau_{average} = 1/\alpha$ and so that the peak in the distribution occurs at $\tau = (n - 1)/(n\alpha)$. As n increases, this distribution becomes progressively more peaked, with the mean and mode in the delay converging. In the limit as n goes to infinity, the distribution tends towards a δ function; our integrodifferential equation then reduces to a delay-differential equation with delay $1/\alpha$.

Equation (5.55a) possesses two equilibria, $N = 0$ and $N = K$. To ascertain the stability of the carrying capacity, we introduce

$$x(t) \equiv N(t) - K \tag{5.58}$$

and linearize,

$$\frac{dx}{dt} = -r \int_0^\infty k(\tau) x(t - \tau) \, d\tau. \tag{5.59}$$

If we try an exponential solution of the form

$$x = x_0 e^{\lambda t}, \tag{5.60}$$

we obtain the characteristic equation

$$\lambda = -r \int_0^\infty k(\tau) e^{-\lambda \tau} \, d\tau. \tag{5.61}$$

In general, this will be a transcendental equation. However, for kernels that are of the form of equation (5.57), the resulting characteristic equation is algebraic.

Let us compare the stability of the carrying capacity for the kernels in Figures 5.19 and 5.20. For

$$k(\tau) = \alpha e^{-\alpha\tau}, \tag{5.62}$$

characteristic equation (5.61) reduces to

$$\lambda = -r\alpha \int_0^\infty e^{-(\alpha+\lambda)\tau} d\tau \tag{5.63}$$

or

$$\lambda = -\frac{r\alpha}{\alpha+\lambda}. \tag{5.64}$$

This may be written more conveniently as

$$\lambda^2 + \alpha\lambda + \alpha r = 0. \tag{5.65}$$

If α and αr are both positive, the Routh–Hurwitz criterion† guarantees us that all roots λ of equation (5.63) have negative real part and that the carrying capacity is asymptotically stable.

If we now turn to the kernel

$$k(\tau) = 4\alpha^2\tau e^{-2\alpha\tau} \tag{5.66}$$

---

† The Routh–Hurwitz criterion (cf. Murray, 1989; Edelstein-Keshet, 1988; Pielou, 1977) states that a necessary and sufficient condition that the equation

$$x^n + a_1 x^{n-1} + \cdots + a_n = 0,$$

(with real coefficients) have only roots with negative real part is that the values of the determinants of the matrices

$$H_1 = (a_1), \quad H_2 = \begin{pmatrix} a_1 & 1 \\ a_3 & a_2 \end{pmatrix}, \quad H_3 = \begin{pmatrix} a_1 & 1 & 0 \\ a_3 & a_2 & a_1 \\ a_5 & a_4 & a_3 \end{pmatrix}, \ldots,$$

$$H_j = \begin{pmatrix} a_1 & 1 & 0 & 0 & \cdot & 0 \\ a_3 & a_2 & a_1 & 1 & \cdot & 0 \\ a_5 & a_4 & a_3 & a_2 & \cdot & 0 \\ \cdot & & & & \cdot & \\ \cdot & & & & \cdot & \\ a_{2j-1} & a_{2j-2} & a_{2j-3} & a_{2j-4} & \cdot & a_j \end{pmatrix}, \ldots, \quad H_n = \begin{pmatrix} a_1 & 1 & 0 & \cdot & \cdot & 0 \\ a_3 & a_2 & a_1 & \cdot & \cdot & \cdot \\ \cdot & & & & & \cdot \\ \cdot & & & & & \cdot \\ \cdot & & & & & \cdot \\ 0 & \cdot & \cdot & \cdot & \cdot & a_n \end{pmatrix}$$

all be positive. Here, the (l, m) term in the matrix H_j is

$$\begin{array}{ll} a_{2l-m}, & \text{for } 0 < 2l - m < n, \\ 1, & \text{for } 2l = m, \\ 0, & \text{for } 2l < m \text{ or } 2l > n + m. \end{array}$$

For quadratic and cubic polynomials, these conditions reduce to $n = 2$:

$$a_1 > 0, \quad a_2 > 0$$

$n = 3$:

$$a_1 > 0, \quad a_3 > 0, \quad a_1 a_2 > a_3.$$

(see Figure 5.20), characteristic equation (5.61),

$$\lambda = -4 r\alpha^2 \int_0^\infty \tau\, e^{-(\lambda + 2\alpha)\tau}\, d\tau, \tag{5.67}$$

reduces to

$$\lambda = -\frac{4 r\alpha^2}{(\lambda + 2\alpha)^2}, \tag{5.68}$$

which may be rewritten

$$\lambda^3 + 4\alpha\lambda^2 + 4\alpha^2\lambda + 4 r\alpha^2 = 0. \tag{5.69}$$

If we apply the Routh–Hurwitz criterion to this cubic polynomial equation, we find that three conditions,

(1) $a_1 > 0$

$$4\alpha > 0, \tag{5.70}$$

(2) $a_3 > 0$

$$4 r\alpha^2 > 0, \tag{5.71}$$

(3) $a_1 a_2 > a_3$

$$16\alpha^3 > 4 r\alpha^2, \tag{5.72}$$

must be satisfied for asymptotic stability of the carrying capacity. The first two conditions are trivially satisfied. The third condition implies that we must have

$$\frac{r}{\alpha} < 4 \tag{5.73}$$

or

$$r\, \tau_{\text{average}} < 4 \tag{5.74}$$

for asymptotic stability of the carrying capacity. Rapid growth or an increase in the average delay (for this kernel) can lead to instability. Descartes's rule of signs† may be used to check for the number of positive and negative real roots. This simple rule is enough to convince us that there are never any positive real eigenvalues and that stability is lost as a complex conjugate pair of eigenvalues crosses the imaginary axis. This will cause oscillations to arise.

The last two examples imply that it is not the average size of the delay

---

† Descartes's rule of signs states that the number of positive real roots of a real algebraic (polynomial) equation is either equal to the number of sign changes in the sequence $1, a_1, \ldots, a_n$ of coefficients (where vanishing terms are disregarded), or is less than this number by a positive even integer. Replacing x with $-x$ yields a similar rule for negative real roots.

per se that determines the onset of instability. Rather, it is the way this delay is distributed. As we increase n and sharpen the delay in family (5.57), it becomes easier to destabilize the carrying capacity. In the limit as n tends towards infinity, the kernel tends towards a delta function and we recapture the instability threshold $r\,\tau_{\text{average}} = \pi/2$ that we derived for the Hutchinson–Wright equation.

Recommended readings

There are many books devoted to the theory of delay-differential equations. These include Bellman and Cooke (1963), El'sgol'ts and Norkin (1973), Hale (1977), and Kolmanovskii and Nosov (1986). Cushing (1977a), MacDonald (1978, 1989), and chapter 6 of Banks (1994) provide useful introductions to delay-differential or to integrodifferential models in biology. Gopalsamy (1992) and Kuang (1993) are advanced treatments on delay-differential equations with applications to population dynamics.

6 Branching processes

In Chapter 3, I introduced birth and death processes as the stochastic analogs of simple differential equations. In a similar spirit, one can ask for a stochastic counterpart to the simple linear difference equation,

$$N_{t+1} = R_0 N_t, \qquad (6.1)$$

that I introduced at the start of Chapter 4.

Earlier, we imagined that each individual left R_0 offspring. Let us now follow the lead of the Reverend H. W. Watson and of Francis Galton (1874), as paraphrased by Harris (1963), and introduce an element of chance into this formula:

Let p_0, p_1, p_2, ... be the respective probabilities that a man has 0, 1, 2, ... sons, let each son have the same probability for sons of his own, and so on. What is the probability that the male line is extinct after r generations, and more generally what is the probability for any given number of descendants in the male line in any given generation?

Watson and Galton were interested in the extinction of family names. But to solve their problem, we must develop a generation-by-generation description of the growth of an arbitrary population. In particular, we must determine the population's size, N_t, in each generation t. The problem is challenging in that N_t is now a random variable with a discrete parameter (time) and a countable state space.

The Galton–Watson process is a simple, discrete, branching process. As stated, this process requires

(1) that we start with a single individual,

$$N_0 = 1; \qquad (6.2)$$

(2) that the number of offspring sired by that individual be a discrete random

variable of given distribution,

$$P(N_1 = n) = p_n, \text{ with } \sum_{n=0}^{\infty} p_n = 1; \tag{6.3}$$

and

(3) that the conditional distribution of N_{t+1}, given $N_t = n$, is the sum of n independent variables, each with the same distribution as N_1.

Condition (3) makes the Galton–Watson process a Markov chain. This condition would be specious if, say, infertility is genetic, so that fewer siblings meant fewer children. Condition (3) also implies that our process is density-independent and that individuals do not interfere with one another.

Probability generating functions played a pivotal role in our analysis of birth and death processes and it should come as no surprise that they will play a crucial role in our analysis of the Galton–Watson process. Indeed, we will make repeated use of the probability generating function

$$F(x) = E(x^{N_1}) = \sum_{n=0}^{\infty} p_n x^n. \tag{6.4}$$

To see why, let me list two new properties of probability generating functions to supplement those that you learned in Chapter 3.

Theorem Let F and G be two independent random variables with probability mass functions f_k and g_k and generating functions $F(x)$ and $G(x)$. For the new random variable $H \equiv F + G$ with probability mass function h_k, the probability generating function $H(x)$ is given by

$$H(x) = F(x) G(x). \tag{6.5}$$

Proof

$$H(x) = \sum_{k=0}^{\infty} h_k x^k \tag{6.6}$$

$$= \sum_{k=0}^{\infty} \left(\sum_{i=0}^{k} f_i g_{k-i} \right) x^k \tag{6.7}$$

$$= \sum_{i=0}^{\infty} \sum_{k=i}^{\infty} f_i g_{k-i} x^k \tag{6.8}$$

$$= \sum_{i=0}^{\infty} f_i x^i \sum_{k=i}^{\infty} g_{k-i} x^{k-i} \tag{6.9}$$

$$= F(x) G(x). \tag{6.10}$$

$\square$

This result is easily extended to the sum of n random variables: the probability generating function for the sum of n variables is simply the corresponding n-fold product of each individual generating function. A more interesting case arises if n is itself the outcome of a random experiment.

Theorem Let S_N be the sum of N independent, identically distributed, random variables with the common probability generating function $G(x)$. If N (≥ 0) is an independent random variable with probability generating function $F(x)$ (so that we are taking the sum of a random number of identical random variables), the probability generating function $H(x)$ of S_N is given by

$$H(x) = F(G(x)). \qquad (6.11)$$

Proof Let p_n be the probability mass function for N. Consider n fixed. Then, by our previous theorem, the generating function for the sum of n independent and identically distributed random variables is $[G(x)]^n$. If we now condition on n, the probability generating function for S_N is simply

$$\sum_{n=0}^{\infty} p_n [G(x)]^n. \qquad (6.12)$$

However, this is just $F(x)$, with x replaced by $G(x)$. $\qquad \square$

Rather than computing each probability of the Galton–Watson process individually, let us instead determine the probability generating function $F_t(x)$ for each generation t. By the first of our requirements for a Galton–Watson process, we have that

$$F_0(x) = x. \qquad (6.13)$$

This simply reiterates that we are starting with an individual.

Our second requirement was that the number of offspring sired by this individual be a random variable with probability mass function p_n. The probability generating function for the first generation is, as a result, given by equation (6.4),

$$F_1(x) = F(x). \qquad (6.14)$$

How about the next generation? Well, at the end of the first generation we have a random number of individuals. Each individual will die and leave a random number of offspring. We must thus sum a random number of random variables, each with probability generating function $F(x)$, to count the second generation. Thus, by our second theorem,

$$F_2(x) = F[F_1(x)] = F \circ F(x). \tag{6.15}$$

We can continue. With each new generation, we iterate with the generating function $F(x)$,

$$F_t(x) = F[F_{t-1}(x)] = F^t(x), \tag{6.16}$$

to obtain a t-fold composition in generation t. In principle, this probability generating function allows us to determine the probability distribution of population sizes in each generation. Only rarely, however, can the t-th iterate be found in a simple explicit form. Even so, equation (6.16) can still be used to find the moments of N_t in each generation in terms of the moments of $F(x)$, and to determine the odds of extinction.

The expected value of N_t is especially easy to compute. We need simply differentiate equation (6.16) at $x = 1$:

$$E(N_t) = \frac{dF_t}{dx}\bigg|_{x=1}. \tag{6.17}$$

Then, by the chain rule,

$$
\begin{align}
E(N_t) &= F'[F_{t-1}(1)] F'_{t-1}(1) \tag{6.18} \\
&= F'(1) F'_{t-1}(1) \tag{6.19} \\
&= R_0 F'_{t-1}(1) \tag{6.20} \\
&= R_0 E(N_{t-1}), \tag{6.21}
\end{align}
$$

where

$$R_0 = F'(1) = E(N_1). \tag{6.22}$$

By repeated application of equation (6.21), we quickly determine that

$$E(N_t) = R_0^t, \tag{6.23}$$

so that the population grows, on average, geometrically with a net reproductive rate equal to the mean number of offspring.

The variance of N_t is only slightly more difficult to compute. We know, from Chapter 3, that

$$\text{Var}(N_t) = \left[\frac{\partial^2 F_t}{\partial x^2} + \frac{\partial F_t}{\partial x} - \left(\frac{\partial F_t}{\partial x}\right)^2\right]_{x=1}. \tag{6.24}$$

The first term is clearly the challenge. We have already seen that

$$F_t(x) = F[F_{t-1}(x)], \tag{6.25}$$

so that

$$F_t'(x) = F'[F_{t-1}(x)] F_{t-1}'(x). \tag{6.26}$$

Differentiating once again gives us

$$F_t''(x) = F'[F_{t-1}(x)] F_{t-1}''(x) + F''[F_{t-1}(x)] [F_{t-1}'(x)]^2, \tag{6.27}$$

so that

$$F_t''(1) = F'(1) F_{t-1}''(1) + F''(1) [F_{t-1}'(1)]^2. \tag{6.28}$$

We may use equation (6.23) to write this more clearly as

$$F_{t+1}''(1) = R_0 F_t''(1) + R_0^{2t} F''(1). \tag{6.29}$$

Problem 6.1 *Linear nonautonomous difference equation*
Show that equation (6.29) has the solution

$$F_t''(1) = \begin{cases} \dfrac{(R_0^t - 1) R_0^t F''(1)}{R_0 (R_0 - 1)}, & R_0 \neq 1, \\ t F''(1), & R_0 = 1. \end{cases} \tag{6.30}$$

Hint   Let $F_t''(1) = R_0^t u_t$.

If $F(x)$ has variance σ^2, equation (6.24) implies that

$$F''(1) = \sigma^2 + R_0 (R_0 - 1). \tag{6.31}$$

By combining equations (6.24), (6.30), and (6.31), we find that

$$\text{Var}(N_t) = \begin{cases} \dfrac{(R_0^t - 1) R_0^t}{R_0 (R_0 - 1)} \sigma^2, & R_0 \neq 1, \\ \sigma^2 t, & R_0 = 1. \end{cases} \tag{6.32}$$

Higher moments can be found similarly.

Determining the chance that a lineage goes extinct – Galton's and Watson's original problem – has a nice geometric solution. The probability of being extinct in generation t is $F_t(0)$. In light of equation (6.16),

$$F_t(0) = F[F_{t-1}(0)] = F^t(0). \tag{6.33}$$

We may determine the chance of being extinct in any generation by recursively iterating the probability generating function with $x = 0$. However,

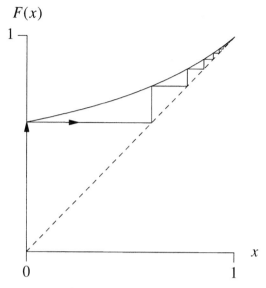

$F(x)$

Fig. 6.1. Subcritical case.

this is the same as cobwebbing (see Chapter 4) along probability generating function $F(x)$ with initial condition zero !

So, what does the probability generating function $F(x)$ look like? Well, since all the probabilities in definition (6.4) are nonnegative, $F(x)$ (and its derivative) must be a nondecreasing function on the closed interval $0 \leq x \leq 1$. Moreover $F(0) = p_0 \leq 1$ and $F(1) = 1$. For the special cases $p_0 = 0$ or $p_0 = 1$, the problem is uninteresting, so I will restrict our attention to $0 < p_0 < 1$. The probability generating function $F(x)$ may now take three, qualitatively different, forms, depending on the size of $R_0 = F'(1)$.

If $R_0 = F'(1) < 1$ (the subcritical case), $F(x)$ remains above the 45° line until it finally intersects this line transversely at $x = 1$ (see Figure 6.1). Cobwebbing now causes $F_t(0)$ to converge to 1: extinction will occur with probability 1. If this extinction is slow and gradual, we may look at the conditional distribution for the population size, conditional on no extinction. After one generation, the generating function for this condition distribution is

$$G(x) = \sum_{n=1}^{\infty} \frac{p_n}{1 - p_0} x^n. \tag{6.34}$$

The denominator may be factored out and the generating function may be rewritten

$$G(x) = \frac{1}{[1 - F(0)]} \sum_{n=1}^{\infty} p_n x^n, \tag{6.35}$$

which simplifies to

$$G(x) = \frac{F(x) - F(0)}{1 - F(0)} \tag{6.36}$$

or

$$G(x) = 1 + \frac{F(x) - 1}{1 - F(0)}. \tag{6.37}$$

By this line of argument, one can also show that

$$G_t(x) = 1 + \frac{F_t(x) - 1}{1 - F_t(0)} \tag{6.38}$$

in generation t. Thus, $G_t(x)$ is determined by $F(x)$. Moreover, by substituting $F(x)$ for x, we obtain

$$G_t[F(x)] = 1 + \frac{F_t[F(x)] - 1}{1 - F_t(0)}, \tag{6.39}$$

which I choose to rewrite as

$$G_t[F(x)] = 1 + \left[\frac{F_{t+1}(x) - 1}{1 - F_{t+1}(0)} \right] \left[\frac{1 - F_{t+1}(0)}{1 - F_t(0)} \right], \tag{6.40}$$

or, better yet, as

$$G_t[F(x)] = 1 + [G_{t+1}(x) - 1] \left[\frac{F_{t+1}(0) - 1}{F_t(0) - 1} \right]. \tag{6.41}$$

For large t, the second term in square brackets tends to R_0 (consider the linearization of the mapping $F(x)$ at $x = 1$). The limit of $G_t(x)$ for large t, $G^*(x)$, must thus satisfy the functional equation

$$G^*[F(x)] = 1 + [G^*(x) - 1] R_0. \tag{6.42}$$

If a population with a subcritical net reproductive rate has not gone extinct after some long period, it is probably in a 'stable' statistically quasistationary state given by this limiting distribution.

If $R_0 = F'(1) = 1$ (the critical case), $F(x)$ has both a tangency and a root at $x = 1$. Extinction, once again, occurs with probability 1, but the approach to extinction is often slow. Indeed, it was shown by Kolmogorov (1938) that if $R = 1$ and $F'''(1) < \infty$, then

$$P(N_t > 0) \sim \frac{2}{t F''(1)}, \tag{6.43}$$

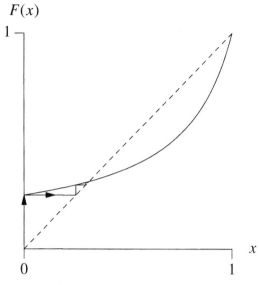

Fig. 6.2. $F(x)$ with $R_0 > 1$.

for large t. Moreover, despite the certainty of extinction, equations (6.21) and (6.32) imply that

$$E(N_t) = 1, \quad \forall t \geq 1, \tag{6.44}$$

$$\text{Var}(N_t) \longrightarrow \infty. \tag{6.45}$$

Hence, if the population has not died out, it may be large.

The final alternative is that $R_0 = F'(1) > 1$ (the supercritical case). Here,

$$F(x) - x \tag{6.46}$$

has a unique solution in the open interval $(0, 1)$. The generating function lies above the 45° line to the left of this root and below the 45° line immediately to the right of this root. It rises back up and intersects the 45° line at $x = 1$ (see Figure 6.2). The fixed point that satisfies (6.46) is easily shown to be asymptotically stable. A trajectory that starts at $x = 0$ rapidly approaches this fixed point. The abscissa of this fixed point is the asymptotic probability of extinction.

In the supercritical case, if a population does not die out, it diverges to infinity. To better describe this growth, consider the random variable consisting of the population size in the t-th generation normed by the expected population size in the t-th generation,

$$W_t \equiv \frac{N_t}{R_0^t}. \tag{6.47}$$

Problem 6.2 *The survival of right whales* (Caswell *et al.*, 1999)
A female right whale may produce zero, one, or two females the following
year. A female at time t produces zero offspring if it dies before $t + 1$, one
offspring (itself) if it survives without reproducing, and two offspring (itself
and its calf) if it survives and reproduces. Let p be the survival probability
and let m be probability of producing a female calf. Thus,

$$p_0 = (1 - p), \tag{6.48a}$$
$$p_1 = p(1 - m), \tag{6.48b}$$
$$p_2 = pm. \tag{6.48c}$$

In 1980, p and m were estimated to be $p = 0.99$ and $m = 0.063$. By 1999,
these parameters had dropped to $p = 0.94$ and $m = 0.038$. Determine the
population growth rate and the extinction probability for 1980 and 1999.

Problem 6.3 *The case of three children*
Suppose that each parent has exactly three children, and that each newborn
child is equally likely to be male or female. Start with a single male parent
($N_0 = 1$) and assume that the number of male progeny can be modeled as
a branching process.

(1) Write down the probability generating function for the number of male offspring after one generation. Use this generating function to compute the
expected number of male offspring after one generation. Sketch the generating
function over the unit interval and on a graph that includes the 45° line.
(2) Calculate the extinction probability for this male lineage.

Problem 6.4 *The unfortunate demise of a male line of descent*
Using numbers from the census of 1920, Lotka (1931a,b, 1998) found that
the probability mass function for the number of male offspring was well
represented by the geometric series

$$p_n = bc^{n-1}, \quad n = 1, 2, \ldots, \tag{6.49}$$

with $b = 0.2126$, $c = 0.5893$, and $p_0 = 0.4825$. Using these figures, determine (a) the probability generating function $F(x)$, (b) the net reproductive
rate R_0, and (c) the probability that a newly created surname will go extinct.

This random variable has several interesting properties. First,

$$E(W_t) = \frac{1}{R_0^t} E(N_t) = 1. \qquad (6.50)$$

Secondly,

$$\text{Var}\,(W_t) = \frac{1}{R_0^{2t}} \text{Var}\,(N_t) \qquad (6.51)$$

so that, for the supercritical case,

$$\text{Var}\,(W_t) = \frac{\sigma^2}{R_0\,(R_0 - 1)} \left(1 - \frac{1}{R_0^t}\right). \qquad (6.52)$$

Finally, and most importantly, note that

$$E(N_{t+1}\,|\,N_t) = R_0\,N_t, \qquad (6.53)$$

which, by the Markov property, can be rewritten

$$E(N_{t+1}\,|\,N_1, N_2, \ldots, N_t) = R_0\,N_t. \qquad (6.54)$$

We may use definition (6.47) to rewrite equation (6.54) in terms of W_t:

$$E(W_{t+1}\,|\,N_1, N_2, \ldots, N_t) = W_t. \qquad (6.55)$$

This last result is important because it implies that W_t is a discrete parameter *martingale*.

Definition A sequence $\{W_t : t \geq 1\}$ is a *martingale* with respect to the sequence $\{N_t : t \geq 1\}$ if, for all $t \geq 1$,

(1) $E(|\,W_t\,|) < \infty$
(2) $E(W_{t+1}\,|\,N_1, N_2, \ldots, N_t) = W_t$.

Martingales are important because, subject to minor conditions on the moments of W_t (satisfied here), they always converge. This is the 'martingale convergence theorem' of Doob. A proof may be found in Grimmett and Stirzaker (1992).

The limiting distribution of W_t is known explicitly in only a few cases. For fractional linear generating functions, the limiting distribution W is an exponential distribution. In general, the Laplace transform $\phi(s)$ of the limiting distribution W satisfies Poincaré's functional equation,

$$\phi(s) = F[\phi(s/R_0)]. \qquad (6.56)$$

Historical hiatus

It was long thought that the theory of branching processes began with the work of Galton (1873) and Watson and Galton (1874). However, it is now appreciated that the French Academician I. J. Bienaymé (1845) anticipated Galton and Watson by some 28 years, as first noted by Heyde and Seneta (1972). Bienaymé was probably stimulated by an empirical study of the duration of noble families by the demographer and statistician L. F. Benoiston de Châteauneuf (1847); he was clearly aware of the importance of the mean number of offspring in determining the probability of extinction. Hence, the Galton–Watson process is increasingly referred to as the Bienaymé–Galton–Watson process. See Kendall (1975) and Heyde and Seneta (1977) for more details.

In 1873, the Swiss botanist Alphonse de Candolle suggested that the extinction of families might have a probabilistic interpretation. Sir Francis Galton (1873) gave Candolle's (1873) suggestion a precise formulation as problem 4001 in the *Educational Times*. However, he received just one answer to his problem – 'from a correspondent who totally failed to perceive its intricacy' (Watson and Galton, 1874). He then turned to a friend, the clergyman and mathematician Rev. H. W. Watson, for help. Watson realized that this problem could be solved by iterating probability generating functions, but then, through an algebraic error, incorrectly concluded that *every* surname must die out.

The Galton–Watson process did not reappear in biology until R. A. Fisher (1922, 1930a,b) and J. B. S. Haldane (1927, 1939) used it to study the rate with which rare mutations vanished from a population. The first complete analysis of the probability of extinction was given by the Danish actuary J. F. Steffensen (1930, 1932). Steffensen's solution followed a challenge by A. K. Erlang. Erlang was motivated by the fact that his mother belonged to a well-known but disappearing Danish family (Jagers, 1975); he was unaware of the British work on this problem. A. J. Lotka (1931a,b, 1998) was the first person to compute the probability of extinction using demographic data. Interest in the Galton–Watson process blossomed after 1940, due largely to its use in modeling nuclear chain reactions (Harris, 1963).

Mathematical meanderings

The Galton–Watson process is a discrete-time branching process in which all individuals live to the same age. In general, semelparous individuals of the same generation may die at different times. The age of reproduction need not be a constant and may itself be a random variable.

Let us imagine that the age of reproduction is a random variable with density function $g(t)$ and distribution function

$$G(t) = \int_0^t g(u)\,du, \tag{6.57}$$

that the probability generating function for the number of offspring is

$$H(x) = \sum_{n=0}^{\infty} q_n x^n,$$ (6.58)

where q_n is the probability of n offspring, and that I may write the probability generating function for the population size $N(t)$ at time t as

$$F(t, x) = E[x^{N(t)}] = \sum_{n=0}^{\infty} p_n(t) x^n.$$ (6.59)

What is the equation that determines $F(t, x)$?

I will start, as usual, with a single individual. If this individual is alive at time t, the probability generating function at t is simply x. This occurs with probability

$$\int_t^{\infty} g(u) \, du = 1 - G(t).$$ (6.60)

The alternative is that this individual died at some time $u \leq t$ and was replaced by n offspring. This occurred with probability $q_n g(u)$, in which case the probability generating function at time t is now the sum of n independent copies of $F(t - u, x)$. Adding together the possibilities yields

$$F(t, x) = x [1 - G(t)] + \int_0^t \sum_{n=0}^{\infty} q_n [F(t - u, x)]^n g(u) \, du$$ (6.61)

or

$$F(t, x) = x [1 - G(t)] + \int_0^t H[F(t - u, x)] g(u) \, du.$$ (6.62)

This process was first described by Bellman and Harris (1948, 1952) and is termed an *age-dependent branching* (or Bellman–Harris) process. If the density function for the age of reproduction is exponentially distributed, the Bellman–Harris process is a Markov process. It is non-Markovian for all other densities.

Problem 6.5 *The exponential case*

Let the density function for the age of reproduction be exponentially distributed:

$$g(t) = \lambda e^{-\lambda t}, \quad t \geq 0.$$ (6.63)

(1) Show that equation (6.62) reduces to the differential equation

$$\frac{\partial F}{\partial t} = \lambda H[F(t, x)] - \lambda F(t, x).$$ (6.64)

(2) Solve this differential equation for binary fission, $H(x) = x^2$. What is the appropriate initial condition? Determine the probability mass function for the population size $N(t)$.

Problem 6.6 *A renewal equation for the mean*
Let

$$M(t) \equiv E[N(t)] \tag{6.65}$$

be the expected value of the population size at time t. Show that $M(t)$ satisfies

$$M(t) = 1 - G(t) + R_0 \int_0^t M(t - u)\, g(u)\, du. \tag{6.66}$$

Equation (6.66) is an integral equation of the *renewal type*. We will consider renewal-type equations in the second half of this book, when we discuss age-structured populations.

Recommended readings

Harris (1963) is the classic reference on branching processes. This book was reprinted in 1989 as a Dover paperback. Bharucha-Reid (1997) covers branching processes within the broader context of Markov processes. Jagers (1975) has a more advanced treatment that considers numerous biological applications. Moran (1962) discusses some of the early uses of branching processes in population genetics and Kendall (1975) provides a useful history of branching processes. Vatutin and Zubkov (1987, 1993) provide a useful survey of recent results.

Section B
INTERACTING POPULATIONS

7 A classical predator–prey model

I will start with the granddaddy of all predator–prey models, the Lotka–Volterra predator–prey model. I will follow Braun's (1978) wonderful description of this model. You should also note that:

(1) This is a bad mathematical model. (It is structurally unstable.)
(2) The Lotka–Volterra predator–prey model is of profound historical interest.

How did this model arise? In the mid-1920s Umberto D'Ancona, an Italian marine biologist, performed a statistical analysis of the fish that were sold in the markets of Trieste, Fiume, and Venice between 1910 and 1923 (D'Ancona, 1926, 1954). Fishing was largely suspended in the upper Adriatic during the First World War, from 1914 to 1918, and D'Ancona showed that this coincided with increases in the relative frequency of some species and decreases in the relative frequency of other species.

Table 7.1 and Figure 7.1 show the relative abundance of selachians in the market of Fiume. 'Selachians' is an old term for sharks and shark-like fish. Selachians are usually predatory. It is thus clear that the frequency of predators increased during the war years and decreased with an increase in fishing. The relative abundance of prey, in turn, followed the opposite pattern. Why did this happen? At the time, Umberto was engaged to Luisa Volterra, an ecologist. Umberto posed this question to Vito Volterra, his future father-in-law and a famous mathematician.

Vito wrote down a simple pair of differential equations to describe this system. Let $N(t)$ be the number (or density) of prey and let $P(t)$ be the number (or density) of predators. The system that Vito Volterra considered is

$$\frac{dN}{dt} = rN - cNP, \tag{7.1a}$$

$$\frac{dP}{dt} = bNP - mP. \tag{7.1b}$$

Table 7.1. *Adriatic catch data*

Year	Proportion of selachians in the market of Fiume (%)
1914	11.9
1915	21.4
1916	22.1
1917	21.2
1918	36.4
1919	27.3
1920	16.0
1921	15.9
1922	14.8
1923	10.7

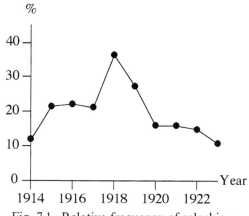

Fig. 7.1. Relative frequency of selachians.

Let us look at each term of this system. The first term on the right-hand side of equation (7.1a) implies that the prey will grow exponentially in the absence of the predator: the prey are limited by the predators. The second term describes the loss of prey due to predators. This loss is assumed to be proportional to both the number of prey and the number of predators, resulting in what is often described as a *mass-action* term. Turning to the right-hand side of equation (7.1b), we see that the loss of prey leads to the production of new predators, and that the predator population decreases exponentially in the absence of prey.

The Lotka–Volterra system, as written, has enough parameters to be unwieldy. I will therefore introduce a change of variables. Let

$$x \equiv \frac{b}{m} N, \quad y \equiv \frac{c}{r} P .$$
(7.2)

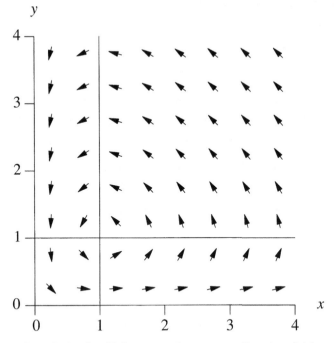

Fig. 7.2. Lotka–Volterra predator–prey direction field.

Equations (7.1a) and (7.1b) can now be rewritten as

$$\frac{dx}{dt} = r\,(1 - y)\,x, \tag{7.3a}$$

$$\frac{dy}{dt} = m\,(x - 1)\,y. \tag{7.3b}$$

The effect of this transformation is to rescale the positive equilibrium so that it is always at $(x, y) = (1, 1)$.

Equations (7.3a) and (7.3b) determine whether the variables x and y increase or decrease at each point of the (x, y) *phase plane*. The ratio of equations (7.3a) and (7.3b),

$$\frac{dy}{dx} = \frac{m\,(x - 1)\,y}{r\,(1 - y)\,x}, \tag{7.4}$$

determines the slope of the vector field at every point of the phase plane. I have drawn the corresponding direction field in Figure 7.2. I have also drawn the predator and prey zero-growth isoclines. The first of these is obtained by setting the right-hand side of equation (7.3b) equal to zero and corresponds to the locus of points where predator numbers stay constant. For this system, the predator zero-growth isoclines are given by $x = 1$ and $y = 0$. Along the prey zero-growth isoclines, prey numbers do not change. There are two prey zero-growth isoclines for this system, $y = 1$ and $x = 0$.

The vectors in Figure 7.2 rotate as we move through the plane. If predator and prey numbers are both low ($x < 1$, $y < 1$), predator numbers decrease but prey numbers increase. If prey numbers are high but predator numbers are low ($x > 1$, $y < 1$), both predators and prey increase. As predator numbers increase ($x > 1$, $y > 1$), prey now begin to decrease. Finally, when predator numbers are high but prey numbers are low ($x < 1$, $y > 1$), both predators and prey decrease.

We can try to pick up additional information by finding the equilibria for this system and by linearizing equations (7.3a) and (7.3b) about each equilibrium point. Equilibria occur at the intersections of the predator and prey zero-growth isoclines. There are two equilibria. One is at $(x, y) = (0, 0)$, the other is at $(x, y) = (1, 1)$. Near $(0, 0)$, we may neglect the nonlinear terms and consider

$$\frac{dx}{dt} \approx r\,x, \tag{7.5a}$$

$$\frac{dy}{dt} \approx -m\,y. \tag{7.5b}$$

The prey increase exponentially fast close to the origin, while the predators decrease (see Figure 7.2). The equilibrium at the origin has one stable direction and one unstable direction and, as such, is referred to as a *saddle point*.

Near the nontrivial equilibrium, we introduce new variables that measure our distance from (1, 1),

$$u \equiv x - 1, \quad v \equiv y - 1. \tag{7.6}$$

As a result,

$$\frac{du}{dt} = -r\,v\,(1 + u), \tag{7.7a}$$

$$\frac{dv}{dt} = m\,u\,(1 + v). \tag{7.7b}$$

Since u and v are small close to the nontrivial equilibrium,

$$\frac{du}{dt} \approx -r\,v, \tag{7.8a}$$

$$\frac{dv}{dt} \approx m\,u. \tag{7.8b}$$

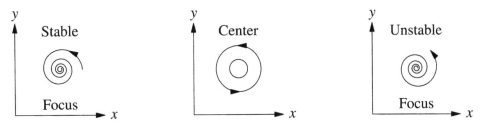

Fig. 7.3. Possible scenarios for (1,1).

It is easy to combine these two equations by noting that

$$\frac{d^2u}{dt^2} = -r\frac{dv}{dt} = -rmu \tag{7.9}$$

so that

$$\frac{d^2u}{dt^2} + rmu = 0. \tag{7.10}$$

Since this a linear, constant-coefficient equation, we may try an exponential solution of the form

$$u(t) = c\,e^{\lambda t}. \tag{7.11}$$

We thus obtain a characteristic equation,

$$\lambda^2 + rm = 0, \tag{7.12}$$

that has two purely imaginary roots,

$$\lambda = \pm\sqrt{rm}\,i. \tag{7.13}$$

What can we conclude from these two purely imaginary roots? We can be certain that the linearized system has simple periodic solutions, since combinations of $\exp(\pm i\sqrt{rm}\,t)$ can be rewritten in terms of sines and cosines. Unfortunately, we cannot conclude that the fully nonlinear system has these same simple periodic solutions. Purely imaginary roots imply that the linearized system is on the cutting edge between instability and asymptotic stability and on the edge between oscillatory solutions that increase in amplitude and those that decrease in amplitude. For the fully nonlinear system, the nonlinear terms that we have neglected are suddenly critical. They may tip the nonlinear system one way or another with regard to stability. Linearization has thus given us an ambiguous answer. The nonlinear system is structurally unstable: small changes in the structure of the equations could lead to changes in stability. Three qualitatively different phase portraits (see

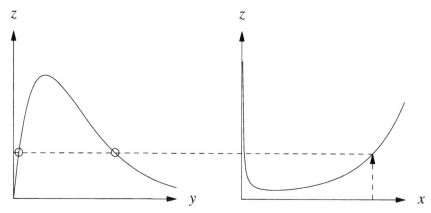

Fig. 7.4. Roots of the first integral.

Figure 7.3) are all consistent (so far) with the information from the linearized system.

Which scenario is correct? Fortunately, equation (7.4) is separable:

$$\frac{dy}{dx} = \frac{m(x-1)y}{r(1-y)x},$$ (7.14)

$$r\frac{(1-y)}{y}\,dy = m\frac{(x-1)}{x}\,dx,$$ (7.15)

$$r(\ln y - y) = m(x - \ln x) + k,$$ (7.16)

$$y^r e^{-ry} = k\frac{e^{mx}}{x^m}.$$ (7.17)

Equation (7.17) cannot be simplified any further. However, we can solve this equation graphically by introducing a new variable z and by considering the two functions

$$z = y^r e^{-ry}, \quad z = k\frac{e^{mx}}{x^m}$$ (7.18)

(see Figure 7.4). For each orbit, (i.e., for each different value of the integration constant k) picking a value of x determines z. Corresponding to each value of z there are either two, one, or zero values of y. The only one of the three scenarios in Figure 7.3 that has no more than two points of intersection between each orbit and each vertical line is the center scenario. We conclude that the nontrivial equilibrium is a center surrounded by a family of periodic orbits, and not a stable or unstable focus.

Figure 7.5 shows a time series corresponding to a typical periodic orbit. The prey cycle 'leads' the predator cycle. A small perturbation to either prey or predator numbers will result in a new periodic orbit with a different

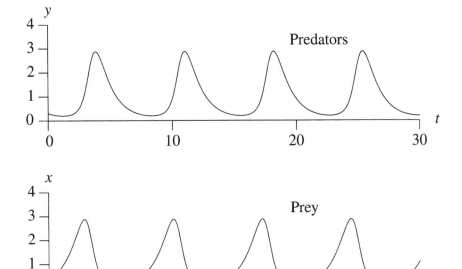

Fig. 7.5. Typical predator–prey cycle.

amplitude. The amplitude of the periodic oscillation is determined by the initial condition.

Let us return, now, to d'Ancona's original question. Why did the relative abundance of predators increase and the relative abundance of prey decrease with a decrease in fishing? To answer this question, Volterra began by integrating equation (7.3a) around a typical cycle of period T:

$$\frac{dx}{dt} = r(1 - y)x, \tag{7.19}$$

$$\int_{t_0}^{t_0+T} \frac{1}{x} dx = r \int_{t_0}^{t_0+T} (1 - y) dt, \tag{7.20}$$

$$\ln\left[\frac{x(t_0 + T)}{x(t_0)}\right] = rT - r \int_{t_0}^{t_0+T} y \, dt. \tag{7.21}$$

Since we are dealing with a periodic orbit of period T, the left-hand side of equation (7.21) reduces to zero and we are left with

$$r \int_{t_0}^{t_0+T} y \, dt = rT \tag{7.22}$$

or

$$\frac{1}{T} \int_{t_0}^{t_0+T} y \, dt = 1. \tag{7.23}$$

However, the left-hand side of this last equation is just the average predator density over the cycle,

$$y_{\text{average}} = 1 ; \tag{7.24}$$

the average number of predators over any periodic orbit is the same as the equilibrium number of predators. One can also show, in a similar way, that the average number of prey over any periodic orbit is the same as the equilibrium number of prey,

$$x_{\text{average}} = 1. \tag{7.25}$$

Let me now introduce the effect of fishing. I will do this for the original, unscaled equations,

$$\frac{dN}{dt} = r N - c N P, \tag{7.26a}$$

$$\frac{dP}{dt} = b N P - m P. \tag{7.26b}$$

The nontrivial equilibrium

$$N_{\text{average}} = \frac{m}{b}, \quad P_{\text{average}} = \frac{r}{c}, \tag{7.27}$$

gives the average number of prey and predators for a cycle of this system. The effect of fishing is to decrease the intrinsic rate of growth of the prey and to increase the mortality rate of the predators by the harvest rate h,

$$\frac{dN}{dt} = (r - h) N - c N P, \tag{7.28a}$$

$$\frac{dP}{dt} = b N P - (m + h) P. \tag{7.28b}$$

For this new set of equations, the nontrivial equilibrium and the time average numbers of prey and predators are simply

$$N_{\text{average}} = \frac{m + h}{b}, \quad P_{\text{average}} = \frac{r - h}{c}. \tag{7.29}$$

The effect of increasing the harvest h then is to increase the average prey level and to decrease the average predator level. The effect of decreasing the harvest rate (d'Ancona's original concern) is just the opposite: to decrease the average numbers of prey and increase the average numbers of predators. This differential response arises even though the predators and prey are being harvested at the same per capita rate.

The results of this analysis are not restricted to fish. Indeed, one can consider this analysis as an admonition against the use of nonspecific insecticides that indiscriminately kill insect pests and insect predators. The long-term effect of such insecticides may be to raise the relative abundance of the pest, with catastrophic consequences (DeBach, 1974).

Recommended readings

Scuda and Ziegler (1978) and Oliveira-Pinto and Conolly (1982) translate and comment on many of Volterra's most important writings.

8 To cycle or not to cycle

The dynamics of the Lotka–Volterra predator–prey model are fascinating. Unfortunately, this model is structurally unstable. A small change in the equations will often eliminate the carefully poised neutrally stable family of periodic orbits that we uncovered. We need, therefore, to look at other predator–prey models. I will analyze these new models using standard phase-plane techniques such as

(1)   linearization,
(2)   the Bendixson–Dulac negative criterion,
(3)   the Hopf bifurcation theorem, and
(4)   the Poincaré–Bendixson theorem.

Since we will look at several systems, it is useful to go through linearization for a general planar system in advance. We will then have the critical results at hand.

Consider an autonomous predator–prey system of the form

$$\frac{dN}{dt} = F(N, P), \tag{8.1a}$$

$$\frac{dP}{dt} = G(N, P), \tag{8.1b}$$

where N is the number of prey and P is the number of predators. The equations for the equilibria (N^*, P^*) are found by setting the right-hand sides equal to zero,

$$F(N^*, P^*) = 0, \quad G(N^*, P^*) = 0. \tag{8.2}$$

To determine the stability of an equilibrium, we introduce new variables that measure the deviation about the equilibrium,

$$x(t) \equiv N(t) - N^*, \quad y(t) \equiv P(t) - P^*.$$

116

We then linearize about the equilibrium,

$$\frac{dx}{dt} = \frac{\partial F}{\partial N}\Bigg|_{(N^*,P^*)} x + \frac{\partial F}{\partial P}\Bigg|_{(N^*,P^*)} y, \tag{8.3a}$$

$$\frac{dy}{dt} = \frac{\partial G}{\partial N}\Bigg|_{(N^*,P^*)} x + \frac{\partial G}{\partial P}\Bigg|_{(N^*,P^*)} y. \tag{8.3b}$$

This last set of equations can be written, more succinctly, as

$$\begin{pmatrix} \dot{x} \\ \dot{y} \end{pmatrix} = \begin{pmatrix} a_{11} & a_{12} \\ a_{21} & a_{22} \end{pmatrix} \begin{pmatrix} x \\ y \end{pmatrix} = \boldsymbol{J}\,x, \tag{8.4}$$

where the a_{ij} are the various partial derivatives. The Jacobian matrix $\boldsymbol{J}$ is called the *community matrix* in ecology. It captures the strength of the interactions in a community at equilibrium.

We now look for solutions of the form

$$x(t) = x_0\,e^{\lambda t}, \quad y(t) = y_0\,e^{\lambda t}. \tag{8.5}$$

With this substitution, equation (8.4) reduces to

$$\lambda x_0 = a_{11} x_0 + a_{12} y_0, \tag{8.6a}$$

$$\lambda y_0 = a_{21} x_0 + a_{22} y_0, \tag{8.6b}$$

or

$$\begin{pmatrix} a_{11} - \lambda & a_{12} \\ a_{21} & a_{22} - \lambda \end{pmatrix} \begin{pmatrix} x_0 \\ y_0 \end{pmatrix} = 0. \tag{8.7}$$

The simplest systematic way of solving equation (8.7) for x_0 and y_0 is to use Cramer's rule:

$$x_0 = \frac{\begin{vmatrix} 0 & a_{12} \\ 0 & a_{22} - \lambda \end{vmatrix}}{\begin{vmatrix} a_{11} - \lambda & a_{12} \\ a_{21} & a_{22} - \lambda \end{vmatrix}}, \tag{8.8a}$$

$$y_0 = \frac{\begin{vmatrix} a_{11} - \lambda & 0 \\ a_{21} & 0 \end{vmatrix}}{\begin{vmatrix} a_{11} - \lambda & a_{12} \\ a_{21} & a_{22} - \lambda \end{vmatrix}}. \tag{8.8b}$$

However, we have a problem. The determinant in each numerator equals zero. Unless the denominators also equal zero, we are forced to accept the trivial solution as the only exponential solution. To avoid this, we will require that

$$\begin{vmatrix} a_{11} - \lambda & a_{12} \\ a_{21} & a_{22} - \lambda \end{vmatrix} = 0. \qquad (8.9)$$

By expanding the determinant, we obtain the characteristic equation

$$\lambda^2 - (a_{11} + a_{22}) \lambda + (a_{11}a_{22} - a_{12}a_{21}) = 0. \qquad (8.10)$$

We can rewrite this as

$$\lambda^2 - p\lambda + q = 0, \qquad (8.11)$$

where $p = \mathrm{Tr}\, J$ and $q = \det J$ are the trace and the determinant of the community matrix. If we are only interested in the stability of the equilibrium, we can apply the Routh–Hurwitz criterion and obtain

$$\mathrm{Tr}\, J < 0, \ \det J > 0, \qquad (8.12)$$

as sufficient conditions for asymptotic stability. However, p and q also determine the detailed character of the flow for the linearized system (see Figure 8.1).

The trace and determinant determine the eigenvalues λ. If the two roots λ are real and negative, the equilibrium is a stable node. If the roots are both positive, we have an unstable node. If they are real and of opposite sign, the equilibrium is a saddle point. If the roots are complex with negative real part, we have a stable focus. If the roots are complex but with positive real part, the equilibrium is an unstable focus. Finally, if the roots are purely imaginary, the linearized system will have a center but the original nonlinear system will have a center or a stable or an unstable focus, depending on the exact nature of the nonlinear terms.

Since the matrix in equation (8.7) is singular, we expect that one scalar equation is degenerate and that the remaining equation determines a line of eigenvectors (x_0, y_0) for each eigenvalue λ. For real eigenvalues, these eigenvectors define the stable and unstable manifolds of the linearized system.

With this linearization in hand, we are now free to examine new predator–prey models. I will begin by reintroducing density dependence, so that the prey grow logistically in the absence of predators,

$$\frac{dN}{dT} = r N \left(1 - \frac{N}{K} \right) - c N P, \qquad (8.13a)$$

$$\frac{dP}{dT} = b N P - m P. \qquad (8.13b)$$

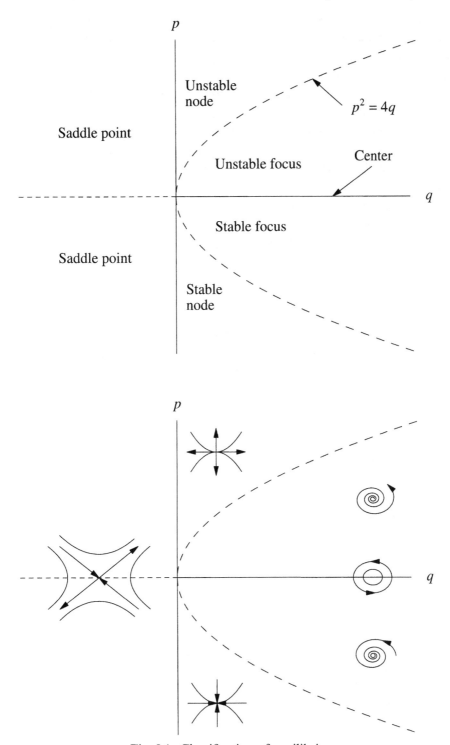

Fig. 8.1. Classification of equilibria.

I have written time as T (rather than t) because I will soon rescale this variable. To ease the analysis, I will nondimensionalize all of the variables. This is best done in steps.

Prey abundance is often measured by comparing the number of prey to their carrying capacity. I will thus introduce the dimensionless variable

$$x \equiv \frac{N}{K} \tag{8.14}$$

and replace each occurrence of N in equations (8.13a) and (8.13b) with $N = Kx$,

$$\frac{dx}{dT} = r x (1 - x) - c x P, \tag{8.15a}$$

$$\frac{dP}{dT} = b K x P - m P. \tag{8.15b}$$

To simplify equation (8.15a) I will now introduce the dimensionless variable

$$y \equiv \frac{c}{r} P, \tag{8.16}$$

and eliminate P,

$$\frac{dx}{dT} = r x (1 - x - y), \tag{8.17a}$$

$$\frac{dy}{dT} = b K x y - m y. \tag{8.17b}$$

Finally, I will rescale time by letting

$$t \equiv r T. \tag{8.18}$$

Note that

$$\frac{d}{dT} = \frac{d}{dt} \frac{dt}{dT} = r \frac{d}{dt}. \tag{8.19}$$

Equations (8.17a) and (8.17b) now take the form

$$\frac{dx}{dt} = x (1 - x - y), \tag{8.20a}$$

$$\frac{dy}{dt} = \frac{bK}{r} y \left(x - \frac{m}{bK} \right). \tag{8.20b}$$

The parameters in equation (8.20b) occur in two clusters. I will rename these

$$\alpha \equiv \frac{m}{bK}, \quad \beta \equiv \frac{bK}{r}. \tag{8.21}$$

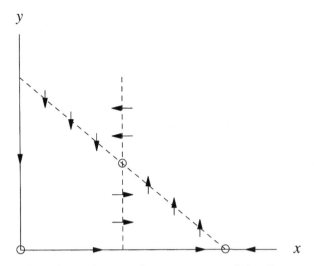

Fig. 8.2. Predator and prey zero-growth isoclines.

With this last substitution, we have

$$\frac{dx}{dt} = x(1 - x - y), \tag{8.22a}$$

$$\frac{dy}{dt} = \beta(x - \alpha)y. \tag{8.22b}$$

The simplified model has three equilibria: $(x^*, y^*) = (0, 0)$, $(x^*, y^*) = (1, 0)$, and $(x^*, y^*) = (\alpha, 1 - \alpha)$. These equilibria occur at the intersections of the prey and predator zero-growth isoclines (see Figure 8.2).

By equation (8.4), the Jacobian or community matrix is just

$$J = \begin{bmatrix} 1 - 2x - y & -x \\ \beta y & \beta(x - \alpha) \end{bmatrix}_{(x^*, y^*)}. \tag{8.23}$$

At $(x^*, y^*) = (0, 0)$,

$$J = \begin{pmatrix} 1 & 0 \\ 0 & -\beta\alpha \end{pmatrix}. \tag{8.24}$$

This matrix has eigenvalues

$$\lambda = 1, -\beta\alpha \tag{8.25}$$

and eigenvectors $(1, 0)$ and $(0, 1)$. The origin is thus a saddle point.

At $(x^*, y^*) = (1, 0)$,

$$J = \begin{bmatrix} -1 & -1 \\ 0 & \beta(1 - \alpha) \end{bmatrix}. \tag{8.26}$$

The eigenvalues are

$$\lambda = -1, \beta(1 - \alpha) \tag{8.27}$$

and the corresponding eigenvectors are $(1, 0)$ and $[1, -1 - \beta(1 - \alpha)]$. If $\alpha > 1$, the predator's per capita death rate (m) exceeds its per capita birth rate (bK). The equilibrium at $(1, 0)$ is now a stable node. If $\alpha < 1$, the equilibrium is an unstable saddle point.

Finally, at $(x^*, y^*) = (\alpha, 1 - \alpha)$,

$$J = \begin{bmatrix} -\alpha & -\alpha \\ \beta(1 - \alpha) & 0 \end{bmatrix} \tag{8.28}$$

with characteristic equation

$$\lambda^2 + \alpha\lambda + \alpha\beta(1 - \alpha) = 0. \tag{8.29}$$

By the Routh–Hurwitz criterion, this equilibrium is stable if $\alpha < 1$ and unstable if $\alpha > 1$. The eigenvalues are given by the quadratic formula,

$$\lambda = \frac{-\alpha \pm \sqrt{\alpha^2 - 4\alpha\beta(1 - \alpha)}}{2}, \tag{8.30}$$

and if we examine the discriminant, we see that we have a node for

$$\alpha > \frac{4\beta}{1 + 4\beta} \tag{8.31}$$

and a focus for

$$\alpha < \frac{4\beta}{1 + 4\beta}. \tag{8.32}$$

Figures 8.3, 8.4 and 8.5 illustrate the behavior of the model for a variety of parameter values. Figure 8.3 has a stable node at $(1, 0)$ and no nontrivial equilibrium in the interior of the first quadrant. Figures 8.4 and 8.5 show a node and a stable focus in the interior of the first quadrant. This model does not appear to have any periodic orbits – in marked contrast to the classical Lotka–Volterra system. The addition of a small amount of prey-density dependence has destroyed the family of periodic orbits that we observed in the last section, reinforcing our conclusion that the classical Lotka–Volterra system is structurally unstable.

Of course, appearances can be deceiving. It would be nice to actually *prove* that equations (8.22a) and (8.22b) do not contain periodic orbits in the first quadrant. Fortunately, several criteria allow us to do just that.

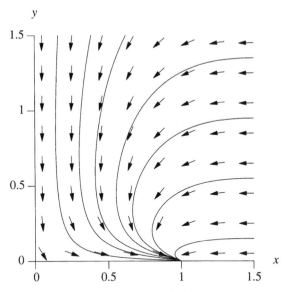

Fig. 8.3. Phase portrait for $\alpha > 1$.

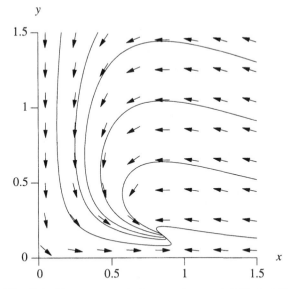

Fig. 8.4. Phase portrait for $\alpha < 1$, $\alpha > 4\beta/(1 + 4\beta)$.

Theorem (Bendixson's negative criterion) Consider the dynamical system

$$\frac{dx}{dt} = F(x, y), \quad \frac{dy}{dt} = G(x, y), \tag{8.33}$$

where F and G are continuously differentiable functions on some simply

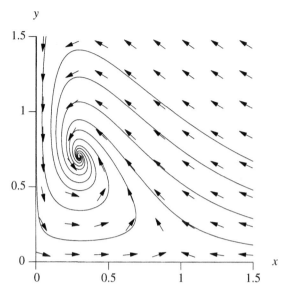

Fig. 8.5. Phase portrait for $\alpha < 1$, $\alpha < 4\beta(1 + 4\beta)$.

connected domain $D \subset \mathfrak{R}^2$. (By simply connected, I mean that the domain has no 'holes' or disjunct portions.) If

$$\nabla \cdot (F, G) \;=\; \frac{\partial F}{\partial x} + \frac{\partial G}{\partial y} \tag{8.34}$$

is of one sign in D, there cannot be a closed orbit contained within D.

Proof The proof is by contradiction. Suppose that we *do* have a closed orbit C, with interior Ω, contained in D that satisfies equations (8.33). Suppose also that the right-hand side of equation (8.34) is of one sign. It follows that

$$\int\!\!\int_{\Omega} \left(\frac{\partial F}{\partial x} + \frac{\partial G}{\partial y} \right) dA \;\neq\; 0. \tag{8.35}$$

However, if we employ Green's theorem in the plane to transform the left-hand side of equation (8.35), we observe that

$$\oint_C F\, dy - G\, dx \;\neq\; 0. \tag{8.36}$$

This last integral may also be rewritten

$$\oint_C \left(F\frac{dy}{dt} - G\frac{dx}{dt} \right) dt \tag{8.37}$$

and, since C has been assumed to satisfy system (8.33), we may take advantage of these equations to simplify equation (8.37),

$$\oint_C (F\,G - G\,F)\,dt = \oint_C 0 \quad dt = 0. \tag{8.38}$$

Equation (8.37) contradicts equation (8.35), and so we must have been mistaken in assuming the existence of a closed periodic orbit (contained within D) that satisfies equations (8.33). If the divergence is of one sign, there cannot be such an orbit. ☐

The French mathematician H. Dulac made the useful observation that the last system is a member of a family of dynamical systems,

$$\frac{dx}{dt} = B(x,\,y)\,F(x,\,y), \tag{8.39a}$$

$$\frac{dy}{dt} = B(x,\,y)\,G(x,\,y), \tag{8.39b}$$

that share the same phase portrait. If one can disprove the existence of a closed orbit for any member of this family, one can disprove the existence of a closed orbit for every member of the family. This leads to a minor, but powerful, extension of Bendixson's negative criterion.

Theorem (Bendixson's–Dulac's negative criterion) Let B be a smooth function on $D \subset \mathfrak{R}^2$ (with all other assumptions as before). If

$$\nabla \cdot (B\,F,\,B\,G) = \frac{\partial\,(B\,F)}{\partial x} + \frac{\partial\,(B\,G)}{\partial y} \tag{8.40}$$

is of one sign in D, then no closed orbit is contained within D.

The Bendixson–Dulac negative criterion does not tell you how to find $B(x,\,y)$. I know of no general method for constructing this function. Nevertheless, you may be lucky enough to find such a function.

EXAMPLE
Consider

$$\frac{dx}{dt} = F(x,\,y) = x\,(1 - x - y), \tag{8.41a}$$

$$\frac{dy}{dt} = G(x,\,y) = \beta\,(x - \alpha)\,y. \tag{8.41b}$$

Let

$$B \equiv \frac{1}{x\,y}. \tag{8.42}$$

Then

$$BF = \frac{(1 - x - y)}{y}, \quad BG = \frac{\beta(x - \alpha)}{x}, \tag{8.43}$$

and

$$\frac{\partial(BF)}{\partial x} + \frac{\partial(BG)}{\partial y} = -\frac{1}{y}. \tag{8.44}$$

The last expression is strictly negative in the interior of the first quadrant. Thus, there cannot be a closed orbit that satisfies equations (8.41a) and (8.41b) and that lies entirely within the interior of the first quadrant. ◇

Compare our last two predator–prey models. The Lotka–Volterra predator–prey system possesses a continuum of periodic orbits. However, this system is structurally unstable. We added a small amount of prey-density dependence and lost all our periodic orbits. Can we build a simple ecological model with robust periodic orbits? If so, what form do these periodic orbits take?

The periodic orbits that we are after are called *limit* cycles. I will start with a toy problem (a contrived nonecological example) so that we can see the essential attributes of limit cycles.

EXAMPLE
Consider

$$\frac{dx}{dt} = \mu x - \omega y - x(x^2 + y^2), \tag{8.45a}$$

$$\frac{dy}{dt} = \omega x + \mu y - y(x^2 + y^2). \tag{8.45b}$$

The origin, $(x^*, y^*) = (0, 0)$, is the only equilibrium. We can determine the stability of this equilibrium from the linearization

$$\begin{pmatrix} \dot{x} \\ \dot{y} \end{pmatrix} = \begin{pmatrix} \mu & -\omega \\ \omega & \mu \end{pmatrix} \begin{pmatrix} x \\ y \end{pmatrix}. \tag{8.46}$$

The characteristic equation for this system is

$$\lambda^2 - 2\mu\lambda + (\mu^2 + \omega^2) = 0 \tag{8.47}$$

and the eigenvalues are

$$\lambda = \mu \pm i\omega \tag{8.48}$$

(see Figure 8.6). As I increase μ through zero, the origin shifts from being a stable focus to being an unstable focus.

Although the local behavior (near the origin) changes at $\mu = 0$, the

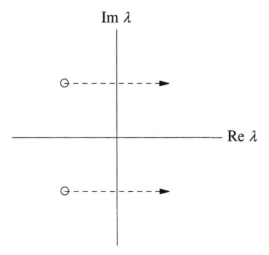

Fig. 8.6. Eigenvalues crossing the imaginary axis.

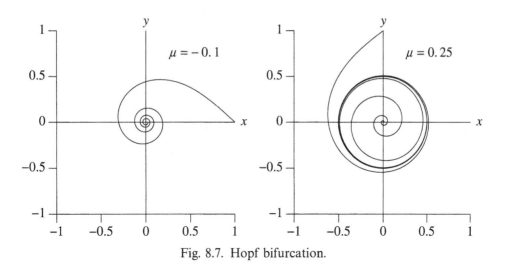

Fig. 8.7. Hopf bifurcation.

global behavior (for large r) does not. For large r, trajectories are drawn in, regardless of the sign of μ. This can be seen by shifting to polar coordinates,

$$\frac{dr}{dt} = r(\mu - r^2), \quad \frac{d\theta}{dt} = \omega. \tag{8.49}$$

In this polar form, all trajectories clearly tend towards the origin for μ negative (see Figure 8.7). For μ positive, trajectories spiral inwards for $r > \sqrt{\mu}$ and outward for $r < \sqrt{\mu}$. There is a single nonzero radius, $r = \sqrt{\mu}$, for which dr/dt is zero. On this circle, $d\theta/dt = \omega$: we continue to move around the circle with constant angular velocity. This simple periodic orbit differs from the periodic orbits for the Lotka–Volterra system. It is an isolated periodic orbit (not part of a continuum of periodic orbits). It also attracts orbits based at nearby initial conditions (hence the name limit cycle). This cycle persists for a range of parameter values.   ◇

Please note that:

(1) Structurally stable planar systems may possess periodic limit cycles.
(2) Limit cycles can be stable or unstable. System (8.49) provides an example of a stable limit cycle. To see an unstable limit cycle, consider

$$\frac{dr}{dt} = r(\mu + r^2), \quad \frac{d\theta}{dt} = \omega. \tag{8.50}$$

The stability in question is not Lyapunov stability. It is *Poincaré* or *orbital stability* (stability in a tube rather than stability in a disk).

(3) You should look for limit cycles to appear or disappear whenever foci change their stability. In our toy example, we observed a *Hopf bifurcation*. A limit cycle was born as the origin changed from a stable to an unstable focus. This limit cycle grew in amplitude as μ increased.

Let me 'flesh out' these ideas. I will start with the issue of stability.

Mathematical meanderings

It is easy enough to extend the definition of Lyapunov stability for an equilibrium point (Chapter 1) to more general solutions such as limit cycles.

Definition A solution $\xi(t)$ is *Lyapunov stable* if, for every initial time t_0 and for all $\varepsilon > 0$, there exists a $\delta > 0$ such that solutions $x(t)$ that start close to $\xi(t_0)$, $\| x(t_0) - \xi(t_0) \| < \delta$, stay close, $\| x(t) - \xi(t) \| < \varepsilon$, for all $t \geq t_0$. For autonomous systems, δ is independent of t_0 and stable solutions are sometimes said to be *uniformly stable*.

Unfortunately, this definition is very restrictive; it requires the perturbed solution to stay within an ε-disk of the true solution at every instant of time. Such *isochronicity* is rare.

EXAMPLE
Consider

$$\frac{dr}{dt} = 0, \quad \frac{d\theta}{dt} = r. \tag{8.51}$$

This system has a continuum of periodic orbits (circles) in addition to the trivial solution. The periodic orbits are not Lyapunov stable. This is because neighboring solutions have different angular velocities and 'lap' each other. Even so, *orbits* do stay close to each other in the phase plane. This suggests that Lyapunov stability is an inadequate criterion for solutions that differ in their angular velocity.   ◇

To correct this problem, we need to introduce another form of stability. To keep things simple, I will assume that we are dealing with an autonomous system that possesses a simple periodic solution $\xi(t)$. This periodic orbit defines a closed path (or orbit) C in the phase (or state) space. I will use $d[x, C]$ to denote the distance between some other point x and this orbit. $d[x, C]$ is the infimum of the distances between x and the points of C.

Definition A periodic solution $\xi(t)$ with closed orbit C is *orbitally* or *Poincaré stable* if, for all $\varepsilon > 0$, there exists a $\delta > 0$ such that every solution $x(t)$ that satisfies $d[x(t_0), C] < \delta$ for time $t = t_0$, satisfies $d[x(t), C] < \varepsilon$ for all $t > t_0$. We say that C is *orbitally asymptotically stable* if it is orbitally stable and if, in addition,

$$\lim_{t \to \infty} d[x(t), C] = 0. \tag{8.52}$$

Loosely speaking, a solution is orbitally stable if orbits that start close in phase space stay close in phase space. Lyapunov stability implies orbital stability, but the converse is not true. The limit cycles in which we are most interested are orbitally asymptotically stable. They are rarely, if ever, Lyapunov stable.

Another item that I should 'flesh out' is the Hopf bifurcation theorem:

Theorem (Hopf) Consider a smooth dynamic system of the form

$$\frac{dx}{dt} = F(x, y, \mu), \tag{8.53a}$$

$$\frac{dy}{dt} = G(x, y, \mu), \tag{8.53b}$$

where x and y are the state variables, μ is a bifurcation parameter, and (x^*, y^*, μ^*) is an equilibrium of the system.

(1) If the Jacobian J, evaluated at (x^*, y^*, μ^*), has a simple pair of imaginary eigenvalues, there is a smooth curve of equilibria $[x(\mu), y(\mu), \mu]$ with $[x(\mu^*), y(\mu^*)] = (x^*, y^*)$. The eigenvalues for the equilibria also vary smoothly with μ.

(2) If, in addition,

$$\frac{d}{d\mu} [Re \, \lambda(\mu)]_{\mu = \mu^*} = d \neq 0, \tag{8.54}$$

then there is a smooth change of coordinates that allows the Taylor expansion of degree three of the dynamical system to be written

$$\frac{dr}{dt} \approx (\mu d + a r^2) r, \tag{8.55a}$$

$$\frac{d\theta}{dt} \approx \omega + c\mu + b r^2. \tag{8.55b}$$

See Guckenheimer and Holmes (1983) for details.

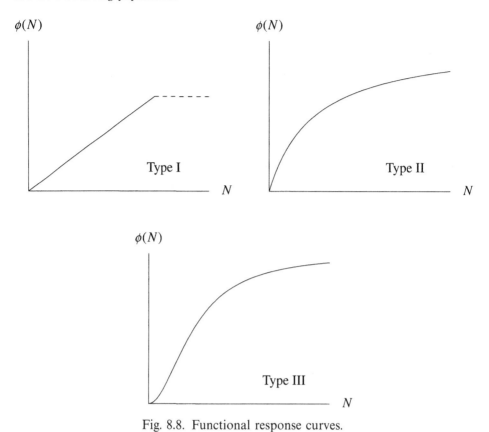

Fig. 8.8. Functional response curves.

There are various proofs of this theorem (Marsden and McCracken, 1976; Glendinning, 1994) and the theorem can also be extended to higher dimension. The significance of equations (8.55a) and (8.55b) is that they are similar to system (8.49), our toy example. As we vary μ, we expect the creation of a periodic orbit. This orbit may be stable or unstable depending on the sign of a.

The coefficients a, b, c, and d in equations (8.55a) and (8.55b) (and the nature of the periodic orbit) can be determined from nonlinear equations (8.53a) and (8.53b). However, determining these parameters is tedious (Glendinning, 1994). We will instead use the Hopf bifurcation theorem as a simple signpost for limit cycles. We will keep our eyes open for limit cycles whenever there is a change in the stability of a focus.

What biological factors create limit cycles? The inclusion of a more realistic *functional response* is one such factor. The functional response (Solomon, 1949) is the rate at which each predator captures prey. Heretofore, the functional response was a linearly increasing function of prey density. However, predators may become satiated. They may also be limited by the 'handling time' of catching and consuming their prey. This limit on the predator's ability can have a profound effect on the dynamics of a predator–prey model.

Holling (1959a,b, 1965, 1966) described three different functional response curves. Some individuals (Crawley, 1992) also argue for a fourth curve.

A type I functional response is a linear relationship between the number of prey eaten per predator per unit time and the prey density. The resulting curve may increase up to some fixed maximum (see Figure 8.8) or it may increase indefinitely.

A type II functional response is a hyperbolic function that saturates because of the time it takes to handle prey. It can be derived in a straightforward way. Let

$$T \equiv \text{total time,} \tag{8.56a}$$

$$T_h \equiv \text{handling time for each prey item,} \tag{8.56b}$$

$$N \equiv \text{number of potential prey,} \tag{8.56c}$$

$$V \equiv \text{number of prey caught (victims).} \tag{8.56d}$$

Imagine that the number of victims is proportional to both the number of potential prey and to the time available for searching,

$$V = \alpha (T - V T_h) N. \tag{8.57}$$

Here, α is the proportionality constant. If we solve for V, we obtain

$$V = \frac{\alpha T N}{1 + \alpha T_h N}. \tag{8.58}$$

For convenience, I will rewrite the type II functional response as

$$\phi(N) = \frac{c N}{a + N}. \tag{8.59}$$

Since $\phi(a) = c/2$, a is referred to as the *half-saturation constant*.

A typical type III functional response is

$$\phi(N) = \frac{c N^2}{a^2 + N^2}. \tag{8.60}$$

This is a sigmoidal curve that has predators foraging inefficiently at low prey densities. This functional response is especially appropriate for predators that must encounter enough prey to form a 'search image'.

Problem 8.1 *More bifurcation diagrams*

Analyze the scalar differential equation

$$\frac{dN}{dt} = r N \left(1 - \frac{N}{K}\right) - \phi(N) P, \tag{8.61}$$

with

(1) $\phi(N) = c N,$
(2) $\phi(N) = c N/(a + N),$
(3) $\phi(N) = c N^2/(a^2 + N^2).$

Treat the number of predators P as a bifurcation parameter. Sketch a bifurcation diagram for each of the three cases.

Some authors have also described a (type IV) functional response that is humped and that declines at high prey densities. This decline may occur because of prey group defense or prey toxicity. See Chapter 9 for details.

What are the consequences of a type II functional response? Consider

$$\frac{dN}{dT} = r N \left(1 - \frac{N}{K}\right) - \frac{c N P}{a + N}, \tag{8.62a}$$

$$\frac{dP}{dT} = \frac{b N P}{a + N} - m P. \tag{8.62b}$$

This system is sometimes referred to as the Rosenzweig–MacArthur system (Rosenzweig and MacArthur, 1963). Introducing the change of variables

$$N = a x, \quad P = r \frac{a}{c} y, \quad T = \frac{1}{r} t, \tag{8.63}$$

reduces this system to

$$\frac{dx}{dt} = x \left(1 - \frac{x}{\gamma}\right) - \frac{x y}{1 + x}, \tag{8.64a}$$

$$\frac{dy}{dt} = \beta \left(\frac{x}{1 + x} - \alpha\right) y, \tag{8.64b}$$

where

$$\alpha \equiv \frac{m}{b}, \quad \beta \equiv \frac{b}{r}, \quad \gamma \equiv \frac{K}{a}. \tag{8.65}$$

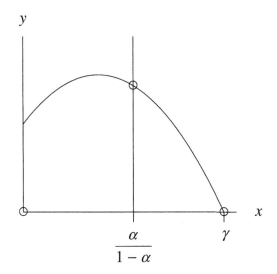

Fig. 8.9. Predator and prey zero-growth isoclines.

If we set $dx/dt = 0$, we obtain

$$x = 0, \quad y = (1 + x) \left(1 - \frac{x}{\gamma} \right) \tag{8.66}$$

as the two prey zero-growth isoclines. Setting $dy/dt = 0$, in turn, gives us

$$y = 0, \quad x = \frac{\alpha}{1 - \alpha}. \tag{8.67}$$

as the two predator zero-growth isoclines. These isoclines are shown in Figure 8.9.

Equilibria occur at the intersections of the predator and prey zero-growth isoclines. There are three equilibria:

$$(x_0, y_0) = (0, 0), \tag{8.68}$$
$$(x_1, y_1) = (\gamma, 0), \tag{8.69}$$

and

$$(x_2, y_2) = \left[x^*, (1 + x^*) \left(1 - \frac{x^*}{\gamma} \right) \right], \tag{8.70}$$

with

$$x^* = \frac{\alpha}{1 - \alpha}. \tag{8.71}$$

We can simplify our analyses of these equilibria by rewriting our rescaled differential equations as

$$\frac{dx}{dt} = f(x)\,[g(x) - y], \tag{8.72a}$$

$$\frac{dy}{dt} = \beta\,[f(x) - \alpha]\,y, \tag{8.72b}$$

where

$$f(x) \equiv \frac{x}{1 + x}, \quad g(x) \equiv (1 + x)\left(1 - \frac{x}{\gamma}\right). \tag{8.73}$$

The Jacobian (or community) matrix can now be written, rather simply, as

$$J = \begin{bmatrix} f(x)\,g'(x) + f'(x)\,g(x) - y\,f'(x) & -f(x) \\ \beta\,f'(x)\,y & \beta\,[f(x) - \alpha] \end{bmatrix}. \tag{8.74}$$

At $(0,0)$, the Jacobian reduces to

$$J = \begin{pmatrix} 1 & 0 \\ 0 & -\alpha\beta \end{pmatrix}. \tag{8.75}$$

This matrix has the eigenvalues

$$\lambda_1 = 1, \quad \lambda_2 = -\alpha\beta. \tag{8.76}$$

Since these eigenvalues are real and of opposite sign, (x_0, y_0) is a saddle point.

At $(\gamma, 0)$, $g(\gamma) = 0$. The Jacobian is now

$$J = \begin{pmatrix} -1 & -f(\gamma) \\ 0 & \beta\,[f(\gamma) - \alpha] \end{pmatrix}. \tag{8.77}$$

The eigenvalues are again given by the diagonal elements,

$$\lambda_1 = -1, \quad \lambda_2 = \beta\left(\frac{\gamma}{1 + \gamma} - \alpha\right). \tag{8.78}$$

This second equilibrium is stable (and a node) if

$$\frac{\gamma}{1 + \gamma} < \alpha \tag{8.79}$$

or, equivalently, if

$$x^* > \gamma. \tag{8.80}$$

If this condition is violated (i.e., if the predator zero-growth isocline is to the left of γ), this second equilibrium is a saddle point.

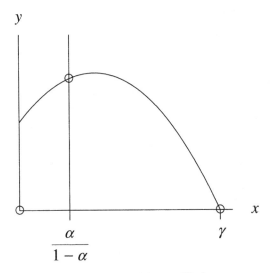

Fig. 8.10. Unstable equilibrium.

The third equilibrium is at $[x^*, g(x^*)]$ with $f(x^*) = \alpha$. The Jacobian is

$$J = \begin{bmatrix} \alpha\, g'(x^*) & -\alpha \\ \beta f'(x^*)g(x^*) & 0 \end{bmatrix} \tag{8.81}$$

and the characteristic equation is

$$\lambda^2 - \alpha g'(x^*)\lambda + \alpha\beta f'(x^*)g(x^*) = 0. \tag{8.82}$$

By the Routh–Hurwitz criterion, the coefficients in λ must be positive for this third equilibrium to be stable. The quantities α, β, and $f'(x)$ are all strictly positive while $g(x^*)$ is positive for $-1 < x^* < \gamma$. The stability of the third equilibrium is thus determined by the sign of $g'(x^*)$. This equilibrium is stable if

$$g'(x^*) < 0, \tag{8.83}$$

and unstable if

$$g'(x^*) > 0. \tag{8.84}$$

The eigenvalues are imaginary if $g'(x^*) = 0$. We have all the makings for a Hopf bifurcation.

The value $g'(x^*)$ is the slope of the (parabolic) prey zero-growth isocline at its intersection with the predator zero-growth isocline. For the isoclines in Figure 8.9, we have a stable equilibrium. For those in Figure 8.10, we have an unstable equilibrium point. A Hopf bifurcation occurs as the predator

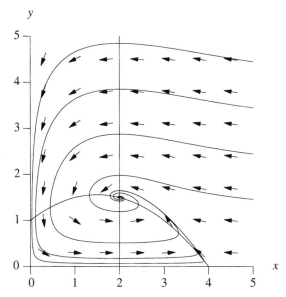

Fig. 8.11. Phase portrait for $g'(x^*) < 0$.

zero-growth isocline passes through the peak of the parabola. Figures 8.11 and 8.12 are typical phase portraits on either side of the Hopf bifurcation.

Increasing the prey's carrying capacity K increases the dimensionless parameter γ. This moves the peak of the prey zero-growth isocline to the right. Large enough increases in K will destabilize the equilibrium point (x_2, y_2) by means of a Hopf bifurcation. Solutions are then drawn to a limit cycle; the predator and prey may then come perilously close to the axes and to local extinction. This effect, wherein a stable predator–prey equilibrium of coexistence is destabilized by increasing the prey's carrying capacity, is called the 'paradox of enrichment' (Rosenzweig, 1971). Rosenzweig argued that one must be extremely careful in enriching ecosystems.

It is possible to construct predator–prey models with more than one limit cycle. Bazykin (1998) and Kuznetsov (1995) have analyzed a variant of the Rosenzweig–MacArthur system that incorporates density-dependent predator mortality and that has as many as two limit cycles and three equilibria coexisting in the interior of the first quadrant.

A more pressing question is whether we can observe other, more exotic, attractors. Can an autonomous planar system of differential equations possess a chaotic attractor? No. We will need a higher (≥ 3) dimension system of differential equations, or a nonautonomous system, to get more interesting behavior. (Contrast this with the behavior of difference equations and

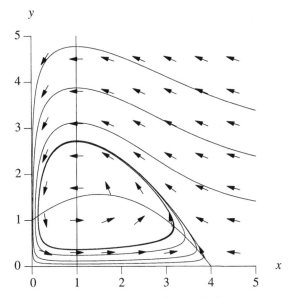

Fig. 8.12. Phase portrait for $g'(x^*) > 0$.

delay-differential equations.) The theory that justifies this statement is due to Poincaré and to Bendixson (1901). The key result of this theory, the Poincaré–Bendixson theorem, can be used to prove the existence of closed (periodic) orbits. In that sense, it is a positive criterion that complements our negative criteria.

Let me first introduce some formalism. We say that the solution, $x = x(t)$, $y = y(t)$, of a planar autonomous system determines a *positive semi-orbit* C^+ if it exists in the interval $t_0 \le t < +\infty$, for some initial time t_0. A point (ξ, η) is an ω or *positive limit point* of C^+ (ω is the last letter of the Greek alphabet) if there exists a sequence of increasing times, $t_n \to +\infty$, such that $x(t_n) \to \xi$, $y(t_n) \to \eta$. The set of ω-limit points of a positive semi-orbit C^+ is the *ω-limit set*, $\omega(C^+)$, of C^+.

Theorem (Poincaré–Bendixson) Consider an autonomous system of differential equations of the form

$$\frac{dx}{dt} = F(x, y), \quad \frac{dy}{dt} = G(x, y), \tag{8.85}$$

where F and G have continuous first partial derivatives and the solutions of this system exist for all t. Suppose that a positive semi-orbit C^+ of this system enters and does not leave some closed bounded domain D and that there are no equilibrium points in D. Then $\omega(C^+)$ is a periodic orbit.

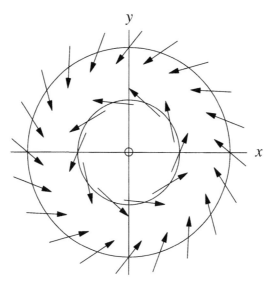

Fig. 8.13. Annular trapping region.

The proof of this theorem is beyond the scope of this chapter. Let me, instead, refer the interested reader to the many recent books on dynamical systems (Wiggins, 1990; Perko, 1991; Farkas, 1994; Glendinning, 1994; Verhulst, 1996) that do contain proofs. The theorem is important because it limits the possible fates of a (positive semi-) orbit. For an autonomous planar system of differential equations, a bounded orbit is either (a) an equilibrium point, (b) asymptotic to an equilibrium point, (c) a periodic orbit, (d) asymptotic to a periodic orbit, or (e) asymptotic to a cycle graph. (I will say more about cycle graphs in the next chapter.) Chaos is not possible in two dimensions.

EXAMPLE
Consider

$$\frac{dx}{dt} = x - y - x(x^2 + y^2), \tag{8.86a}$$

$$\frac{dy}{dt} = x + y - y(x^2 + y^2). \tag{8.86b}$$

This system has a single equilibrium point, an unstable focus at the origin. Consider an annular domain D with inner radius $0 < r_i < 1$ and outer radius $r_o > 1$. It is easy to show that the vector field for equations (8.86a) and (8.86b) points everywhere outward along the inner circle and everywhere inward along the outer circle (Figure 8.13). Since the annulus does not contain any equilibrium points, it must, by the Poincaré–Bendixson theorem, contain a periodic orbit. This system is merely equations (8.45a) and (8.45b) with $\mu = 1$ and $\omega = 1$. It possesses a limit cycle of radius 1. $\qquad\qquad\Diamond$

Recommended readings

Perko (1991) and Verhulst (1996) provide useful introductions to phase-plane methods. Several topics from this section are treated in greater detail in Farkas's (1994) excellent book on periodic motions. Coleman (1983) discusses the implications of the Poincaré–Bendixson theorem for population ecology.

9 Global bifurcations in predator–prey models[†]

The bifurcations in Chapter 8 were *local*. Each bifurcation could be detected by examining the changes that took place in *any* small neighborhood of a single equilibrium point. Even the Hopf bifurcation, which gave rise to a limit cycle, involved a local change in the stability of a focus. There are bifurcations, however, where we need to examine the flow over a wide expanse of phase space. These *global* bifurcations have been found in several predator–prey models. We examine two of these models in this chapter. For each model, a global bifurcation leads to the creation or destruction of a limit cycle.

The first model (Freedman and Wolkowicz, 1986),

$$\frac{dN}{dT} = r N \left(1 - \frac{N}{K}\right) - \phi(N) P, \tag{9.1a}$$

$$\frac{dP}{dT} = b \phi(N) P - m P, \tag{9.1b}$$

is similar in appearance to the Rosenzweig–MacArthur system. N is the number of prey, P is the number of predators, and T is time. In the absence of the predator, the prey grows logistically with intrinsic rate of growth r and carrying capacity K. The predator possesses a constant per capita mortality rate m, consumes the prey with functional response $\phi(N)$, and converts consumed prey into new predators with efficiency b. N and P are nonnegative variables; r, K, b, and m are positive parameters.

This model differs from earlier models in its functional response. We are now interested in a 'type IV' (Crawley, 1992) functional response. This is a functional response in which the predator's per capita rate of predation decreases at sufficiently high prey density, due to either prey interference or prey toxicity. Examples abound. Freedman and Wolkowicz (1986) cite

---

[†] This chapter was coauthored by Stéphane Rey.

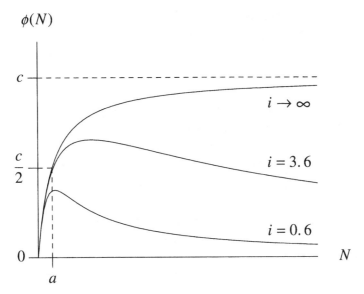

Fig. 9.1. Type IV functional response.

the example of musk ox that are more successful at fending off wolves when in herds than when alone. Collings (1997) considers spider mites that produce webbing. At high densities, the webbing interferes with the mites' predators. Veldkamp and Jannasch (1972) and Van Gemerden (1974) have even measured a type IV functional response for the uptake of hydrogen sulfide by purple sulfur bacteria. These anaerobic bacteria take up hydrogen sulfide for photosynthesis, but may suffer ill effects at high hydrogen sulfide concentrations.

We are interested in the behavior of equations (9.1a) and (9.1b) for a type IV functional response of the form

$$\phi(N) = \frac{cN}{\frac{N^2}{i} + N + a},$$
(9.2)

where c, i, and a are positive parameters. This functional form was introduced by Haldane (1930) in enzymology. It was then used by Andrews (1968) and Yano (1969) as a substrate uptake function. In the limit of large i, the right-hand side of equation (9.2) reduces to a type II functional response (see Figure 9.1). The parameters c and a can be interpreted as the maximum per capita predation (consumption) rate and the half-saturation constant in the absence of any inhibitory effect. The parameter i, in turn, is a direct measure of the predator's immunity from, or tolerance of, the prey. As i decreases, the predator's foraging efficiency decreases.

To reduce the number of parameters, we introduce the dimensionless variables

$$x \equiv \frac{N}{a}, \; y \equiv \frac{c}{r a} P, \; t \equiv r T. \tag{9.3}$$

Our system reduces to

$$\frac{dx}{dt} = x \left(1 - \frac{x}{\gamma} \right) - \frac{x y}{\frac{x^2}{\alpha} + x + 1}, \tag{9.4a}$$

$$\frac{dy}{dt} = \frac{\beta \delta x y}{\frac{x^2}{\alpha} + x + 1} - \delta y, \tag{9.4b}$$

with

$$\alpha \equiv \frac{i}{a}, \; \beta \equiv \frac{bc}{m}, \; \gamma \equiv \frac{K}{a}, \; \delta \equiv \frac{m}{r}. \tag{9.5}$$

Note that α is proportional to the immunity and that γ is proportional to the carrying capacity of the prey.

We will begin, as usual, with a stability analysis of the equilibria. This analysis will reveal a great deal about the dynamics of our model. It will not, however, show everything. To simplify this analysis, we introduce the functions

$$f(x) \equiv \frac{x}{\frac{x^2}{\alpha} + x + 1}, \tag{9.6a}$$

$$g(x) \equiv \left(1 - \frac{x}{\gamma} \right) \left(\frac{x^2}{\alpha} + x + 1 \right), \tag{9.6b}$$

$$h(x) \equiv \beta f(x) - 1 = \frac{-\frac{x^2}{\alpha} + (\beta - 1) x - 1}{\frac{x^2}{\alpha} + x + 1}, \tag{9.6c}$$

and rewrite equations (9.4a) and (9.4b) as

$$\frac{dx}{dt} = f(x) [g(x) - y], \tag{9.7a}$$

$$\frac{dy}{dt} = \delta h(x) y. \tag{9.7b}$$

After setting the right-hand side of equation (9.7a) equal to zero, we obtain two prey zero-growth isoclines: the y-axis and the curve $y = g(x)$ (see Figure 9.2). The curve $y = g(x)$ passes through the points $(0, 1)$ and $(\gamma, 0)$. When $\gamma > 1$, $g'(0) > 0$ and $g'(\gamma) < 0$, so that $g(x)$ has a local maximum between $x = 0$ and $x = \gamma$, at a point that we designate M.

If we now set the right-hand side of equation (9.7b) equal to zero, we obtain the predator zero-growth isoclines. The x-axis is one such isocline. Other

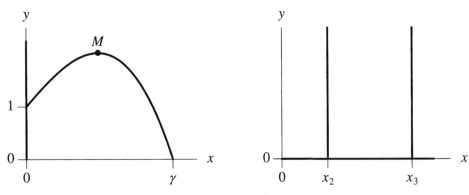

Fig. 9.2. Zero-growth isoclines.

isoclines occur at the roots of $h(x)$. These roots occur at the intersections of the functional response $f(x)$ with the horizontal line of ordinate $1/\beta$. If $\beta < 1$ or $\alpha < 4/(\beta - 1)^2$, $h(x)$ does not have any positive roots. In contrast, if $\beta > 1$ and $\alpha > 4/(\beta - 1)^2$, then $h(x)$ has two positive real roots,

$$x_3, x_2 = \frac{\alpha}{2}\left[(\beta - 1) \pm \sqrt{(\beta - 1)^2 - \frac{4}{\alpha}}\right], \tag{9.8}$$

and two predator zero-growth isoclines, $x = x_2$ and $x = x_3$ (see Figure 9.2).

The sign of $h(x)$ will play an important role in our stability analyses. If $\beta < 1$ or $\alpha < 4/(\beta - 1)^2$, then $h(x) < 0$ for all $x \geq 0$. If $\beta > 1$ and $\alpha > 4/(\beta - 1)^2$, then $h(x)$ is positive between its two roots,

$$h(x) > 0, \quad x_2 < x < x_3, \tag{9.9}$$

and negative outside of this interval,

$$h(x) < 0, \quad 0 < x < x_2 \text{ or } x > x_3. \tag{9.10}$$

Inequalities (9.9) and (9.10) imply that

$$h'(x_2) > 0, \quad h'(x_3) < 0. \tag{9.11}$$

Equilibria occur at the intersections of the predator and prey zero-growth isoclines. Table 9.1 lists these equilibria, along with conditions for their existence. At E_0, both species are extinct. At E_1, the prey is at its carrying capacity and the predator is extinct. Finally, at E_2 and E_3, the predator and prey coexist.

Table 9.1. *Equilibria*

Equilibrium	Coordinates	Conditions for existence
E_0	$(0,0)$	
E_1	$(\gamma,0)$	
E_2	$[x_2, g(x_2)]$	$\beta > 1,\ \alpha > 4/(\beta-1)^2,\ \gamma > x_2$
E_3	$[x_3, g(x_3)]$	$\beta > 1,\ \alpha > 4/(\beta-1)^2,\ \gamma > x_3$

The Jacobian for equations (9.7a) and (9.7b) is

$$J(x, y) = \begin{pmatrix} f'(x)\,[g(x) - y] + f(x)\,g'(x) & -f(x) \\ \delta\,h'(x)\,y & \delta\,h(x) \end{pmatrix}. \qquad (9.12)$$

We may determine the nature and stability of each equilibrium in terms of its community matrix (the Jacobian evaluated at that equilibrium point). If the eigenvalues of a community matrix all have negative real part, the corresponding equilibrium is asymptotically stable.

At E_0, the Jacobian reduces to

$$J(0, 0) = \begin{pmatrix} 1 & 0 \\ 0 & -\delta \end{pmatrix}. \qquad (9.13)$$

The eigenvalues $\lambda_1 = 1$ and $\lambda_2 = -\delta$ are real and of opposite sign: E_0 is an unstable saddle point.

At E_1, the community matrix is

$$J(\gamma, 0) = \begin{pmatrix} -1 & -f(\gamma) \\ 0 & \delta\,h(\gamma) \end{pmatrix}. \qquad (9.14)$$

This matrix has $\lambda_1 = -1$ and $\lambda_2 = \delta\,h(\gamma)$ as its eigenvalues. In light of our discussion regarding $h(x)$, we may conclude that E_1 is a stable node for $\beta < 1$ or $\alpha < 4/(\beta - 1)^2$; that it is an unstable saddle point for $\beta > 1$, $\alpha > 4/(\beta - 1)^2$, and $x_2 < \gamma < x_3$; and that it returns to being a stable node for $\beta > 1$, $\alpha > 4/(\beta - 1)^2$ and either $0 < \gamma < x_2$ or $\gamma > x_3$.

At E_2, the community matrix

$$J[x_2, g(x_2)] = \begin{pmatrix} \dfrac{1}{\beta}\,g'(x_2) & -\dfrac{1}{\beta} \\ \delta\,h'(x_2)\,g(x_2) & 0 \end{pmatrix} \qquad (9.15)$$

has the characteristic equation

$$\lambda^2 - \frac{1}{\beta}\,g'(x_2)\,\lambda + \frac{\delta}{\beta}\,h'(x_2)\,g(x_2) = 0, \qquad (9.16)$$

Table 9.2. *Regions of parameter space*

Case	Conditions			
I_1	$\beta < 1$, or	$\alpha < 4/(\beta - 1)^2$		
I_2	$\beta > 1$,	$\alpha > 4/(\beta - 1)^2$,	$\gamma < x_2$	
II	$\beta > 1$,	$\alpha > 4/(\beta - 1)^2$,	$x_2 < \gamma < x_3$,	$g'(x_2) < 0$
III	$\beta > 1$,	$\alpha > 4/(\beta - 1)^2$,	$\gamma > x_3$,	$g'(x_2) < 0$
IV	$\beta > 1$,	$\alpha > 4/(\beta - 1)^2$,	$x_2 < \gamma < x_3$,	$g'(x_2) > 0$
V	$\beta > 1$,	$\alpha > 4/(\beta - 1)^2$,	$\gamma > x_3$,	$g'(x_2) > 0$

with $h'(x_2) > 0$ and $g(x_2) > 0$. By Descartes's rule of signs, E_2 is a node, a focus, or a center. By the Routh–Hurwitz criterion, this equilibrium is asymptotically stable for $g'(x_2) < 0$ and unstable for $g'(x_2) > 0$. Just to the right of vertex M of $g(x)$, E_2 is a stable focus. Just to the left of this vertex, it is an unstable focus. By the Hopf bifurcation theorem, we expect a limit cycle close to $g'(x_2) = 0$.

Finally, at E_3, the Jacobian reduces to

$$J[x_3, g(x_3)] = \begin{pmatrix} \dfrac{1}{\beta} g'(x_3) & -\dfrac{1}{\beta} \\ \delta\, h'(x_3)\, g(x_3) & 0 \end{pmatrix}, \tag{9.17}$$

with characteristic equation

$$\lambda^2 - \frac{1}{\beta} g'(x_3)\lambda + \frac{\delta}{\beta} h'(x_3)\, g(x_3) = 0, \tag{9.18}$$

where $h'(x_3) < 0$, and $g(x_3) > 0$. Descartes's rule of signs implies that the two eigenvalues are real and of opposite sign, regardless of the sign of $g'(x_3)$. The equilibrium E_3 is an unstable saddle point.

Figure 9.3 depicts five cases for the existence and stability of equilibria. Solid circles indicate stable nodes or foci; open circles indicate unstable nodes or foci; crosses indicate saddle points. Each case occurs in a distinct portion of the (α, β, γ) parameter space (see Table 9.2). (The parameter δ has no effect on the location or stability of equilibria.) I have divided case I, with two equilibria, into two subcases: I_1 and I_2 have the same equilibria but different zero-growth isoclines.

Each region of parameter space is separated from its neighbors by one or more surfaces; a local bifurcation occurs as one crosses each surface. The equations for the surfaces are obtained by rewriting the conditions in Table 9.2 in terms of the parameters α, β, and γ, using equations (9.8). The

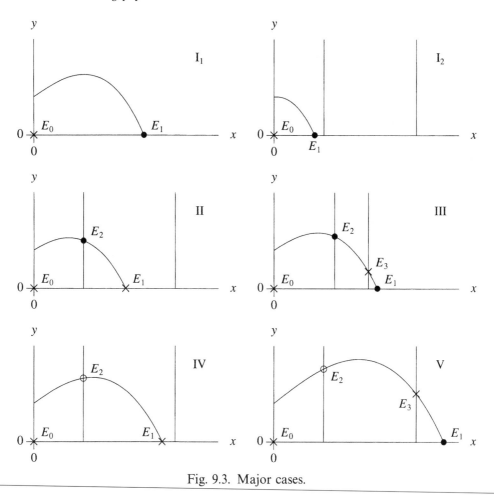

Fig. 9.3. Major cases.

surface for the Hopf bifurcation, for example, reduces to

$$\gamma = \frac{-2 + \frac{\alpha}{2}(3\beta - 1)\left[(\beta - 1) - \sqrt{(\beta - 1)^2 - \frac{4}{\alpha}}\right]}{\beta - \sqrt{(\beta - 1)^2 - \frac{4}{\alpha}}}. \tag{9.19}$$

Problem 9.1 *A Hopf bifurcation*
Derive equation (9.19).

Figure 9.4 shows the $\beta = 2$ slice of the (α, β, γ) parameter space. This slice is typical for all $\beta > 1$. However, this bifurcation diagram is based on

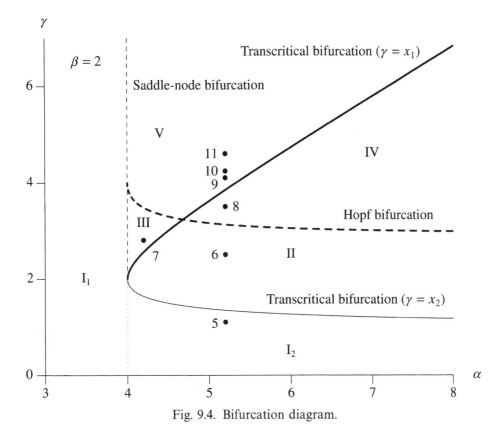

Fig. 9.4. Bifurcation diagram.

local analyses about equilibria. Does it capture all the interesting dynamics and bifurcations in this system? The simplest way to answer this question is to simulate equations (9.4a) and (9.4b) in each region of the bifurcation diagram. We have done this for points 5 to 11 of Figure 9.4 for $\delta = 2.5$.

In regions I, II, and III, the dynamics of numerical simulations are totally dominated by equilibria. In region I (Figure 9.5), the carrying capacity of the prey is too low to sustain the predators; orbits converge to equilibrium E_1. As we enter region II, there is a transcritical bifurcation: E_1 and E_2 exchange stability. Equilibrium E_1 is now an unstable saddle point whereas E_2 is a stable node or focus (Figure 9.6). Finally, as we enter region III, there is another transcritical bifurcation. Equilibria E_1 and E_3 exchange stability, leaving E_1 as a stable node and E_3 as an unstable saddle point (Figure 9.7). In region III, we have bistability: equilibria E_1 and E_2 are both stable. The corresponding domains of attractions are separated by the stable manifolds (the separatrices) of E_3. The predator now either survives or goes extinct, depending on the initial conditions.

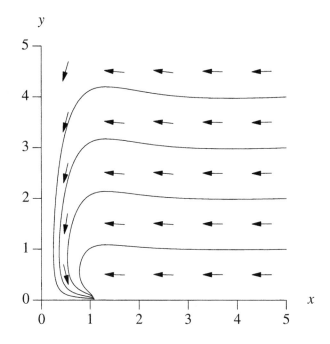

Fig. 9.5. Phase portrait for $\alpha = 5.2$, $\beta = 2.0$, $\gamma = 1.1$, $\delta = 2.5$.

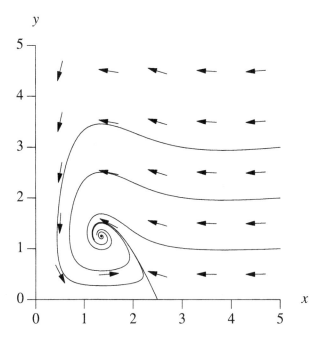

Fig. 9.6. Phase portrait for $\alpha = 5.2$, $\beta = 2.0$, $\gamma = 2.5$, $\delta = 2.5$.

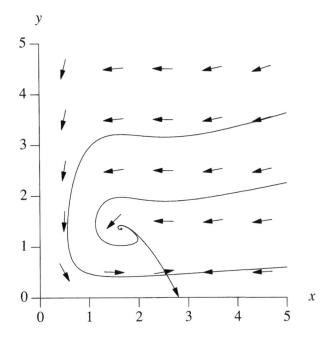

Fig. 9.7. Phase portrait for $\alpha = 4.2$, $\beta = 2.0$, $\gamma = 2.8$, $\delta = 2.5$.

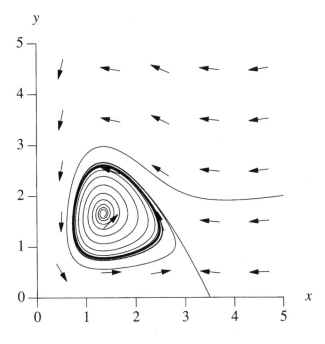

Fig. 9.8. Phase portrait for $\alpha = 5.2$, $\beta = 2.0$, $\gamma = 3.5$, $\delta = 2.5$.

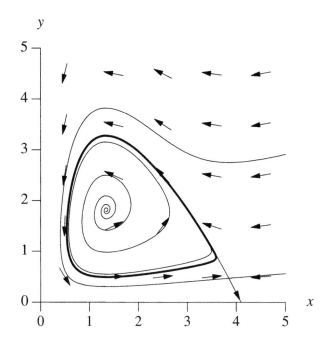

Fig. 9.9. Phase portrait for $\alpha = 5.2$, $\beta = 2.0$, $\gamma = 4.1$, $\delta = 2.5$.

Our simulations also show that a stable limit cycle is born in a supercritical Hopf bifurcation as we cross into regions IV or V. In region IV, E_2 is unstable and is surrounded by a stable limit cycle that attracts positive orbits (Figure 9.8). For higher γ, in region V (Figure 9.9), the limit cycle and equilibrium E_1 are both stable and the stable manifold of saddle point E_3 separates the phase plane into two domains of attraction. Now, something fascinating happens. As we increase γ, the limit cycle runs into the saddle point and disappears in a global bifurcation. (Our linearized stability analyses failed to pick up this bifurcation!) The last vestige of the limit cycle is a *homoclinic orbit* with beginning and end at E_3 (Figure 9.10). After the disappearance of the limit cycle, all orbits tend toward E_1 (Figure 9.11).

A global bifurcation in which a limit cycle runs into a saddle point, forms a homoclinic orbit, and disappears is known as a *homoclinic* (Farkas, 1994; Kuznetsov, 1995), a *saddle separatrix loop* (Hoppensteadt and Izhikevich, 1997), or an *Andronov–Leontovich* bifurcation (Andronov and Leontovich, 1939). Homoclinic bifurcations are examples of dangerous bifurcations. Dangerous bifurcations are discontinuous, with the sudden, blue-sky disappearance of the attractor and a fast dynamic jump to a distant unrelated attractor (Thompson *et al.*, 1994). The homoclinic bifurcation in our example was

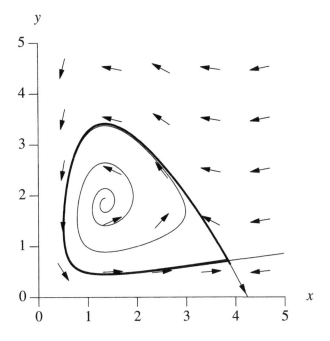

Fig. 9.10. Phase portrait for $\alpha = 5.2$, $\beta = 2.0$, $\gamma = 4.24$, $\delta = 2.5$.

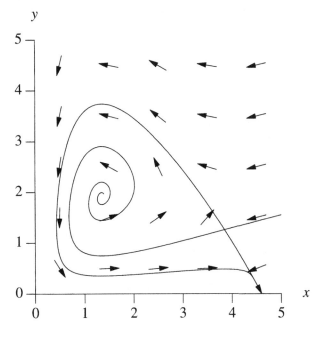

Fig. 9.11. Phase portrait for $\alpha = 5.2$, $\beta = 2.0$, $\gamma = 4.6$, $\delta = 2.5$.

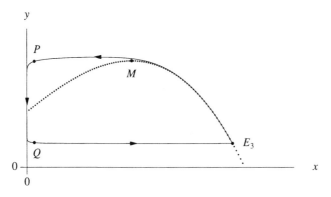

Fig. 9.12. Homoclinic orbit for $\alpha = 5.2$, $\beta = 2.0$,
$\gamma = 4.077267$, and $\delta = 0.02$.

catastrophic for the predator; the predator population collapsed immediately after the bifurcation. This homoclinic bifurcation occurred as we increased γ, or, equivalently, increased the carrying capacity K for fixed a, and is thus an example of the paradox of enrichment (Freedman and Wolkowicz, 1986).

Mathematical meanderings

Can we locate our homoclinic bifurcation in parameter space? To do so, we must search for homoclinic orbits, typically by integrating our differential equations numerically. Sometimes, the predator–prey limit cycle acts like a *relaxation oscillator*, with rapid changes in the prey population interspersed between other slower changes. In these instances, we may approximate the homoclinic orbit analytically.

We'll look for homoclinic orbits for small δ. Equations (9.7a) and (9.7b) define a vector field in the phase plane with slope

$$\frac{dy}{dx} = \frac{\delta\, h(x)\, y}{f(x)\, [g(x) - y]}. \tag{9.20}$$

In the case of small δ, the resulting flow is nearly horizontal, corresponding to large changes in prey number for small changes in predator number, once we're away from the prey zero-growth isoclines $x = 0$ and $y = g(x)$. If we examine the Jacobian at E_3 for small δ, we find that the unstable eigenvector has a slope close to $g'(x_3)$ and that the stable eigenvector is nearly horizontal. To a first approximation then (see Figure 9.12), a homoclinic orbit must follow the curve $y = g(x)$ until it reaches the vertex $M = (x_M, y_M)$. It then shoots quickly to the left until it gets close to the y-axis. At the point $P = (\sqrt{\delta}, y_P)$, the slope is still small, $dy/dx = O(\sqrt{\delta})$, by equation (9.20), so that P is still on the horizontal portion of the orbit. We can thus approximate y_P by y_M. Between P and $Q = (\sqrt{\delta}, y_Q)$, x is small ($x \leq \sqrt{\delta}$) and we may approximate equation (9.20) by

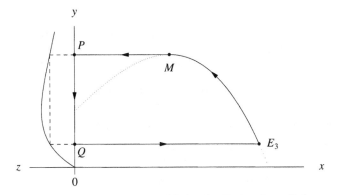

Fig. 9.13. Homoclinic orbit in the limit of small δ.

$$\frac{dy}{dx} = \frac{-\delta\, y}{x(1-y)}. \tag{9.21}$$

This differential equation is separable. Using the coordinates of P as initial conditions, we find that

$$x = \sqrt{\delta}\left(\frac{y_M\, e^{-y_M}}{y\, e^{-y}}\right)^{\frac{1}{\delta}}. \tag{9.22}$$

Since the coordinates of Q must also satisfy this equation, it follows that

$$y_M\, e^{-y_M} = y_Q\, e^{-y_Q}. \tag{9.23}$$

At Q, the orbit shoots horizontally to the right, since $dy/dx = O(\sqrt{\delta})$ at Q. For a homoclinic orbit, this orbit must now hit the equilibrium $E_3 = (x_3, y_3)$. We thus require that $y_Q = y_3$ so that

$$y_M\, e^{-y_M} = y_3\, e^{-y_3} \tag{9.24}$$

or

$$g(x_M)\,e^{-g(x_M)} = g(x_3)\,e^{-g(x_3)} \tag{9.25}$$

for a homoclinic orbit. This approximate condition becomes exact in the limit of small δ. Figure 9.13 shows a simple geometric interpretation of equation (9.24) in the limit of small δ. By bouncing off the function $z = y\exp(-y)$ in Figure 9.13 as shown (dashed line) one may uniquely determine the point Q in terms of the point P.

The function $g(x)$ is given by equation (9.6b). The coordinate x_3 is given by positive case of equation (9.8). The coordinate of x_M is easily shown, using elementary calculus, to be

$$x_M = \frac{(\gamma - \alpha) + \sqrt{(\gamma - \alpha)^2 + 3\alpha(\gamma - 1)}}{3}. \tag{9.26}$$

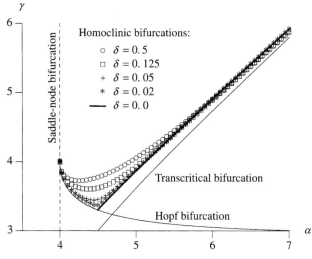

Fig. 9.14. Homoclinic bifurcations.

Equations (9.6b), (9.8), (9.25), and (9.26), together, give an implicit equation for γ in terms of α and β that can be solved using a root finder. Figure 9.14 shows the locus of points that satisfies this equation for $\beta = 2$ as a heavy solid line. For high α, this curve lies just above the curve for the transcritical bifurcation of equilibria E_1 and E_3. As α decreases, the gap between these two bifurcation curves widens. At $\alpha \approx 4.5$, the curve of homoclinic bifurcations intersects the Hopf bifurcation curve. Further to the left and in the limit of small δ, we imagine that a Hopf bifurcation is immediately followed by the homoclinic destruction of a limit cycle.

By combining a root finder and a numerical ordinary differential equation integrator, we were also able to find the locus of points corresponding to a homoclinic bifurcation for various nonzero values of δ. These are shown in Figure 9.14 (for $\beta = 2$). These curves lie below the $\delta = 0$ homoclinic bifurcation curve for high values of α and above it for low values of α. All of these curves coalesce at a special point,

$$(\alpha, \gamma) = \left[\frac{4}{(\beta - 1)^2}, \frac{4}{\beta - 1} \right], \tag{9.27}$$

known as a Bogdanov–Takens (double-zero) codimension-2 bifurcation point. This point occurs at the intersection of the saddle-node (or fold), Hopf, and homoclinic bifurcation curves (Kuznetsov, 1995) and corresponds, in this model, to a double equilibrium point at the maximum M of the prey zero-growth isocline $y = g(x)$.

The obvious effect of introducing a type IV functional response was to create an Allee effect for the predator. For high values of the rescaled prey carrying capacity γ and/or for low values of the rescaled immunity α (see Figure 9.4), a saddle point (E_3) appeared by means of a transcritical bifurcation. The predator population had to lie above this saddle point to persist. Otherwise, the predators were swamped by the prey, with too few predators to bring prey numbers and prey interference under control. For even higher values of γ and/or lower values of α (see Figure 9.14), the predator–prey limit cycle approached the separatrices of a single saddle point, formed a homoclinic loop, and disappeared. This led to the collapse of the predator population.

We can also generate a global bifurcation by introducing an Allee effect in the prey. Consider, for example, the system

$$\frac{dN}{dT} = r N \left(\frac{N}{K_0} - 1 \right) \left(1 - \frac{N}{K} \right) - c N P, \tag{9.28a}$$

$$\frac{dP}{dT} = b N P - m P, \tag{9.28b}$$

(Conway and Smoller, 1986). We will assume that the parameters r, K, c, b, and m are positive and that $0 < K_0 < K$. By introducing the dimensionless variables,

$$x \equiv \frac{N}{K}, \quad y \equiv \frac{c}{r} P, \quad t \equiv r T, \tag{9.29}$$

we may reduce equations (9.28a) and (9.28b) to a system with three parameters,

$$\frac{dx}{dt} = x \left[\left(\frac{x}{\gamma} - 1 \right) (1 - x) - y \right] = x [g(x) - y], \tag{9.30a}$$

$$\frac{dy}{dt} = \beta (x - \alpha) y, \tag{9.30b}$$

where

$$\alpha \equiv \frac{m}{b K}, \quad \beta \equiv \frac{b K}{r}, \quad \gamma \equiv \frac{K_0}{K}. \tag{9.31}$$

The x-axis and the vertical line $x = \alpha$ are the predator zero-growth isoclines. The y-axis and the parabola

$$y = g(x) = \left(\frac{x}{\gamma} - 1 \right) (1 - x) \tag{9.32}$$

are the prey zero-growth isoclines. These isoclines are shown in Figure 9.15.

The equilibria for this system are listed in Table 9.3. Both species are absent at E_0. This equilibrium can easily be shown to be a stable node.

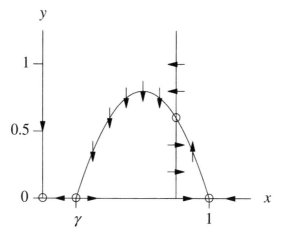

Fig. 9.15. Zero-growth isoclines with Allee effect.

Table 9.3. *Equilibria*

Equilibrium	Coordinates
E_0	$(0, 0)$
E_1	$(\gamma, 0)$
E_2	$(1, 0)$
E_3	$[\alpha, g(\alpha)]$

E_1, with no predators and γ prey, is a saddle point for $\alpha > \gamma$. Its unstable manifold is the x-axis; its stable manifold extends into the first quadrant. This equilibrium is the Allee threshold for the prey. E_2 has the prey at their carrying capacity and the predators extinct. This equilibrium is an unstable saddle point for $\alpha < 1$.

At E_3, predator and prey coexist. The stability of this equilibrium is determined by the community matrix

$$J = \begin{bmatrix} x\,g'(x) + g(x) - y & -x \\ \beta\,y & \beta\,(x - \alpha) \end{bmatrix}_{[\alpha,\,g(\alpha)]} \tag{9.33}$$

or

$$J = \begin{bmatrix} \alpha\,g'(\alpha) & -\alpha \\ \beta\,g(\alpha) & 0 \end{bmatrix}. \tag{9.34}$$

The community matrix has the characteristic equation

$$\lambda^2 - \alpha\,g'(\alpha)\,\lambda + \alpha\beta\,g(\alpha) = 0. \tag{9.35}$$

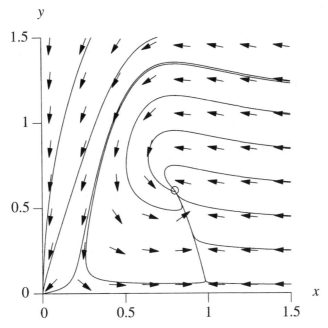

Fig. 9.16. Phase portrait for $\alpha = 0.8$, $\beta = 1$, $\gamma = 0.2$.

For $\gamma < \alpha < 1$, the stability of E_3 is determined by the sign of $g'(\alpha)$, the sign of the slope of the prey zero-growth isocline at E_3. If this slope is negative, then, by the Routh–Hurwitz criterion, the equilibrium is stable. If this slope is positive, the equilibrium is unstable. If the slope is zero, characteristic equation (9.35) has purely imaginary eigenvalues. As we move the predator zero-growth isocline $x = \alpha$ from just to the right to just to the left of the peak of the prey-zero growth isocline (by decreasing α), we go from a stable focus to an unstable focus and expect to see a Hopf bifurcation.

Figures 9.16 through 9.18 highlight this Hopf bifurcation. Figure 9.16 shows E_3 as a stable node. This attractor coexists with another stable node at the origin. The two basins of attractions are separated by a separatrix, the stable manifold of the saddle point at $(\gamma, 0)$. Orbits above the separatrix go to the origin (with extirpation of both species). Orbits below the separatrix are drawn towards E_3. Figure 9.17 shows E_3 as a stable focus just before the Hopf bifurcation. Figure 9.18 shows the phase portrait just after the Hopf bifurcation. The stable focus has turned into an unstable focus and is now surrounded by a stable limit cycle.

Keep in mind that the Hopf bifurcation theorem is a local result. This theorem predicts the existence of a limit cycle in the neighborhood of a critical value of α. However, it makes no statement and sheds no light on the

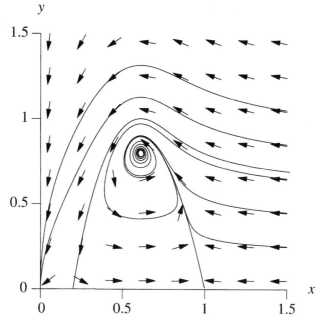

Fig. 9.17. Phase portrait for $\alpha = 0.61$, $\beta = 1$, $\gamma = 0.2$.

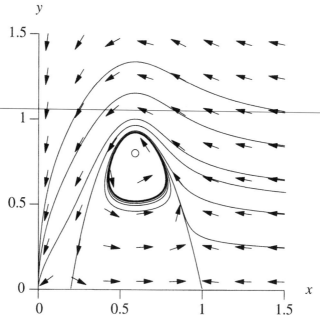

Fig. 9.18. Phase portrait for $\alpha = 0.59$, $\beta = 1$, $\gamma = 0.2$.

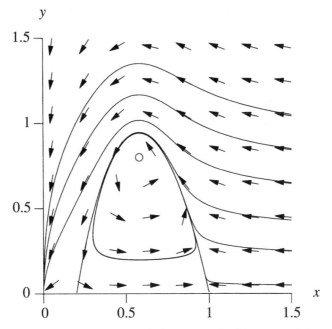

Fig. 9.19. Phase portrait for $\alpha = 0.58$, $\beta = 1$, $\gamma = 0.2$.

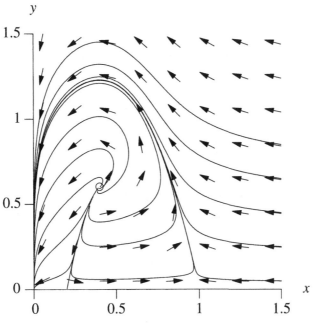

Fig. 9.20. Phase portrait for $\alpha = 0.4$, $\beta = 1$, $\gamma = 0.2$.

long-term existence of this limit cycle. And indeed, for this model, the limit cycle runs into trouble as we continue to decrease α. Figures 9.18 and 9.19 show that the limit cycle (a) gets larger, and (b) approaches the separatrix, as we continue to decrease α. Eventually, the limit cycle hits the separatrix and disappears in a global bifurcation. The last vestige of the limit cycle is a *cycle graph* consisting of the two *heteroclinic orbits* that connect the saddle points E_1 and E_2. One orbit lies along the abscissa. The other is only present when the unstable manifold emanating out of E_2 coincides with stable manifold entering E_1. After the disappearance of the limit cycle (Figure 9.20), all orbits tend towards the origin. In this example, both the predator and the prey go extinct.

Recommended readings

Kuznetsov (1995) and Hoppensteadt and Izhikevich (1997) give interesting and complementary introductions to the theory of global bifurcations.

10 Chemostat models

The fascinating single-species models of Chapters 1 through 6 and the equally interesting predator–prey models of Chapters 8 and 9 can be an experimentalist's nightmare. The chief problem lies with the carrying capacity K. In principle, this is a fine parameter. It is the size (or density) at which a population's birth rate equals its death rate. It defines the capacity of the environment to 'carry' organisms. This is all fine and good, but how do you measure the carrying capacity for an organism – especially if the carrying capacity is unstable? Is there some way of predicting the asymptotic behavior of a population short of measuring the birth and death rates for a range of densities?

Many organisms are limited by their inability to procure or assimilate enough of some essential environmental nutrient. We will follow the strategy of constructing a *resource-based model* for these organisms. Such models have experimental advantages over the models that we considered earlier. As a result, they are of increasing importance in the ecological literature.

Resource-based models have their origins in models for growth in a chemostat. A chemostat is a common laboratory apparatus for the culture and growth of microorganisms. You pump sterile growth medium into the chemostat at a constant rate and you keep the volume within the chemostat constant by letting excess medium (and microbes) flow out through a siphon. By modeling the uptake of some essential limiting nutrient (substrate), you can predict the effective carrying capacity for the species. Limnologists and oceanographers have taken these simple resource-based models and used them to predict the nutrient, phytoplankton, and zooplankton densities in lakes and oceans.

To see how the carrying capacity can arise from a simple resource-based model, let us start with a simple model for the concentration of substrate and the density of heterotroph in a chemostat. Picture, if you will, a bacterium that uses glucose for growth. We pump the substrate in at a constant rate and we let the substrate and the heterotroph drain out of the chemostat at a rate proportional to their concentrations. If the heterotroph takes up substrate with a simple mass-action functional response, we then have

$$\frac{dS}{dT} = D(S_i - S) - \frac{\mu_1}{Y_1} S H, \tag{10.1a}$$

$$\frac{dH}{dT} = \mu_1 S H - D H, \tag{10.1b}$$

where S is the concentration of the substrate, H is the density of heterotroph, S_i is the concentration of the inflowing substrate, D is the 'dilution rate', μ_1 is a growth constant for the heterotroph, and Y_1 is the yield of heterotroph per unit mass of substrate. S, H, and T are nonnegative variables; S_i, D, μ_1, and Y_1 are positive parameters.

There are the usual advantages to analyzing dimensionless equations. We therefore rescale all concentrations by S_i. In addition, we rescale the heterotroph by its yield constant. Finally, we treat the reciprocal of the dilution rate as a natural measure of time:

$$x \equiv \frac{S}{S_i}, \quad y \equiv \frac{H}{Y_1 S_i}, \quad t \equiv D T. \tag{10.2}$$

After some algebra, this yields

$$\frac{dx}{dt} = 1 - x - A x y, \tag{10.3a}$$

$$\frac{dy}{dt} = A x y - y, \tag{10.3b}$$

where

$$A \equiv \frac{\mu_1 S_i}{D}. \tag{10.4}$$

I will take a roundabout route to analyzing these equations. By adding equations (10.3a) and (10.3b) we obtain

$$\frac{dx}{dt} + \frac{dy}{dt} = 1 - x - y. \tag{10.5}$$

If we treat the sum of $x(t)$ and $y(t)$ as a single variable, equation (10.5) reduces to a simple ordinary differential that has the solution

$$x(t) + y(t) = 1 + (x_0 + y_0 - 1) e^{-t}, \tag{10.6}$$

where x_0 and y_0 are the initial concentrations of rescaled substrate and heterotroph. It follows that $x + y \to 1$ as $t \to \infty$. We may, in other words, study the asymptotic behavior of equations (10.3a) and (10.3b) along the line

$$x + y = 1. \tag{10.7}$$

Since we are primarily interested in the growth of the heterotroph, let us use equation (10.7) to eliminate x from equation (10.3b). This leaves us with

$$\frac{dy}{dt} = A(1 - y)y - y \tag{10.8}$$

or

$$\frac{dy}{dt} = (A - 1)y \left[1 - \frac{y}{\left(\frac{A-1}{A} \right)} \right]. \tag{10.9}$$

The right-hand side of equation (10.9) is a simple quadratic function of y. Equation (10.9) can thus be identified with the logistic equation. Note that

$$\lim_{t \to \infty} y(t) = \frac{A - 1}{A} = 1 - \frac{D}{\mu_1 S_i}. \tag{10.10}$$

Thus, in terms of the original dimensional units, the effective 'carrying capacity' of the heterotroph is simply

$$K = Y_1 \left(S_i - \frac{D}{\mu_1} \right). \tag{10.11}$$

The carrying capacity is defined in terms of quantities, Y_1, S_i, D, and μ_1, that are all easy to measure. This carrying capacity increases with the inflowing substrate concentration, yield coefficient, or growth constant, and decreases with the dilution rate.

This approach can be extended to more complicated models in which we include a holozoic predator that eats either a heterotroph or an autotroph (e.g., protozoa that eat bacteria or populations of zooplankton that eat phytoplankton). Two predator–prey–substrate models that have received great attention (Canale, 1970; Cunningham and Nisbet, 1983) are the double mass-action model

$$\frac{dS}{dT} = D(S_i - S) - \frac{\mu_1}{Y_1} S H, \tag{10.12a}$$

$$\frac{dH}{dT} = \mu_1 S H - D H - \frac{\mu_2}{Y_2} H P, \tag{10.12b}$$

$$\frac{dP}{dT} = \mu_2 H P - D P, \tag{10.12c}$$

and the double-Monod model

$$\frac{dS}{dT} = D\left(S_i - S\right) - \frac{\mu_1}{Y_1}\frac{S\,H}{K_1 + S}, \tag{10.13a}$$

$$\frac{dH}{dT} = \mu_1 \frac{S\,H}{K_1 + S} - D\,H - \frac{\mu_2}{Y_2}\frac{H\,P}{K_2 + H}, \tag{10.13b}$$

$$\frac{dP}{dT} = \mu_2 \frac{H\,P}{K_2 + H} - D\,P. \tag{10.13c}$$

P is the concentration of the holozoic predator; Y_2 is the yield coefficient for the predator; μ_2 is the predator's growth coefficient, for the double mass-action model, or its maximum growth rate, for the double-Monod model; K_1 and K_2 are the half-saturation constants for the heterotroph and for the predator. The remaining variables and parameters have the same meanings that they did at the start of this chapter. The double-Monod model differs from the double mass-action model in having a type II functional response (also called Michaelis–Menten or Monod kinetics) both for the heterotroph and for the holozoic predator.

Both of these models may be simplified by introducing the change of variables

$$x \equiv \frac{S}{S_i}, \quad y \equiv \frac{H}{Y_1\,S_i}, \quad z \equiv \frac{P}{Y_1\,Y_2\,S_i}, \quad t \equiv DT. \tag{10.14}$$

The double mass-action model simplifies to

$$\frac{dx}{dt} = 1 - x - A\,x\,y, \tag{10.15a}$$

$$\frac{dy}{dt} = A\,x\,y - y - B\,y\,z, \tag{10.15b}$$

$$\frac{dz}{dt} = B\,y\,z - z, \tag{10.15c}$$

with

$$A \equiv \frac{\mu_1\,S_i}{D}, \quad B \equiv \frac{\mu_2\,S_i\,Y_1}{D}. \tag{10.16}$$

The double-Monod model reduces to

$$\frac{dx}{dt} = 1 - x - \frac{A\,x\,y}{a + x}, \tag{10.17a}$$

$$\frac{dy}{dt} = \frac{A\,x\,y}{a + x} - y - \frac{B\,y\,z}{b + y}, \tag{10.17b}$$

$$\frac{dz}{dt} = \frac{B\,y\,z}{b + y} - z, \tag{10.17c}$$

with

$$A \equiv \frac{\mu_1}{D}, \quad a \equiv \frac{K_1}{S_i}, \quad B \equiv \frac{\mu_2}{D}, \quad b \equiv \frac{K_2}{Y_1 S_i}.$$

Let us consider the double mass-action model in more detail. If we add the three equations in this system, we see that

$$\frac{dx}{dt} + \frac{dy}{dt} + \frac{dz}{dt} = 1 - x - y - z. \tag{10.18}$$

If we treat the sum of the three concentrations as a single variable, equation (10.18) is easy to integrate and yields

$$x(t) + y(t) + z(t) = 1 + (x_0 + y_0 + z_0 - 1) e^{-t}. \tag{10.19}$$

Thus, after some transient, equations (10.15a), (10.15b), and (10.15c) will satisfy

$$x(t) + y(t) + z(t) = 1; \tag{10.20}$$

trajectories will lie on a two-dimensional triangular 'simplex' in the first octant of the (x, y, z) phase space. If we consider trajectories on this simplex, the double mass-action model reduces to

$$\frac{dy}{dt} = A (1 - y - z) y - y - B y z, \tag{10.21a}$$

$$\frac{dz}{dt} = B y z - z. \tag{10.21b}$$

Equations (10.21a) and (10.21b) can be analyzed by phase-plane techniques, and it will probably come as no surprise, in light of our earlier analyses, that this system has three equilibria, corresponding to 'washout' of both the predator and the prey,

$$x_1 = 1, \quad y_1 = 0, \quad z_1 = 0; \tag{10.22}$$

washout of just the predator,

$$x_2 = \frac{1}{A}, \quad y_2 = 1 - \frac{1}{A}, \quad z_2 = 0; \tag{10.23}$$

or coexistence of the predator, heterotroph, and substrate,

$$x_3 = 1 - \frac{A}{A + B}, \quad y_3 = \frac{1}{B}, \quad z_3 = \frac{A}{A + B} - \frac{1}{B}. \tag{10.24}$$

It is easy enough to show that the first equilibrium is a stable node (on the simplex) if

$$A < 1 \tag{10.25}$$

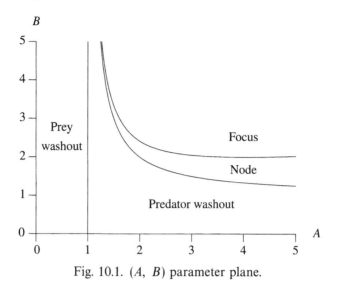

Fig. 10.1. (A, B) parameter plane.

and an unstable saddle point if this inequality is reversed. Similarly, given that $A > 1$, the second equilibrium is a stable node if

$$B < \frac{A}{A - 1} \tag{10.26}$$

and an unstable saddle point if this inequality is reversed. Finally, given that $A > 1$ and that $B > A/(A - 1)$, the third equilibrium is a stable node if

$$B < \frac{A}{2(\sqrt{A} - 1)} \tag{10.27}$$

and a stable focus if this inequality is reversed. The transitions are summarized in Figure 10.1 An experimentalist can cause transitions by changing either the dilution rate or the inflowing substrate concentration. Some of the orbits that are found on simplex (10.20) are shown in Figures 10.2 through 10.6. The outcomes are similar to those for a Lotka–Volterra system with prey-density dependence.

Problem 10.1 *Mass-action chemostat*
Derive inequalities (10.25), (10.26), and (10.27).

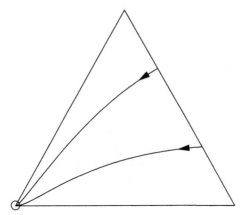

Fig. 10.2. $A = 0.5, \quad B = 1.0.$

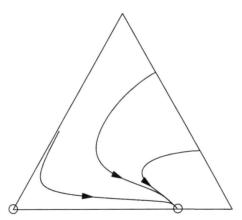

Fig. 10.3. $A = 4.0, \quad B = 1.0.$

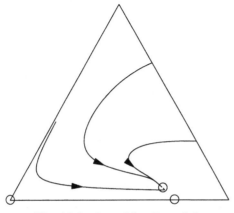

Fig. 10.4. $A = 4.0, \quad B = 1.5.$

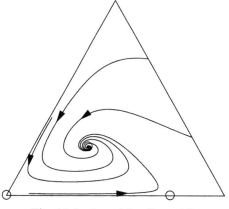

Fig. 10.5. $A = 4.0$, $B = 4.0$.

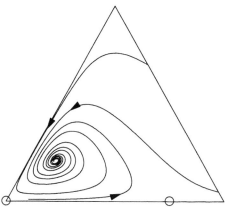

Fig. 10.6. $A = 4.0$, $B = 8.0$.

Mathematical meanderings

All the models that we've considered have been autonomous: there was no forcing and time did not appear explicitly in the equations. However, most organisms live in diurnally or seasonally forced environments. We, as humans, are certainly aware of the profound differences between day and night and between summer and winter. Fortunately, we do a good job of correcting for these differences. Even so, we may suffer the serious effects of this forcing. There is evidence, for example, that seasonal variation in contact rates drives the dynamics of childhood disease epidemics (Schaffer, 1985; Olsen and Schaffer, 1990). For many organisms, seasonal or diurnal periodicity in competition coefficients plays a pivotal role in the coexistence of competitors (Van Gemerden, 1974; Cushing, 1980; Dhondt and Eyckerman, 1980; Ebenhoh, 1988). Seasonal pulses in the production of resources can also lead to the episodic reproduction referred to in Chapter 4.

Periodic forcing can have an especially strong effect on predators and prey. Periodically forced predator–prey models exhibit a striking variety of periodic (Cushing, 1977b, 1982; Bardi 1981) and chaotic (Inoue and Kamifukumoto, 1984; Leven *et al.*, 1987; Schaffer, 1988; Allen, 1990; Doveri *et al.*, 1992; Kot *et al.*, 1992; Pavlou and Kevrekidis, 1992; Sabin and Summers, 1992; Rinaldi and Muratori, 1993; Rinaldi *et al.*, 1993) solutions. The same factors that lead to fascinating synamics also, unfortunately, lead to analytic intractability. As a result, more recent studies of periodically forced predator–prey models have relied on numerical methods.

One way to reduce the analytic intractability of forced systems is to consider periodically pulsed systems. One can find examples of impulsive forcing and impulsive parametric excitation in fields as disparate as mechanical engineering (Bernussou, 1977; Hsu, 1977, 1987) and pharmacokinetics (Mazumdar, 1989). Models with impulsive forcing can often be rewritten as simple discrete-time mappings or as difference equations. An early ecological example of this approach is the work by Ebenhoh (1988); piecewise-linear growth functions and periodic pulsing were combined to produce a simple model for competing algae.

The double mass-action model may be modified to include periodic pulsing of substrate (Funasaki and Kot, 1993). It then takes the form

$$\frac{dS}{dT} = -DS - \frac{\mu_1}{Y_1} SH, \tag{10.28a}$$

$$\frac{dH}{dT} = \mu_1 SH - DH - \frac{\mu_2}{Y_2} HP, \tag{10.28b}$$

$$\frac{dP}{dT} = \mu_2 HP - DP, \tag{10.28c}$$

$$S\left(\frac{\tau^+}{D}\right) = S\left(\frac{n\tau^-}{D}\right) + \tau S_i, \tag{10.28d}$$

with $n = 0, 1, 2, \ldots$. Differential equations (10.28a), (10.28b), and (10.28c) apply between pulses. Equation (10.28d) describes the actual pulsing. τ/D is the period of the pulsing; τS_i is the amount of limiting substrate pulsed each τ/D units of time. DS_i units of substrate are added, on average, per unit of time. This system may be rescaled, as before, leaving us with the dimensionless system

$$\frac{dx}{dt} = -x - Axy, \tag{10.29a}$$

$$\frac{dy}{dt} = Axy - y - Byz, \tag{10.29b}$$

$$\frac{dz}{dt} = Byz - z, \tag{10.29c}$$

$$x(n\tau^+) = x(n\tau^-) + \tau. \tag{10.29d}$$

Let us start with a simpler system. With no holozoic predator, the dimensionless system reduces to

$$\frac{dx}{dt} = -x - A x y, \tag{10.30a}$$

$$\frac{dy}{dt} = A x y - y, \tag{10.30b}$$

$$x(n\tau^+) = x(n\tau^-) + \tau. \tag{10.30c}$$

This nonlinear system possesses simple periodic solutions. Can we derive these periodic solutions?

If we add equations (10.30a) and (10.30b),

$$\frac{dx}{dt} + \frac{dy}{dt} = -(x + y), \tag{10.31}$$

we can integrate and solve for the total concentration in the chemostat between pulses,

$$x(t) + y(t) = (x_{n\tau} + y_{n\tau}) e^{-(t-n\tau)}, \tag{10.32}$$

$n\tau < t < (n + 1)\tau$, with $x_{n\tau}$ and $y_{n\tau}$ the initial concentration of substrate and bacteria at time $n\tau$. Equation (10.32) allows us to decouple the substrate–bacterium equations (10.30a) and (10.30b),

$$\frac{dx}{dt} = -[A(x_{n\tau} + y_{n\tau}) e^{-(t-n\tau)} + 1] x + A x^2, \tag{10.33a}$$

$$\frac{dy}{dt} = [A(x_{n\tau} + y_{n\tau}) e^{-(t-n\tau)} - 1] y - A y^2. \tag{10.33b}$$

The latter equations may, in turn, be solved as Bernoulli equations:

$$x(t) = \frac{x_{n\tau}(x_{n\tau} + y_{n\tau}) e^{-(t-n\tau)}}{x_{n\tau} + y_{n\tau} e^{A(x_{n\tau}+y_{n\tau})[1-e^{-(t-n\tau)}]}}, \tag{10.34a}$$

$$y(t) = \frac{(x_{n\tau} + y_{n\tau}) y_{n\tau} e^{A(x_{n\tau}+y_{n\tau})[1-e^{-(t-n\tau)}]} e^{-(t-n\tau)}}{x_{n\tau} + y_{n\tau} e^{A(x_{n\tau}+y_{n\tau})[1-e^{-(t-n\tau)}]}}. \tag{10.34b}$$

Equations (10.34a) and (10.34b) apply between pulses. After each new pulse of substrate, we may write

$$x_{(n+1)\tau} = \frac{x_{n\tau}(x_{n\tau} + y_{n\tau}) e^{-\tau}}{x_{n\tau} + y_{n\tau} e^{A(x_{n\tau}+y_{n\tau})(1-e^{-\tau})}} + \tau, \tag{10.35a}$$

$$y_{(n+1)\tau} = \frac{(x_{n\tau} + y_{n\tau}) y_{n\tau} e^{A(x_{n\tau}+y_{n\tau})(1-e^{-\tau})} e^{-\tau}}{x_{n\tau} + y_{n\tau} e^{A(x_{n\tau}+y_{n\tau})(1-e^{-\tau})}}. \tag{10.35b}$$

Equations (10.35a) and (10.35b) are difference equations that describe the substrate and bacterial concentrations at each periodic pulse in terms of the concentrations at the previous pulse. These equations allow us to stroboscopically sample the chemostat at its forcing period. The limiting behavior of equations (10.35a) and (10.35b), coupled with between-pulse equations (10.34a) and (10.34b), determines the asymptotic behavior of the substrate and bacteria (in the absence of protozoa) within the pulsed chemostat.

The above difference equations may be analyzed directly. However, if you add equations (10.35a) and (10.35b), you will observe that the total concentration satisfies a very simple difference equation:

$$x_{(n+1)\tau} + y_{(n+1)\tau} = e^{-\tau}(x_{n\tau} + y_{n\tau}) + \tau. \tag{10.36}$$

The last equation is linear. After solving this equation exactly,

$$x_{n\tau} + y_{n\tau} = \frac{\tau}{1 - e^{-\tau}} + \left(x_0 + y_0 - \frac{\tau}{1 - e^{-\tau}}\right) e^{-n\tau}. \tag{10.37}$$

we may take the limit of a large number of pulses. In this limit, the total concentration of substrate and bacteria approaches a constant,

$$\lim_{n \to \infty} (x_{n\tau} + y_{n\tau}) = \frac{\tau}{1 - e^{-\tau}}, \tag{10.38}$$

and stroboscopic equations (10.35a) and (10.35b) decouple, leaving us with an autonomous pair of first-order difference equations:

$$x_{(n+1)\tau} = \frac{\tau e^{-\tau} x_{n\tau}}{\tau e^{A\tau} - (e^{A\tau} - 1)(1 - e^{-\tau}) x_{n\tau}} + \tau, \tag{10.39a}$$

$$y_{(n+1)\tau} = \frac{\tau e^{(A-1)\tau} y_{n\tau}}{\tau + (e^{A\tau} - 1)(1 - e^{-\tau}) y_{n\tau}}. \tag{10.39b}$$

Equations (10.39a) and (10.39b) each possess two equilibria. The bacterium goes extinct at the first equilibrium,

$$x_e^* = \frac{\tau}{1 - e^{-\tau}}, \quad y_e^* = 0, \tag{10.40}$$

but survives at the second equilibrium,

$$x_s^* = \frac{\tau e^{A\tau}}{e^{A\tau} - 1}, \tag{10.41a}$$

$$y_s^* = \frac{\tau [e^{(A-1)\tau} - 1]}{(e^{A\tau} - 1)(1 - e^{-\tau})}, \tag{10.41b}$$

We may determine the stability of the equilibria by evaluating the slope λ of each different equation at each of its equilibria. $|\lambda| < 1$ implies asymptotic stability. After evaluating the derivatives of (10.39a) and (10.39b) at equilibrium (10.40), we find

$$\lambda_e = e^{(A-1)\tau}. \tag{10.42}$$

Evaluating these same derivatives at equilibrium coordinates (10.41a) and (10.41b) produces

$$\lambda_s = e^{-(A-1)\tau}. \tag{10.43}$$

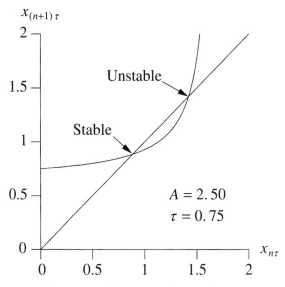

Fig. 10.7. Substrate stroboscopic map.

For $A < 1$, the first equilibrium is stable. As stroboscopic samples of the substrate and of the heterograph ($x_{n\tau}$ and $y_{n\tau}$) approach (10.40) the heterograph is lost or washes out the chemostat. Since equations (10.34a) and (10.34b) still describe the substrate and bacterial concentrations between pulses, trajectories approach the periodic solution

$$x_e(t) = \frac{\tau}{1 - e^{-\tau}} e^{-(t - n\tau)}, \quad y_e(t) = 0, \tag{10.44}$$

$n\tau < t < (n + 1)\tau$.

For $A > 1$ (see Figures 10.7 and 10.8), the second equilibrium is stable. Trajectories now approach the periodic solution

$$x_s(t) = \frac{\tau e^{-(t - n\tau)}}{1 - e^{-\tau} + (e^{-\tau} - e^{-A\tau}) e^{A\tau [1 - e^{-(t-n\tau)}]/(1 - e^{-\tau})}}, \tag{10.45a}$$

$$y_s(t) = \frac{\tau e^{-(t - n\tau)} (e^{-\tau} - e^{-A\tau})}{(1 - e^{-\tau})^2 e^{-A\tau [1 - e^{-(t-n\tau)}]/(1 - e^{-\tau})} + (1 - e^{-\tau})(e^{-\tau} - e^{-A\tau})}, \tag{10.45b}$$

$n\tau < t < (n + 1)\tau$. See Figure 10.9. There is also a transcritical bifurcation at $A = 1$, as equilibria and periodic solutions pass through each other and exchange stability. You should also note that the solutions $x_e(t)$ and $x_s(t)$ are discontinuous for t as a multiple of τ.

Periodic solutions dominate the behavior of the pulsed mass-action chemostat without a holozoic predator. This system has two periodic solutions as its attractors. These two solutions may be written in closed form, as shown by equations (10.44), (10.45a), and (10.45b).

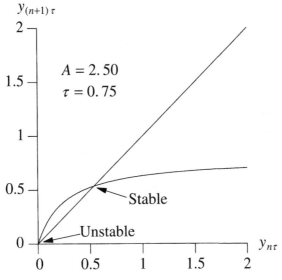

Fig. 10.8. Heterotroph stroboscopic map.

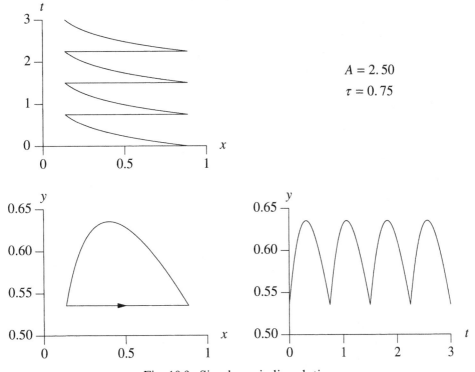

Fig. 10.9. Simple periodic solution.

For $A > 1$, the second periodic solution is stable to small perturbations in substrate or bacterial concentration. Is this periodic solution also stable to small perturbations in protozoa? Can a holozoic protozoan successfully invade a pulsed mass-action chemostat containing substrate and bacteria?

Equations (10.45a) and (10.45b), together with $z_s(t) = 0$, are one solution to the full unsimplified substrate–bacterium–protozoan system. To determine whether protozoa can invade the pulsed chemostat, we introduce small perturbations $u(t)$, $v(t)$, and $w(t)$,

$$x(t) = x_s(t) + u(t), \tag{10.46a}$$

$$y(t) = y_s(t) + v(t), \tag{10.46b}$$

$$z(t) = z_s(t) + w(t), \tag{10.46c}$$

and linearize equations (10.29a), (10.29b), and (10.29c) with respect to the perturbations. This results in a linear system of equations

$$\frac{du}{dt} = -[A\, y_s(t) + 1]\, u - A\, x_s(t)\, v, \tag{10.47a}$$

$$\frac{dv}{dt} = A\, y_s(t)\, u + [A\, x_s(t) - 1]\, v - B\, y_s(t)\, w, \tag{10.47b}$$

$$\frac{dw}{dt} = [B\, y_s(t) - 1]\, w, \tag{10.47c}$$

with periodic coefficients.

Difficulties arise in solving most linear differential equations with periodic coefficients. These equations are usually studied using Floquet theory (Richards, 1983; Zwillinger, 1992). In our case, mercifully, we need only solve the decoupled equation (10.47c) to determine whether protozoa increase or decrease after being introduced. In addition, since the concentration $y_s(t)$ is periodic with period τ, integrating over one period,

$$w_{(n+1)\tau} = w_{n\tau}\, e^{\int_{n\tau}^{(n+1)\tau} [B\, y_s(t) - 1]\, dt}, \tag{10.48}$$

produces

$$w_{(n+1)\tau} = w_{n\tau}\, e^{\left[B\, \frac{(A-1)}{A} - 1\right]\tau}. \tag{10.49}$$

This last equation describes the growth of the protozoa over one period.

The geometric rate of growth, the Floquet multiplier $\mu = \exp(\lambda\tau)$ with characteristic exponent $\lambda = \{[B\,(A-1)/A] - 1\}$, is critical to our analysis. For $B < A/(A-1)$, $\lambda < 0$ and $\mu < 1$ so that efflux predominates over growth and the invasion fails. Trajectories now approach the (x, y) plane and periodic solution (10.44), for $A < 1$, or the periodic solutions (10.45a) and (10.45b), for $A > 1$.

For $A > 1$ and $B = A/(A-1)$, we have a transcritical bifurcation. An unstable saddle-like limit cycle from outside the first octant passes through the periodic orbit with coordinates (10.45a) and (10.45b). The two periodic orbits trade stability so that the invading orbit enters the first octant as a stable limit cycle for $B > A/(A-1)$. The existence of a periodic solution in the positive

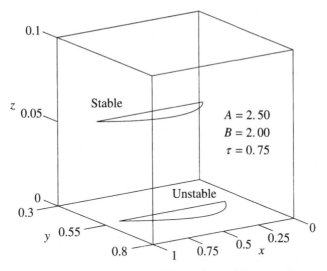

Fig. 10.10. Transcritical bifurcation of limit cycles.

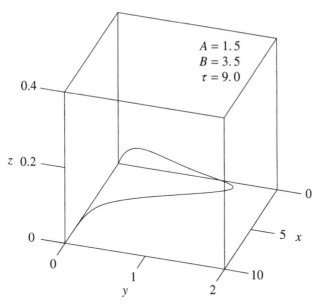

Fig. 10.11. τ-periodic limit cycle.

octant for $A > 1$ and small $B > A/(A - 1)$ can be proved by following Butler and Waltman (1981) and Waltman (1983). See also Hsu *et al.* (1978). Other details can be ascertained numerically. Thus, for B just above $A/(A - 1)$, we observe a stable limit cycle inside the first octant (see Figure 10.10) and the periodic coexistence of substrate, bacteria, and protozoa. The protozoan has successfully invaded the chemostat.

For $A > 1$ and for small enough B in excess of $A/(A - 1)$, the substrate,

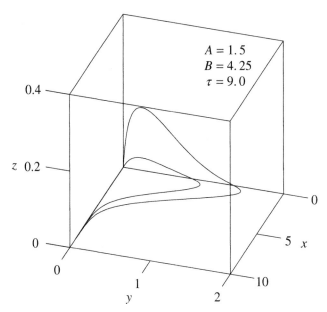

Fig. 10.12. 2τ-periodic limit cycle.

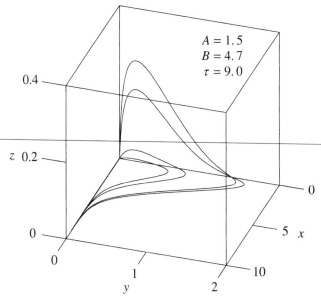

Fig. 10.13. 4τ-periodic limit cycle.

bacteria, and protozoa coexist on a limit cycle of period τ. An example of this simple cycle can be found in Figure 10.11. After each pulse of substrate, there is a surge in the number of bacteria. This is followed by a second surge in the number of protozoa.

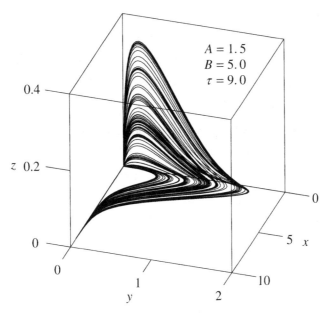

$$A = 1.5$$
$$B = 5.0$$
$$\tau = 9.0$$

Fig. 10.14. Chaotic strange attractor.

For sufficiently large B, this periodic solution loses its stability. Increases in B lead to a cascade of period-doubling bifurcations (Figures 10.12 and 10.13) and to the appearance of chaotic strange attractors. Figure 10.14 illustrates a strange attractor. There are large and erratic fluctuations in the number of bacteria and protozoa (Figure 10.15) on this attractor. The substrate exhibits smaller, but equally erratic, variation in its behavior.

I have documented the role of B by stroboscopically sampling the concentration of bacteria for different values of B. I numerically integrated equations (10.29a), (10.29b), and (10.29c) for 1250 cycles of pulsing and 2000 values of B. For each B, I have plotted the last 250 measurements of bacterial concentration. Since I am sampling the system stroboscopically at the forcing period, periodic solutions of period τ appear as fixed points, periodic solutions of period 2τ appear as pairs of points, and so on. The bifurcation diagram (Figure 10.16) shows a wealth of phenomena including: (1) the invasion of the protozoan at $B = 3$, (2) the first period-doubling at $B \approx 4.06$, (3) a cascade of period doublings, (4) chaotic solutions, and (5) periodic windows within the chaotic regime (e.g., the 3-cycle at $B \approx 5.2$).

The bifurcation diagram in (Figure 10.16) arises from a simple mapping embedded within our pulsed system. Various methods allow us to uncover this mapping. For fixed B, one can construct a time-τ or stroboscopic map, or a Poincaré or first return map (Parker and Chua, 1989). In our case, it is easier to construct a peak-to-peak map.

In Figure 10.17, I take the strange attractor of Figures 10.14 and 10.15 and plot each peak in protozoa as a function of the preceding peak. This simple procedure uncovers a simple unimodal map that is the deterministic heart of

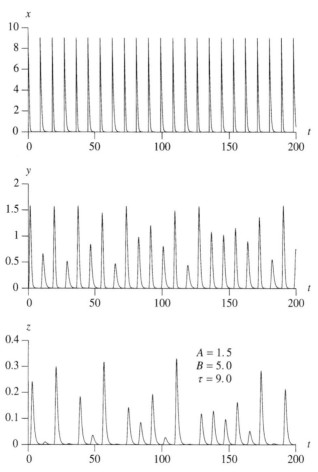

Fig. 10.15. Chaotic time series.

the chaotic behavior of the strange attractor. The seemingly erratic surges in protozoa fall out along this simple curve. This peak-to-peak map is quite similar to the Ricker curve that we studied in Chapter 4 – one is easily beguiled into discussing density dependence, scramble competition, or overcompensation. Increasing B steepens the mapping. Comparable changes occur with an increase in the pulse period τ, or with an increase in the uptake constant A. There is a tradeoff between the parameters B and τ: for fixed A, large τ and small B, small τ and large B, or intermediate τ and intermediate B are each capable of fomenting instability and chaos.

Although I have emphasized the double mass-action model and its pulsed analog, there has also been a keen interest in the double-Monod

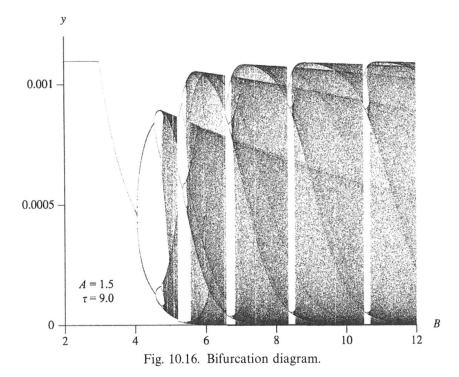

Fig. 10.16. Bifurcation diagram.

model and its periodically forced counterpart (Canale, 1970; Cunning-ham and Nisbet, 1983; Kot *et al.*, 1992; Pavlou and Kevrekidis, 1992). In the double-Monod model, the predator and the prey each exhibit a type II functional response (Monod kinetics) rather than mass-action kinetics. This difference in functional response is of great import, since it permits the existence of a limit cycle with no periodic forcing. The effect of forcing, then, is to introduce a second, typically incommensurate, frequency. The forced double-Monod model follows a rather complicated route to chaos with quasiperiodicity leading to phase-locking and only later a strange attractor. Stroboscopic sections reveal circle maps for the quasiperiodic regimes, and noninvertible maps of the interval for the chaotic regime.

Periodic forcing and pulsing are not the only mechanisms that cause chaos in predator–prey models. The Poincaré–Bendixson theorem precludes chaotic dynamics for an autonomous system in the plane, but the added complexity that one attains by adding more species or more dimensions to an autonomous system is often enough to set off chaotic dynamics. For example, Hastings and Powell (1991) studied a simple three-species food

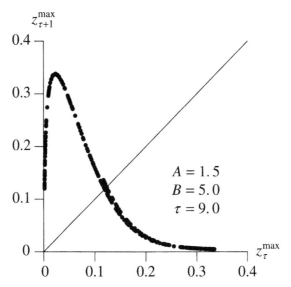

Fig. 10.17. Peak-to-peak map for a strange attractor.

chain model of the form

$$\frac{dx}{dt} = x(1 - x) - f_1(x) y, \tag{10.50a}$$

$$\frac{dy}{dt} = f_1(x) y - f_2(y) z - d_1 y, \tag{10.50b}$$

$$\frac{dz}{dt} = f_2(y) z - d_2 z, \tag{10.50c}$$

with

$$f_i(u) = \frac{a_i u}{1 + b_i u}, \tag{10.51}$$

and showed that this system possesses striking strange attractors. Kuznetsov and Rinaldi (1995) recently reexamined this system; they describe a method for geometrically constructing strange attractors. And even a three-species mass-action (Lotka–Volterra) model for a predator and its two prey can exhibit a strange attractor if one allows for competition between the prey (Gilpin, 1979; Schaffer *et al.*, 1986).

Recommended readings

Smith and Waltman (1995) provide an excellent survey of the dynamics of microbial chemostat systems.

11 Discrete-time predator–prey models

Difference equations appeared in the previous chapter as stroboscopic, peak-to-peak, or Poincaré maps for a predator–prey model with periodically pulsed substrate. Predator–prey models can also be formulated as discrete-time mappings *ab initio*. This is appropriate, and convenient, when organisms have discrete, nonoverlapping generations.

Let N_t be the number of potential prey or hosts in a population and let P_t be the number of predators or parasitoids.† A discrete-time model for these two species may easily mimic the continuous-time models of Chapters 7 and 8. For example, consider a model in which a fraction $1/(1 + aP_t)$ of all hosts survive long enough to reproduce (with net reproductive rate of R_0), and in which each parasitized host gives rise to b new parasitoids:

$$N_{t+1} = \frac{R_0 N_t}{1 + a P_t}, \tag{11.1a}$$

$$P_{t+1} = b \left(1 - \frac{1}{1 + a P_t} \right) N_t. \tag{11.1b}$$

The change of variables

$$x_t \equiv a b N_t, \quad y_t \equiv a P_t, \tag{11.2}$$

---

† I should begin by reminding you of the definition of a parasitoid. Predators, of course, kill their prey, typically for food. Parasites live in or on a host and draw food, shelter, or other requirements from that host, often without killing it. Female parasitoids, in turn, typically search for and kill, but do not consume, their hosts. Rather, they oviposit on, in, or near the host and use the host as a source of food and shelter for future developing young. If you ever saw the movie *Alien* you probably have a good image of parasitoids. However, unlike the large parasitoids of that movie, most parasitoids are small—insect-sized. There are close to 50 000 described species of hymenopteran (of the order containing wasps and bees) parasitoids, about 15 000 described species of dipteran (of the order containing flies) parasitoids, and some 3000 described species in other orders. There are even hyperparasitoids (parasitoids on other parasitoids). For these species, the production of new parasitoids is tightly coupled to the functional response.

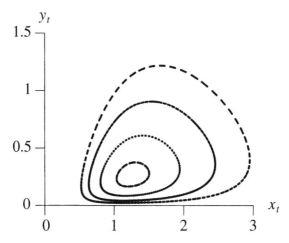

Fig. 11.1. Neutrally stable quasiperiodic solutions.

reduces this system to

$$x_{t+1} = \frac{R_0\, x_t}{1 + y_t}, \tag{11.3a}$$

$$y_{t+1} = \frac{x_t\, y_t}{1 + y_t}. \tag{11.3b}$$

These equations possess a continuum of neutrally stable quasiperiodic solutions (see Figure 11.1) that are reminiscent of the orbits of the classical Lotka–Volterra model described in Chapter 7.

Discrete-time predator–prey models can also exhibit dynamics that are far wilder than that of their continuous-time counterparts. To give you some feel for the range of dynamics that can occur, I will begin with the ecological discretization of a continuous-time predator–prey model. Then I will introduce you to some standard host–parasitoid models.

Earlier, we considered

$$\frac{dN}{dt} = r N \left(1 - \frac{N}{K} \right) - c N P, \tag{11.4a}$$

$$\frac{dP}{dt} = b N P - m P. \tag{11.4b}$$

This system differs from the Lotka–Volterra predator–prey model in that prey that lack predators grow logistically. You will recall that this system possesses three equilibria, corresponding to (a) extinction of both the predator and the prey, (b) extinction of the predator and survival of the prey at its carrying capacity, or (c) survival of both the predator and the prey. The last equilibrium is either a stable node or a stable focus.

If reproduction is discrete, we may replace the derivatives in equations (11.4a) and (11.4b) with the divided differences

$$\frac{dN}{dt} \approx \frac{N_{t+\Delta t} - N_t}{\Delta t}, \quad \frac{dP}{dt} \approx \frac{P_{t+\Delta t} - P_t}{\Delta t}. \tag{11.5}$$

Setting the generation time Δt to 1 and forcing all predators to die after one generation ($m = 1$) produces a system,

$$N_{t+1} = N_t + r N_t \left(1 - \frac{N_t}{K}\right) - c N_t P_t, \tag{11.6a}$$

$$P_{t+1} = b N_t P_t, \tag{11.6b}$$

that was first studied by Maynard Smith (1968). After rescaling,

$$x_t \equiv \frac{N_t}{K}, \quad y_t \equiv \frac{c P_t}{b K}, \quad a \equiv b K, \tag{11.7}$$

this system simplifies to

$$x_{t+1} = (1 + r) x_t - r x_t^2 - a x_t y_t, \tag{11.8a}$$

$$y_{t+1} = a x_t y_t. \tag{11.8b}$$

This mapping has the undesirable property that it may generate negative numbers of individuals. At the same time, this mapping is easier to analyze than similar models that do preserve first-quadrant invariance.

The equilibria for this system appear as fixed points of the mapping,

$$x_{t+1} = x_t = x^*, \tag{11.9a}$$

$$y_{t+1} = y_t = y^*. \tag{11.9b}$$

It is easy to show that there are three equilibria, (x_i^*, y_i^*), $i = 0, 1, 2$, corresponding to extinction of both the predator and the prey,

$$(x_0^*, y_0^*) = (0, 0); \tag{11.10}$$

extinction of the predator,

$$(x_1^*, y_1^*) = (1, 0); \tag{11.11}$$

and coexistence of the predator and prey,

$$(x_2^*, y_2^*) = \left[\frac{1}{a}, \frac{r(a - 1)}{a^2}\right]. \tag{11.12}$$

The stability of each equilibrium is determined by the eigenvalues λ of the Jacobian,

$$J = \begin{bmatrix} (1 + r) - 2 r x_t - a y_t & -a x_t \\ a y_t & a x_t \end{bmatrix}, \tag{11.13}$$

evaluated at the equilibrium. We are dealing with linear difference (not differential) equations. As a result, solutions grow like λ^t (rather than $e^{\lambda t}$).

For the trivial equilibrium at $(0, 0)$, the Jacobian reduces to

$$J = \begin{pmatrix} 1 + r & 0 \\ 0 & 0 \end{pmatrix}. \tag{11.14}$$

The eigenvalues are simply

$$\lambda = (1 + r), \, 0. \tag{11.15}$$

Perturbations along the ordinate decay rapidly. Perturbations along the abscissa grow if $r > 0$. The origin is thus a saddle point for positive r.

For the equilibrium point with prey but no predators, the Jacobian reduces to

$$J = \begin{pmatrix} 1 - r & -a \\ 0 & a \end{pmatrix}. \tag{11.16}$$

This matrix has eigenvalues $(1 - r)$ and a and eigenvectors $(1, 0)$ and $[1, (1 - r - a)/a]$. This equilibrium is unstable if $a > 1$. Even if the predator is inefficient, $a < 1$, and dies out, this equilibrium is still unstable for $r > 2$. This should not surprise you: equation (11.8a), with no predators, reduces to the logistic difference equation of Chapter 4. For $a < 1$ and $2 < r < 3$ the equilibrium at $(1, 0)$ is supplanted by stable periodic or chaotic attractors along the prey axis.

For the equilibrium at $[1/a, r(a - 1)/a^2]$, the Jacobian reduces to

$$J = \begin{pmatrix} 1 - \dfrac{r}{a} & -1 \\ r - \dfrac{r}{a} & 1 \end{pmatrix}, \tag{11.17}$$

with characteristic equation

$$\lambda^2 + \left(\frac{r}{a} - 2 \right) \lambda + \left(1 + r - 2\frac{r}{a} \right) = 0. \tag{11.18}$$

The roots of this quadratic equation are eigenvalues. We are not so much interested in these eigenvalues' exact values as in whether they lead to stability or instability. We need a discrete-time analog to the Routh–Hurwitz condition. There is such a condition; it is called the Jury test.

We can derive the Jury conditions for a quadratic equation by elementary means. Consider the equation form

$$P(\lambda) \equiv \lambda^2 + a_1 \lambda + a_2 = 0. \tag{11.19}$$

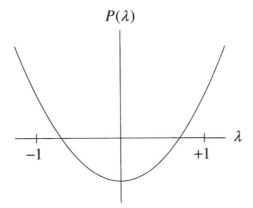

Fig. 11.2. Concave-up parabola.

$P(\lambda)$ is a concave-up parabola. Suppose that this parabola has two real roots. These roots must lie in the interval $-1 < \lambda < 1$ (see Figure 11.2) for our equilibrium to be asymptotically stable. Equivalently, we require

$$P(+1) > 0 \tag{11.20}$$

and

$$P(-1) > 0. \tag{11.21}$$

If the roots

$$\lambda = \frac{-a_1 \pm \sqrt{a_1^2 - 4a_2}}{2}, \tag{11.22}$$

are complex, the modulus of the roots is

$$|\lambda| = \sqrt{\lambda \bar{\lambda}}. \tag{11.23}$$

Since

$$\lambda \bar{\lambda} = \frac{1}{4}\left(-a_1 + \sqrt{a_1^2 - 4a_2}\right)\left(-a_1 - \sqrt{a_1^2 - 4a_2}\right) = a_2, \tag{11.24}$$

the positive coefficient a_2 must satisfy

$$a_2 < 1 \tag{11.25}$$

for asymptotic stability.

Inequalities (11.20), (11.21), and (11.25) constitute the Jury test for a quadratic equation. This test is a necessary and sufficient condition for all roots (real and complex) to be of modulus less than 1. These inequalities also shed light on how stability is lost. If we tune a parameter in such a way

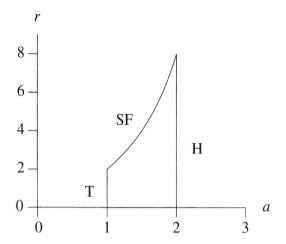

Fig. 11.3. Region of stability. (Reprinted from
Mathematical Biosciences, **110**, Neubert, M. G. and Kot,
M. The subcritical collapse of predator populations in
discrete-time predator–prey models, pp. 45–66. © 1992,
with permission from Elsevier Science.)

that inequality (11.20) is violated, but inequalities (11.21) and (11.25) hold
true, we may expect that stability has been lost because an eigenvalue has
passed through $+1$. Similar inferences can be made with regards to the other
inequalities.

For characteristic equation (11.18), the three Jury conditions reduce to

$$r \left(1 - \frac{1}{a} \right) > 0, \tag{11.26a}$$

$$4 - 3 \frac{r}{a} + r > 0, \tag{11.26b}$$

$$r \left(1 - \frac{2}{a} \right) > 0. \tag{11.26c}$$

We may conclude that the nontrivial equilibrium with both predators and
prey is asymptotically stable if

$$1 < a < 2, \quad 0 < r < \frac{4a}{3 - a} \tag{11.27}$$

(see Figure 11.3).

Boundary T in Figure 11.3 corresponds to a transcritical bifurcation. As the
equilibrium (x_2^*, y_2^*) moves from the fourth quadrant into the first quadrant,
it passes through and exchanges stability with (x_1^*, y_1^*) (see Figure 11.4). A
sufficiently large increase in the predator's birth rate or the prey's carrying
capacity allows the predator to invade the system.

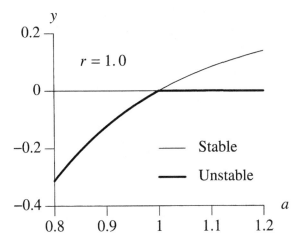

Fig. 11.4. Transcritical bifurcation. (Reprinted from
Mathematical Biosciences, **110**, Neubert, M. G. and Kot,
M. The subcritical collapse of predator populations in
discrete-time predator–prey models, pp. 45–66. © 1992,
with permission from Elsevier Science.)

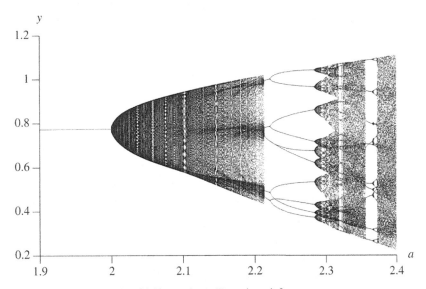

Fig. 11.5. Hopf bifurcation. (Reprinted from
Mathematical Biosciences, **110**, Neubert, M. G. and Kot,
M. The subcritical collapse of predator populations in
discrete-time predator–prey models, pp. 45–66. © 1992,
with permission from Elsevier Science.)

As we continue to increase a, we observe a Hopf bifurcation (see Figure 11.5). This bifurcation occurred in our study of the lagged logistic map in Chapter 5. To the left of H, (x_2^*, y_2^*) is a stable focus. To the right of H, we observe an unstable focus and attracting invariant circle. This bifurcation can be viewed as an example of Rosenzweig's (1971) paradox of enrichment: we destabilize the predator–prey equilibrium by increasing the prey's carrying capacity. There are two exceptions to this Hopf bifurcation. They occur at $r = 4$ and $r = 6$, where there are strong $1:4$ and $1:3$ resonances (Lauwerier, 1986; Lauwerier and Metz, 1986). At $r = 4$, the eigenvalues of (x_2^*, y_2^*) are fourth roots of unity and (x_2^*, y_2^*) loses its stability to a stable 4-cycle. At $r = 6$, the eigenvalues are third roots of unity and one can observe an unstable 3-cycle. For larger values of a, the invariant circle generated by the Hopf bifurcation deforms and breaks apart, giving rise to a strange attractor. Details and figures can be found in Neubert and Kot (1992).

Finally, as we increase r and cross SF, we force an eigenvalue to pass through $\lambda = -1$. You might expect equilibrium (x_2^*, y_2^*) to go unstable and give rise to a stable 2-cycle. However, the bifurcation behavior of (x_2^*, y_2^*) is more complicated (see Figure 11.6). I will focus on this behavior for small $a \approx 1.1$. As r increases, the equilibrium at (x_1^*, y_1^*) undergoes the series of period-doubling bifurcations that is the hallmark of the logistic difference equation. The first such period doubling gives rise to a 2-cycle,

$$\hat{X} = [\hat{X}_1, \hat{X}_2], \tag{11.28a}$$

$$\hat{X}_1 = \left(\frac{r + 2 - \sqrt{r^2 - 4}}{2r}, 0 \right), \tag{11.28b}$$

$$\hat{X}_2 = \left(\frac{r + 2 + \sqrt{r^2 - 4}}{2r}, 0 \right). \tag{11.28c}$$

The order 2^n cycles of this period-doubling sequence reside on the prey axis. Since the equilibrium X_1 was already a saddle point, they are all unstable.

While this is happening (see Figure 11.7), the stable fourth-quadrant 2-cycle,

$$\tilde{X} = [\tilde{X}_1, \tilde{X}_2] = [(\tilde{x}_1, \tilde{y}_1), (\tilde{x}_2, \tilde{y}_2)], \tag{11.29a}$$

$$\tilde{x}_1 = \frac{r - a(r + 2) + \sqrt{r(a + 1)(4a - 3r + ar)}}{2a(a - r)}, \tag{11.29b}$$

$$\tilde{y}_1 = \frac{1 + r(1 - \tilde{x}_1)}{a} - \frac{1}{a^3 \tilde{x}_1^2}, \tag{11.29c}$$

$$\tilde{x}_2 = \frac{r - a(r + 2) - \sqrt{r(a + 1)(4a - 3r + ar)}}{2a(a - r)}, \tag{11.29d}$$

$$\tilde{y}_2 = \frac{1 + r(1 - \tilde{x}_2)}{a} - \frac{1}{a^3 \tilde{x}_2^2}, \tag{11.29e}$$

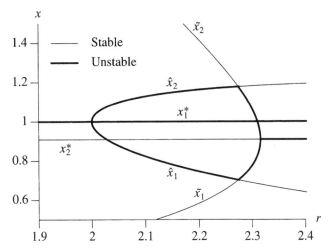

Fig. 11.6. Bifurcations for increasing *r*. (Reprinted from *Mathematical Biosciences*, **110**, Neubert, M. G. and Kot, M. The subcritical collapse of predator populations in discrete-time predator–prey models, pp. 45–66. © 1992, with permission from Elsevier Science.)

moves into the first quadrant. This 2-cycle passes through and exchanges stability with $\hat{X}$ in a transcritical bifurcation. Both the equilibrium (x_2^*, y_2^*) and the prey-axis 2-cycle $\hat{X}$ are now stable (Figure 11.8). At one attractor, the predator coexists with the prey; at the other, the predator is absent. The stable manifolds of the saddle-like 2-cycle $\tilde{X}$ separate the two basins of attraction. Finally, as we cross SF, the unstable 2-cycle $\tilde{X}$ collides with the equilibrium (x_2^*, y_2^*). The 2-cycle $\tilde{X}$ is destroyed and equilibrium (x_2^*, y_2^*) loses its stability in a *subcritical* flip bifurcation (see Figure 11.9). The prey-axis 2-cycle $\hat{X}$ is now the only stable attractor.

For larger values of *a*, larger values of *r* are required to violate inequality (11.26b); (x_1^*, y_1^*) will undergo further period doublings before (x_2^*, y_2^*) loses its stability. The prey-axis attractor will then be of higher order $(4, 8, 16, \ldots)$. For high enough values of *r*, the dynamics on the prey axis will be chaotic. However, if $r > 3$, the logistic map is no longer invariant on the interval $(0, 4/3)$ and trajectories will diverge to minus infinity.

There are many highlights in the study of discrete-time planar mappings. Equilibria can lose their stability to transcritical, flip (period-doubling), or Hopf bifurcations. There are supercritical and subcritical flip and Hopf bifurcations that can lead to chaotic dynamics. Some bifurcations are safe, with the continuous growth of new attractor paths; others are dangerous, with fast dynamic jumps to distant unrelated attractors.

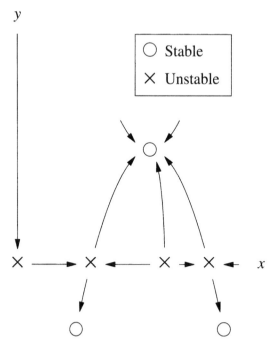

Fig. 11.7. Before the transcritical bifurcation. (Reprinted from *Mathematical Biosciences*, **110**, Neubert, M. G. and Kot, M. The subcritical collapse of predator populations in discrete-time predator–prey models, pp. 45–66. © 1992, with permission from Elsevier Science.)

Simple, discrete-time, predator–prey models such as equations (11.8a) and (11.8b) are especially good for showing what *can* happen. However, any discussion of discrete-time $(+, -)$ interactions must rapidly turn to host–parasitoid models. That is where the action is in ecology. There has been extensive experimental work on parasitoids (Hassell, 1978; Godfray, 1994). Parasitoids are also of great economic importance in regulating agricultural pests (DeBach, 1974; Van Driesche and Bellows, 1996). It is worth noting that discrete-time host–parasitoid models are older than continuous-time predator–prey models; these models have been around since at least the 1920s. Pride of place goes to the Nicholson–Bailey model (Nicholson, 1933; Nicholson and Bailey, 1935).

The framework for the Nicholson–Bailey model (and for most host–parasitoid models) is a pair of equations of the form

$$N_{t+1} = \lambda N_t f(N_t, P_t), \tag{11.30a}$$

$$P_{t+1} = c\,[1 - f(N_t, P_t)]\,N_t, \tag{11.30b}$$

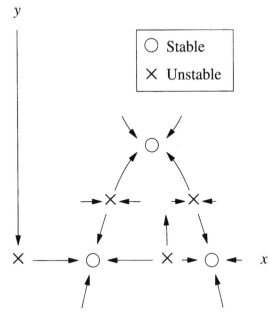

Fig. 11.8. After the transcritical bifurcation. (Reprinted from *Mathematical Biosciences*, **110**, Neubert, M. G. and Kot, M. The subcritical collapse of predator populations in discrete-time predator–prey models, pp. 45–66. © 1992, with permission from Elsevier Science.)

where N_t and P_t are numbers of hosts and parasitoids in generation t, $f(N_t, P_t)$ is the fraction of hosts that avoid being parasitized, λ is the net reproductive rate of the hosts, and c is, loosely speaking, the clutch size of the parasitoids.

There are several assumptions that lead to the final form of the Nicholson–Bailey model (Hassell, 1978; Edelstein-Keshet, 1988). The foremost is that the number of encounters N_e between potential hosts and parasitoids is proportional to the product of their densities,

$$N_e = a N_t P_t. \tag{11.31}$$

Nicholson named the proportionality constant a the *area of discovery*. The dimension of a will depend on those of N_t and P_t. If N_t and P_t are total population sizes, a will be dimensionless. However, if N_t and P_t are expressed as numbers per unit area, a will have units of area. The area of discovery is thought to be a species-specific parameter.

A second assumption is that the encounters are distributed randomly among the available hosts. Randomly here means according to a Poisson

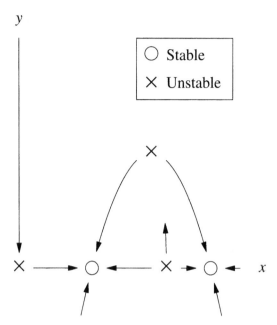

Fig. 11.9. After the subcritical flip bifurcation.
(Reprinted from *Mathematical Biosciences*, **110**, Neubert,
M. G. and Kot, M. The subcritical collapse of predator
populations in discrete-time predator–prey models,
pp. 45–66. © 1992, with permission from Elsevier
Science.)

process. Let $p_n(t)$ be the probability that a given host has been found n times
by time t. Assume that the probability of an encounter in some small time
interval Δt is equal to Δt times the average encounter rate μ, and that the
probability of more than one encounter is $o(\Delta t)$. Thus, for $n > 0$,

$$p_n(t + \Delta t) = \mu \, \Delta t \, p_{n-1}(t) + (1 - \mu \, \Delta t) \, p_n(t) + o(\Delta t). \qquad (11.32)$$

It follows that

$$
\begin{aligned}
\frac{dp_n}{dt} &= \lim_{\Delta t \to 0} \frac{p_n(t + \Delta t) - p_n(t)}{\Delta t} \\
&= \mu \, [p_{n-1}(t) - p_n(t)], \quad n > 0.
\end{aligned}
\qquad (11.33)
$$

The equation for $n = 0$ is even simpler since $n = 0$ at time $t + \Delta t$ if
and only if $n = 0$ at time t and no new encounters have occurred in Δt. It
follows that

$$\frac{dp_0}{dt} = -\mu \, p_0(t). \qquad (11.34)$$

If we start the process off with no encounters,

$$p_0(0) = 1, \tag{11.35}$$

equation (11.34) implies that

$$p_0(t) = e^{-\mu t}. \tag{11.36}$$

We may now chain our way up equations (11.33) to obtain

$$p_n(t) = \frac{e^{-\mu t} (\mu t)^n}{n!}. \tag{11.37}$$

For our discrete-time model, the period of vulnerability to parasitism is exactly 1 and equation (11.37) simplifies to

$$p_n = \frac{e^{-\mu} \mu^n}{n!}. \tag{11.38}$$

In light of our first assumption, μ, the average number of encounters, takes the value

$$\mu = \frac{N_e}{N_t} = a P_t. \tag{11.39}$$

It follows that

$$p_0 = e^{-\mu} = e^{-a P_t} \tag{11.40}$$

is the probability that a particular host will escape parasitism. As the parasitoid population increases, it will be less and less likely that a host will escape parasitism.

Our final assumption is that only the first encounter between a host and a parasitoid is important. In particular, we will assume that a host that has been parasitized will bear exactly c parasitoid progeny, regardless of the number of encounters.

These three assumptions, together, lead us to the Nicholson–Bailey equations:

$$N_{t+1} = \lambda N_t e^{-a P_t}, \tag{11.41a}$$
$$P_{t+1} = c N_t (1 - e^{-a P_t}). \tag{11.41b}$$

Let us round up the usual suspects. We need to find the equilibria for this system and to ascertain their stability. There are two equilibria, the trivial equilibrium,

$$(N_0^*, P_0^*) = (0, 0), \tag{11.42}$$

corresponding to extinction of both species, and the equilibrium of coexistence,

$$(N_1^*, P_1^*) = \left(\frac{1}{ac} \frac{\lambda}{\lambda - 1} \ln \lambda, \frac{1}{a} \ln \lambda \right). \tag{11.43}$$

The stability of each equilibrium is determined by the eigenvalues of the Jacobian,

$$J = \begin{bmatrix} \lambda e^{-aP_t} & -a\lambda N_t e^{-aP_t} \\ c(1 - e^{-aP_t}) & ac N_t e^{-aP_t} \end{bmatrix}, \tag{11.44}$$

at that equilibrium.

For the trivial equilibrium at $(0, 0)$, the Jacobian reduces to

$$J = \begin{pmatrix} \lambda & 0 \\ 0 & 0 \end{pmatrix}, \tag{11.45}$$

with eigenvalues λ and 0. If $\lambda > 1$, the origin is an unstable saddle point.

At the second equilibrium, the Jacobian reduces to

$$J = \begin{bmatrix} 1 & -\frac{1}{c} \frac{\lambda}{(\lambda - 1)} \ln \lambda \\ c \left(1 - \frac{1}{\lambda} \right) & \frac{1}{(\lambda - 1)} \ln \lambda \end{bmatrix}. \tag{11.46}$$

If we let Λ represent an eigenvalue, the characteristic equation for Jacobian (11.46) reduces to

$$P(\Lambda) = \Lambda^2 - \left[1 + \frac{\ln \lambda}{(\lambda - 1)} \right] \Lambda + \frac{\lambda \ln \lambda}{(\lambda - 1)} = 0. \tag{11.47}$$

As usual, we require that all of the Λ values have magnitude less than 1 for asymptotic stability.

We may determine whether characteristic equation (11.47) implies asymptotic stability by examining Jury conditions (11.20), (11.21), and (11.25). For $\lambda > 1$,

$$P(1) = 1 - \left[1 + \frac{\ln \lambda}{(\lambda - 1)} \right] + \frac{\lambda \ln \lambda}{(\lambda - 1)} = \ln \lambda > 0 \tag{11.48}$$

and

$$P(-1) = 1 + \left[1 + \frac{\ln \lambda}{(\lambda - 1)} \right] + \frac{\lambda \ln \lambda}{(\lambda - 1)} = 2 + \frac{(\lambda + 1)}{(\lambda - 1)} \ln \lambda > 0. \tag{11.49}$$

So far, so good. However, when we turn to the third Jury condition, equation (11.25), we find that we must show that

$$\frac{\lambda \ln \lambda}{(\lambda - 1)} < 1 \tag{11.50}$$

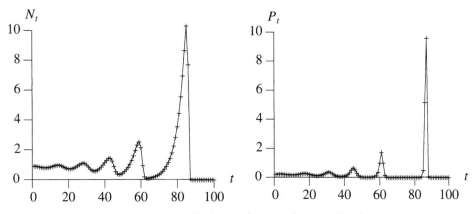

Fig. 11.10. Oscillations of increasing amplitude.

or, equivalently, that

$$S(\lambda) = \lambda - 1 - \lambda \ln \lambda > 0. \tag{11.51}$$

This inequality is not satisfied since $S(1) = 0$ and

$$S'(\lambda) = -\ln \lambda \leq 0. \tag{11.52}$$

The equilibrium of coexistence is thus unstable. Simulations (see Figure 11.10) of the Nicholson–Bailey model produce oscillations of ever-increasing amplitude.

The Nicholson–Bailey model is the canonical model for host–parasitoid interactions. And yet this model is intrinsically unstable; it generates increasingly large oscillations. These oscillations, when coupled with stochasiticity, drive the host and/or the parasitoid extinct. What is one to make of this? Nicholson and Bailey thought it an accurate depiction of nature. They believed that the interaction between host and parasitoid is intrinsically unstable and that the effect of increasing oscillations is to break up the parent population into local populations that each wax, wane, go extinct, and are recolonized. They believed spatial heterogeneity to be the factor that keeps host–parasitoid systems intact. This viewpoint gained support from the laboratory experiments of Huffaker (1958). Huffaker studied the interactions between the six-spotted mite *Eotetranychus sexmaculatus* and the predatory mite *Typhlodromus occidentalis*. The six-spotted mite feeds on oranges, and by setting out oranges and rubber balls on a tray, Huffaker could control the food supply, dispersal, and spatial heterogeneity of the system. In Huffaker's simplest experiments, the predator would quickly discover the prey and reduce the prey to such low levels that the predators would soon starve and go

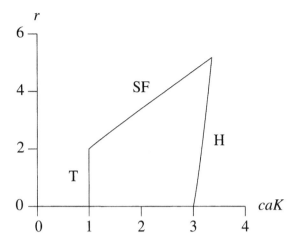

Fig. 11.11. Stability region in parameter plane.
(Reprinted from *Mathematical Biosciences*, **110**, Neubert,
M. G. and Kot, M. The subcritical collapse of predator
populations in discrete-time predator–prey models,
pp. 45–66. © 1992, with permission from Elsevier
Science.)

extinct. As Huffaker increased the spatial heterogeneity of his system, he en-
hanced population persistence. Ultimately, with 120 partly covered oranges,
a maze of Vaseline barriers (to control dispersal), and various other devices,
Huffaker obtained three oscillations of the predator and the prey before the
predator finally went extinct from a shortage of prey. From this perspective,
stability and persistence are found at the regional level.

Others have sought stability and persistence at the local level. The easiest
way to stabilize the Nicholson–Bailey model is to introduce self-limitation
(density dependence) of the prey population (Beddington *et al.*, 1975). The
resulting model,

$$N_{t+1} = N_t \exp\left[r\left(1 - \frac{N_t}{K}\right) - a P_t\right], \tag{11.53a}$$

$$P_{t+1} = c N_t (1 - e^{-a P_t}), \tag{11.53b}$$

is similar in its behavior to equations (11.8a) and (11.8b) There is an equilibrium of coexistence that can, once again, lose its stability by means of
transcritical, subcritical flip, or Hopf bifurcations (see Figure 11.11).

However, many scientists firmly believe that the persistence of successful
biological control arises from a direct stabilizing effect of the parasitoid itself.
This has spurred a vigorous search for the biological attributes of parasitoids

that stabilize host–parasitoid interactions (Mills and Getz, 1996). A more realistic functional response, such as a type II or type III functional response, is typically destabilizing (Hassell, 1978). Nonetheless, some mechanisms do lead to greater stability. These include the aggregation of parasitoid attacks, switching between hosts (which may provide refuges for some hosts), and density-dependent sex ratios. Determining which, if any, of these factors are important for natural populations is extremely difficult (Murdoch, 1994). Understanding what makes for successful biological control remains one of the most important and one of the most elusive goals of theoretical ecology.

Recommended readings

The monograph by Hassell (1978) is an invaluable reference. Edelstein-Keshet (1988) provides a useful introduction to discrete-time host–parasitoid models.

12 Competition models

Predation is the most dynamic interaction. Many ecologists, however, put greater weight on competition. For many years, competition was thought to play a predominant role in structuring ecological communities. This paramountcy has not held up and most ecologists now take a pluralistic approach towards interactions. Even so, there is no question of the overall importance of competition.

There is a classical model of competition due to Lotka (1932) and Volterra (1926). The Lotka–Volterra competition model is an *interference* competition model: two species are assumed to diminish each other's per capita growth rate by direct interference.

We begin by assuming that two species, with populations of size N_1 and N_2, each grow logistically in the absence of the other. Each species has a per capita growth rate that decreases linearly with population size,

$$\frac{1}{N_1}\frac{dN_1}{dt} = r_1\left(1 - \frac{N_1}{K_1}\right), \tag{12.1a}$$

$$\frac{1}{N_2}\frac{dN_2}{dt} = r_2\left(1 - \frac{N_2}{K_2}\right), \tag{12.1b}$$

and each species has its own intrinsic rate of growth and its own carrying capacity. We now add competition. We assume that the effect of interspecific competition is similar to that of intraspecific crowding. Each individual of the second species causes a decrease in the per capita growth of the first species, and vice versa. Because the two species are different, heterospecific individuals may have a stronger effect or a weaker effect on the per capita growth rate than conspecific individuals. To parametrize this effect, we introduce a pair of competition coefficients, α_{12} and α_{21}, that describe the strength of the effect of species 2 on species 1 and of species 1 on species 2

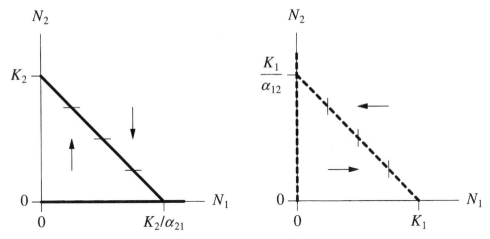

Fig. 12.1. Zero-growth isoclines.

relative to that of conspecific individuals,

$$\frac{1}{N_1}\frac{dN_1}{dt} = r_1 \left[1 - \frac{(N_1 + \alpha_{12} N_2)}{K_1} \right], \tag{12.2a}$$

$$\frac{1}{N_2}\frac{dN_2}{dt} = r_2 \left[1 - \frac{(N_2 + \alpha_{21} N_1)}{K_2} \right]. \tag{12.2b}$$

This is the Lotka–Volterra competition model. It may be rewritten, more succinctly, as

$$\frac{dN_1}{dt} = \frac{r_1}{K_1} N_1 (K_1 - N_1 - \alpha_{12} N_2), \tag{12.3a}$$

$$\frac{dN_2}{dt} = \frac{r_2}{K_2} N_2 (K_2 - N_2 - \alpha_{21} N_1). \tag{12.3b}$$

The complete characterization of the dynamics of equations (12.3a) and (12.3b) revolves around the orientation of the zero-growth isoclines. (The reasons for this will become clearer in Chapter 13.) The N_2 zero-growth isoclines are

$$N_2 = 0 \tag{12.4a}$$

and

$$N_2 = K_2 - \alpha_{21} N_1 \tag{12.4b}$$

(see Figure 12.1). Below the graph given by equation (12.4b), N_2 increases; above this line, N_2 decreases. The N_1 zero-growth isoclines, in turn, are given by

$$N_1 = 0 \tag{12.5a}$$

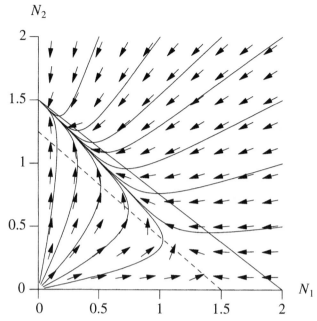

Fig. 12.2. Phase portrait for $\alpha_{12} > \frac{K_1}{K_2}$, $\alpha_{21} < \frac{K_2}{K_1}$.

and

$$N_1 = K_1 - \alpha_{12} N_2. \tag{12.5b}$$

Below the line given by (12.5b), N_1 increases; above this, N_1 decreases. One of the isoclines (12.4b) and (12.5b) may lie entirely above the other. Alternatively, these two zero-growth isoclines may cross. There are four cases, depending on the relative positions of the x and y intercepts of these two zero-growth isoclines. Each of the four cases corresponds to a qualitatively different phase portrait. Let us consider each phase portrait in turn.

If each intercept of the line given by (12.4b) is greater than the corresponding intercept of that for (12.5b), so that

$$\alpha_{12} > \frac{K_1}{K_2}, \quad \alpha_{21} < \frac{K_2}{K_1}, \tag{12.6}$$

N_2 rapidly excludes N_1 (see Figure 12.2). Thus, if species 2 has a relatively large effect on species 1 and species 1 has a relatively small effect on species 2, we expect that species 1 will go extinct and that species 2 will approach its carrying capacity.

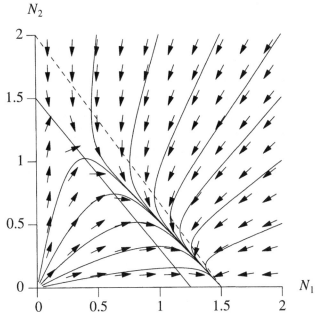

Fig. 12.3. Phase portrait for $\alpha_{12} < \frac{K_1}{K_2}$, $\alpha_{21} > \frac{K_2}{K_1}$.

If the inequalities are reversed,

$$\alpha_{12} < \frac{K_1}{K_2}, \; \alpha_{21} > \frac{K_2}{K_1}, \tag{12.7}$$

so that species 2 has a small effect of species 1 and species 1 has a large effect on species 2, the competitive outcome is also reversed: species 1 approaches its carrying capacity and species 2 goes extinct (see Figure 12.3).

If

$$\alpha_{12} > \frac{K_1}{K_2}, \; \alpha_{21} > \frac{K_2}{K_1}, \tag{12.8}$$

the N_2 zero growth isocline crosses the N_1 zero growth from above. In this case, interspecific effects are large for both species. The two equilibria $(K_1, 0)$ and $(0, K_2)$, corresponding to the exclusion of one or the other species, are now both stable nodes (see Figure 12.4). One or the other of the species will go extinct, depending on the initial conditions. There is a saddle point that lies between the two nodes. The stable manifolds of this saddle point form the boundaries for the two domains of attraction.

Finally, when interspecific competition is relatively weak,

$$\alpha_{12} < \frac{K_1}{K_2}, \; \alpha_{21} < \frac{K_2}{K_1}, \tag{12.9}$$

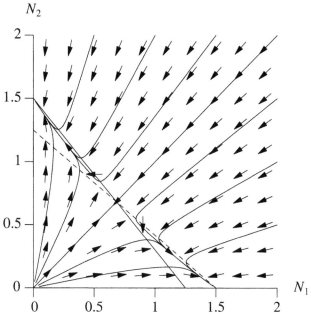

Fig. 12.4. Phase portrait for $\alpha_{12} > \frac{K_1}{K_2}$, $\alpha_{21} > \frac{K_2}{K_1}$.

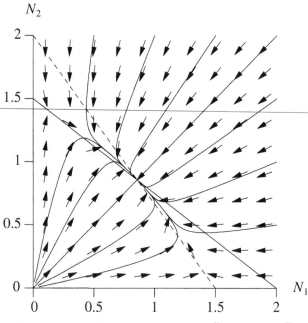

Fig. 12.5. Phase portrait for $\alpha_{12} < \frac{K_1}{K_2}$, $\alpha_{21} < \frac{K_2}{K_1}$.

the equilibria $(K_1, 0)$ and $(0, K_2)$ are unstable saddle points and trajectories are drawn towards a stable node in the interior of the first quadrant (see Figure 12.5).

How can we be sure that there are no limit cycles lurking about? Recall the Bendixson–Dulac negative criterion. Let

$$B(N_1, N_2) \equiv \frac{1}{N_1 N_2}. \tag{12.10}$$

Since the divergence

$$\frac{\partial (B \dot{N}_1)}{\partial N_1} + \frac{\partial (B \dot{N}_2)}{\partial N_2} = -\frac{r_1}{K_1} \frac{1}{N_2} - \frac{r_2}{K_2} \frac{1}{N_1} \tag{12.11}$$

is strictly negative in the interior of the first quadrant, we can be sure that there are no closed orbits contained entirely within the first quadrant.

For three of the four cases that we have considered, one species successfully excludes the other. Only in case 4, where interspecific effects were weak relative to intraspecific effects, did the two competing species coexist. This forms the basis for *Gause's Principle* (Gause, 1934, 1935) or the *Principle of Competitive Exclusion* (Hardin, 1960). This principle states that, if two species are too similar, they cannot coexist. As stated, this principle is somewhat vague. It begs the question: just how similar can two species be and still coexist? There have, in fact, been numerous attempts to develop a more formal theory of limiting similarity, but most of the biological conclusions that have been reached depend on both the specific model that is analyzed and the method of analysis (Abrams, 1983; Schoener, 1986).

Problem 12.1 *Competition with unlimited growth*
Analyze the competition model

$$\frac{dN_1}{dT} = r_1 N_1 \left[1 - \frac{(N_1 + \alpha_{12} N_2)}{K_1} \right], \tag{12.12a}$$

$$\frac{dN_2}{dT} = r_2 N_2 \left(1 - \alpha_{21} \frac{N_1}{K_2} \right). \tag{12.12b}$$

In this model, only one of the species, N_1, has a finite carrying capacity.

Problem 12.2 *Discrete-time competition*

Consider two species that have discrete, nonoverlapping generations and whose populations grow according to

$$x_{t+1} = \frac{\lambda_1 x_t}{1 + a_1(x_t + \alpha_{12} y_t)},$$ (12.13a)

$$y_{t+1} = \frac{\lambda_2 y_t}{1 + a_2(y_t + \alpha_{21} x_t)}.$$ (12.13b)

Analyze this system of difference equations.

One major problem with the Lotka–Volterra competition model is that the competition coefficients α_{ij} are often quite difficult to measure. Frequently, one must grow N_1 and N_2 together in order to measure these coefficients. There is an element of circularity here, as one is forced to perform a competition experiment in order to be able to predict the outcome of a competition experiment. If two species compete with regards to their efficiency at exploiting a common resource, rather than by direct interference, we can instead construct an *exploitation* competition model and make predictions on the basis of the characters of the individual species rather than on the basis of their mutual interactions. (The distinction between interference and exploitation competition is usually attributed to Elton and Miller (1954) and Park (1954).) Interestingly, the behavior of an exploitation competition model may depend on whether the resource is biotic or abiotic.

Problem 12.3 *Nontransitive competition*

Analyze the behavior of three competing species whose dynamics are governed by the system (May and Leonard, 1975)

$$\frac{dx}{dt} = x(1 - x - \alpha y - \beta z),$$ (12.14a)

$$\frac{dy}{dt} = y(1 - \beta x - y - \alpha z),$$ (12.14b)

$$\frac{dz}{dt} = z(1 - \alpha x - \beta y - z),$$ (12.14c)

where

$$0 < \beta < 1 < \alpha. \tag{12.15}$$

Hint The Jacobian of this system is a circulant matrix.

Let us begin by analyzing a model with an abiotic resource. Consider a simple, two-species model for competition in a chemostat (Powell, 1958; Stewart and Levin, 1973; Hsu *et al.*, 1977; Hsu, 1978; Waltman, 1983; Smith and Waltman, 1995). An abiotic substrate is pumped into the chemostat at a constant rate and two species each take up this substrate with a type II, Michaelis–Menten, or Monod functional response,

$$\frac{dS}{dT} = D\,(S_i - S) - \frac{1}{Y_1}\frac{m_1\,S\,N_1}{K_1 + S} - \frac{1}{Y_2}\frac{m_2\,S\,N_2}{K_2 + S}, \tag{12.16a}$$

$$\frac{dN_1}{dT} = \frac{m_1\,S\,N_1}{K_1 + S} - D\,N_1, \tag{12.16b}$$

$$\frac{dN_2}{dT} = \frac{m_2\,S\,N_2}{K_2 + S} - D\,N_2. \tag{12.16c}$$

The variables are S, the substrate concentration, N_1, the concentration of the first species, N_2, the concentration of the second species, and T, time. Each species has its own maximum growth rate, m_1 or m_2, yield coefficient, Y_1 or Y_2, and half-saturation constant, K_1 or K_2. D is the dilution rate and S_i is the inflowing substrate concentration. After introducing the nondimensionalization

$$s \equiv \frac{S}{S_i}, \quad x \equiv \frac{N_1}{Y_1\,S_i}, \quad y \equiv \frac{N_2}{Y_2\,S_i}, \quad t \equiv D\,T, \tag{12.17}$$

equations (12.16a), (12.16b), and (12.16c) simplify to

$$\frac{ds}{dt} = 1 - s - \frac{A_1\,s\,x}{a_1 + s} - \frac{A_2\,s\,y}{a_2 + s}, \tag{12.18a}$$

$$\frac{dx}{dt} = \frac{A_1\,s\,x}{a_1 + s} - x, \tag{12.18b}$$

$$\frac{dy}{dt} = \frac{A_2\,s\,y}{a_2 + s} - y, \tag{12.18c}$$

with

$$A_1 \equiv \frac{m_1}{D}, \quad a_1 \equiv \frac{K_1}{S_i}, \quad A_2 \equiv \frac{m_2}{D}, \quad a_2 \equiv \frac{K_2}{S_i}. \tag{12.19}$$

By adding equations (12.18a), (12.18b), and (12.18c), we obtain

$$\frac{ds}{dt} + \frac{dx}{dt} + \frac{dy}{dt} = 1 - s - x - y. \tag{12.20}$$

If we treat the sum of $s(t)$, $x(t)$, and $y(t)$ as a single variable, equation (12.20) reduces to a single first-order differential equation that has the solution

$$s(t) + x(t) + y(t) = 1 + (s_0 + x_0 + y_0 - 1)\, e^{-t}, \tag{12.21}$$

where s_0, x_0, and y_0 are the initial concentrations of the rescaled substrate and of the two competitors. It follows that $s + x + y \to 1$ as $t \to \infty$. We can therefore study the asymptotic behavior on the plane

$$s + x + y = 1. \tag{12.22}$$

Since we are more interested in the competitors than in the substrate, we can also use equation (12.22) to eliminate s from equations (12.18b) and (12.18c). This leaves us with

$$\frac{dx}{dt} = \frac{A_1\,(1 - x - y)\,x}{a_1 + 1 - x - y} - x, \tag{12.23a}$$

$$\frac{dy}{dt} = \frac{A_2\,(1 - x - y)\,y}{a_2 + 1 - x - y} - y \tag{12.23b}$$

as a system of two equations for the asymptotic concentrations of the two competing species.

Equations (12.23a) and (12.23b) can be analyzed in the usual manner. This system possesses a trivial equilibrium at $(0, 0)$. In the neighborhood of this equilibrium,

$$\frac{dx}{dt} \approx \left(\frac{A_1}{a_1 + 1} - 1 \right) x, \tag{12.24a}$$

$$\frac{dy}{dt} \approx \left(\frac{A_2}{a_2 + 1} - 1 \right) y. \tag{12.24b}$$

Since we want both species to be viable in the absence of competition, we will require that both right-hand coefficients be positive or, equivalently, that

$$\frac{a_1}{A_1 - 1} < 1, \quad \frac{a_2}{A_2 - 1} < 1. \tag{12.25}$$

With this detail in hand, we can now look at the zero-growth isoclines of equations (12.23a) and (12.23b).

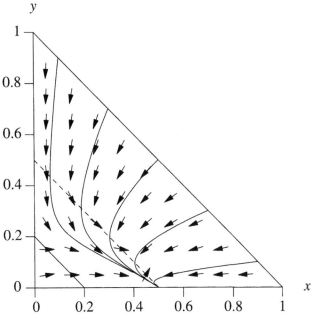

Fig. 12.6. Phase portrait for $\frac{a_1}{A_1 - 1} < \frac{a_2}{A_2 - 1}$.

The x zero-growth isoclines are

$$x = 0 \qquad (12.26)$$

and

$$x + y = 1 - \frac{a_1}{A_1 - 1}. \qquad (12.27)$$

The y zero-growth isoclines, in turn, are just

$$y = 0 \qquad (12.28)$$

and

$$x + y = 1 - \frac{a_2}{A_2 - 1}. \qquad (12.29)$$

Since isoclines (12.27) and (12.29) are parallel straight lines that never cross, there are three possible outcomes. If

$$\frac{a_1}{A_1 - 1} < \frac{a_2}{A_2 - 1}, \qquad (12.30)$$

x excludes y (see Figure 12.6). If

$$\frac{a_2}{A_2 - 1} < \frac{a_1}{A_1 - 1}, \qquad (12.31)$$

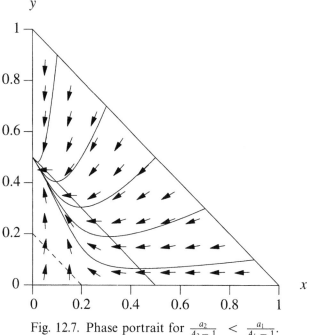

Fig. 12.7. Phase portrait for $\frac{a_2}{A_2 - 1} < \frac{a_1}{A_1 - 1}$.

y excludes x (see Figure 12.7). Finally, in the unlikely event that

$$\frac{a_1}{A_1 - 1} = \frac{a_2}{A_2 - 1}, \tag{12.32}$$

there is a line of equilibria and x and y coexist.

Since isoclines (12.27) and (12.29) never cross, there are fewer outcomes for this resource-based competition model than there were for classical Lotka–Volterra competition. Typically, one species excludes the other, independent of initial conditions. In particular,

$$\frac{a_w}{A_w - 1} < \frac{a_1}{A_1 - 1}, \tag{12.33}$$

where A_w and a_w are the rescaled maximum growth rate and the half-saturation constant of the winner and A_1 and a_1 are the corresponding parameters for the loser. Only in the unlikely event that the inequality is replaced by equality can the two species coexist.

Inequality (12.33) can occur for many reasons. Both species could have the same half-saturation constant, with the winner having a higher maximum growth rate (see Figure 12.8). Alternatively, both species could have the same maximum growth rate, with the winner having a lower half-saturation constant (see Figure 12.9). In both of these examples, the winner's uptake

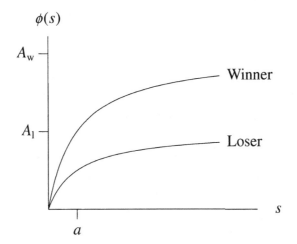

Fig. 12.8. Functional responses: $a = a_w = a_l$ and $A_w > A_l$.

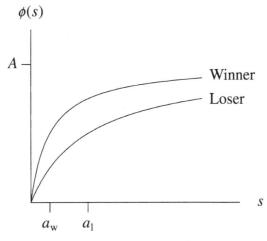

Fig. 12.9. $A = A_w = A_l$ and $a_w < a_l$.

function lies strictly above that of the loser. In general, the two uptake functions may also cross (see Figure 12.10), with one species doing better at low substrate concentrations and the other species doing better at high substrate concentrations. In this case, one must use inequality (12.33) in order to predict the winner.

These predictions concerning competition in a chemostat on an abiotic substrate have received robust experimental support. Hansen and Hubbell (1980) performed a series of experiments, using strains of the bacteria *Escherichia coli* and *Pseudomonas aeruginosa*, in which they compared

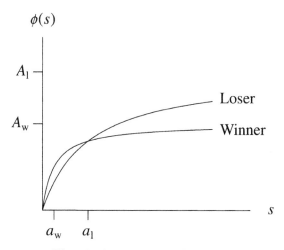

Fig. 12.10. $A_w < A_1$ and $a_w < a_1$.

predicted and actual outcomes. In particular, they performed an experiment in which $A_1 \approx A_2$, but $a_1 < a_2$; another in which $a_1 \approx a_2$, but $A_1 > A_2$; and a third in which the inequality in (12.33) gave way to equality. In each case, exclusion or coexistence was correctly predicted. In addition, there was good agreement between the predicted trajectories and the actual time series (though the experimental time series did show a greater tendency for damped oscillations). This degree of agreement between theory and experiment is quite spectacular by ecological standards.

The above analysis and experiments suggest that the coexistence of two species in an exploitation competition setting is even less likely than their coexistence in the presence of interference. One might even conjecture that a single resource can support no more than one species. We have, however, limited ourselves to an *abiotic* substrate. Many species take biotic prey. Can two predators, both living on the same biotic prey, coexist?

Consider the system (Koch, 1974; Hsu *et al.*, 1978; Waltman, 1983)

$$\frac{dN}{dT} = r N \left(1 - \frac{N}{K_0} \right) - \frac{1}{Y_1} \frac{m_1 N P_1}{K_1 + N} - \frac{1}{Y_2} \frac{m_2 N P_2}{K_2 + N}, \quad (12.34a)$$

$$\frac{dP_1}{dT} = \frac{m_1 N P_1}{K_1 + N} - d_1 P_1, \quad (12.34b)$$

$$\frac{dP_2}{dT} = \frac{m_2 N P_2}{K_2 + N} - d_2 P_2, \quad (12.34c)$$

for a biotic prey, N, and its two predators, P_1 and P_2. This system is no longer in the chemostat. The parameters r and K_0 are the intrinsic rate of growth and the carrying capacity of the prey and d_1 and d_2 are now the per capita death rates of the two predators. The change of variables

$$x \equiv \frac{N}{K_0}, \quad y \equiv \frac{P_1}{Y_1 K_0}, \quad z \equiv \frac{P_2}{Y_2 K_0}, \quad t \equiv r T \qquad (12.35)$$

allows us to reduce this system to

$$\frac{dx}{dt} = x(1 - x) - \frac{A_1 \, x \, y}{a_1 + x} - \frac{A_2 \, x \, z}{a_2 + x}, \qquad (12.36a)$$

$$\frac{dy}{dt} = \frac{A_1 \, x \, y}{a_1 + x} - D_1 \, y, \qquad (12.36b)$$

$$\frac{dz}{dt} = \frac{A_2 \, x \, z}{a_2 + x} - D_2 \, z, \qquad (12.36c)$$

where

$$A_1 \equiv \frac{m_1}{r}, \quad a_1 \equiv \frac{K_1}{K_0}, \quad D_1 \equiv \frac{d_1}{r}, \qquad (12.37a)$$

$$A_2 \equiv \frac{m_2}{r}, \quad a_2 \equiv \frac{K_2}{K_0}, \quad D_2 \equiv \frac{d_2}{r}. \qquad (12.37b)$$

To guarantee that this system is truly competitive, we will also require that

$$\frac{A_1}{a_1 + 1} > D_1, \quad \frac{A_2}{a_2 + 1} > D_2. \qquad (12.38)$$

Can the two competing predators coexist at an equilibrium? If there were such a rest point, the right-hand side of equation (12.36b) would have to be zero, so that

$$y = 0 \qquad (12.39)$$

or

$$x = \frac{a_1 D_1}{A_1 - D_1}. \qquad (12.40)$$

Similarly, the right-hand side of equation (12.36c) would have to be zero, so that

$$z = 0 \qquad (12.41)$$

or

$$x = \frac{a_2 D_2}{A_2 - D_2}. \qquad (12.42)$$

The two expressions for x, equations (12.40) and (12.42), are not, in general, the same. An equilibrium point in the interior of the first octant with y and z both present is all but impossible.

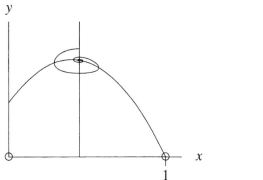
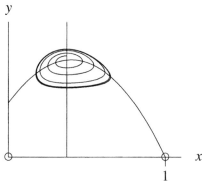

Fig. 12.11. Predator and prey zero-growth isoclines and
a Hopf bifurcation.

At the same time, we have needlessly limited our discussion to equilibria.
Consider the same system without the second predator,

$$\frac{dx}{dt} = x(1 - x) - \frac{A_1 \, x \, y}{a_1 + x},$$ (12.43a)

$$\frac{dy}{dt} = \frac{A_1 \, x \, y}{a_1 + x} - D_1 \, y.$$ (12.43b)

This set of equations should look familiar. It is simply a reparametrized
version of system (8.62). In Chapter 8, we showed that this system undergoes
a Hopf bifurcation as the predator zero-growth isocline passes through the
peak of the prey zero-growth isocline (see Figure 12.11). Equations (12.36a),
(12.36b), and (12.36c) may, as a result, possess a family of limit cycles in
the (x, y) plane. These limit cycles are stable to perturbations in the (x, y)
plane. The more interesting question is whether they are also stable to
perturbations in the z direction. If they are not, and if the corresponding
equilibria or limit cycles in the (x, z) plane are also unstable, it is possible
that the two predators may coexist in the interior of the first octant on a
periodic orbit.

To test this possibility, let us rewrite equations (12.36a), (12.36b), and
(12.36c) in a way that highlights functional dependencies:

$$\frac{dx}{dt} = x f(x, y, z),$$ (12.44a)

$$\frac{dy}{dt} = y g(x),$$ (12.44b)

$$\frac{dz}{dt} = z h(x).$$ (12.44c)

The exact forms of $f(x, y, z)$, $g(x)$, and $h(x)$ are clear from our original equations. If we linearize this system about the periodic (x, y) limit cycle $[x_p(t), y_p(t), 0]$, the appropriate Jacobian is just

$$
J = \begin{pmatrix} x\frac{\partial f}{\partial x} + f & \frac{\partial f}{\partial y} & \frac{\partial f}{\partial z} \\ y\frac{\partial g}{\partial x} & g & 0 \\ z\frac{\partial h}{\partial x} & 0 & h \end{pmatrix}_{[x_p(t), y_p(t), 0]}
\tag{12.45}
$$

or

$$
J = \begin{pmatrix} x\frac{\partial f}{\partial x} + f & \frac{\partial f}{\partial y} & \frac{\partial f}{\partial z} \\ y\frac{\partial g}{\partial x} & g & 0 \\ 0 & 0 & h \end{pmatrix}_{[x_p(t), y_p(t), 0]}.
\tag{12.46}
$$

This Jacobian is reducible and we may decouple the behavior in the z direction:

$$
\frac{dz}{dt} \approx h[x_p(t)]\, z = \left[\frac{A_2\, x_p(t)}{a_2 + x_p(t)} - D_2 \right] z.
\tag{12.47}
$$

As a result,

$$
z(t) \approx z_0\, e^{\int_0^t \frac{A_2\, x_p(s)}{a_2 + x_p(s)} - D_2\, ds}.
\tag{12.48}
$$

Since $x_p(t)$ is periodic, it is sufficient to consider the relative change of the perturbation over one period τ,

$$
\mu = z_0\, e^{\int_0^\tau \frac{A_2\, x_p(s)}{a_2 + x_p(s)} - D_2\, ds}.
\tag{12.49}
$$

This relative change is also known as a *Floquet multiplier*. If a_2 is large, $\mu < 1$, so that small perturbations in the z direction decay. If a_2 is sufficiently small, $\mu > 1$ and these same perturbations grow.

The previous analysis suggests that a good way to look for coexistence for the two predators is to start with a stable limit cycle in the (x, y) plane and to gradually decrease a_2 until this limit cycle is no longer stable. Figure 12.12 shows the results of one such experiment. For $A_1 = 0.05$, $A_2 = 0.2$, $D_1 = 0.03$, $D_2 = 0.05$, $a_1 = 0.1$, and $a_2 = 1.5$, there is a stable limit cycle in the (x, y) plane (not shown). As we decrease a_2, an unstable limit cycle passes through and exchanges stability with the (x, y) limit cycle in a transcritical bifurcation of limit cycles. As we continue to decrease a_2, this new limit cycle rises up into the first octant. For sufficiently small a_2, the new limit cycle undergoes a second transcritical bifurcation. Solutions are then drawn to a stable limit cycle in the (x, z) plane. Coexistence of the two predators occurs on the limit cycles in the interior of the first octant.

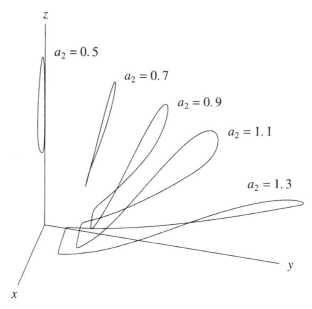

Fig. 12.12. Limit cycles of coexistence.

It is worth noting that, in this example, $a_1 < a_2$ and $A_1/D_1 < A_2/D_2$. That is, the species with the lower half-saturation constant also has the smaller birth-rate-to-death-rate ratio. This is the rule for this system (Hsu *et al.*, 1978): for the two predators to coexist one predator must do better at low prey densities, while the other does better at high prey densities. However, this condition is not sufficient for coexistence. If the prey carrying capacity K_0 is too low, the predator with the higher birth-rate-to-death-rate ratio is sure to win, while if the carrying capacity K_0 is too high, the predator with the lower half-saturation constant wins. Despite these caveats, it is still remarkable that two predators are able to coexist on a single prey item in a constant environment.

In the last example, two competitors relied on some delicate nonlinear dynamics in order to coexist. Most competing organisms have methods for avoiding competitive exclusion that go beyond the mechanisms that I have discussed. For example, one can follow Emlen (1984) and ask what constitutes a single resource. If two predators coexist because one feeds on the larval stage and the other feeds on the adult stage of some species (Haigh and Maynard Smith, 1972), do they, in fact, share one resource? What if one chipmunk prefers the stems of dandelions and the other prefers the heads and seeds (Carleton, 1965)? In these two examples, the competing predators coexist by partitioning aspects of the prey. Competition is also avoided,

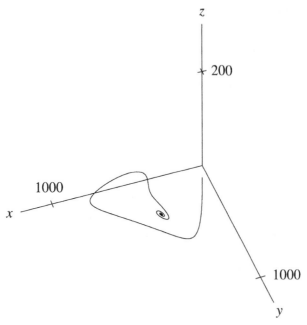

Fig. 12.13. Stable focus for $\alpha_{31} = 0.002$.

in many instances, by partitioning of space and time. The classic example of the former is contained in MacArthur's (1958) study of the segregation of five sympatric species of *Dendroica* warblers on the basis of height and positions in trees. Many species also live in temporally or spatially varying environment, where competitors may trade off the competitive advantage. For example, one species of purple sulfur bacteria may do better early in the day while the other does better late in the day (Van Gemerden, 1974), or one species of titmouse may do better in the breeding season while the other does better in winter (Dhondt and Eyckerman, 1980). Disturbance may also prevent dominance of one species over another (Connell, 1978).

Finally, it is important to remember that competition rarely occurs in isolation and that other interactions may ameliorate competition. Paine (1966, 1969) has demonstrated that *Pisaster ochraceus*, a starfish of the rocky shores of the Pacific coast of North America, functions as a *keystone predator*. It has an indirect beneficial effect on a suite of inferior competitors by depressing the abundance of a superior competitor.

Predation can have an equally profound effect on models of competition. Gilpin (1979) highlighted some of the interesting effects that can occur by

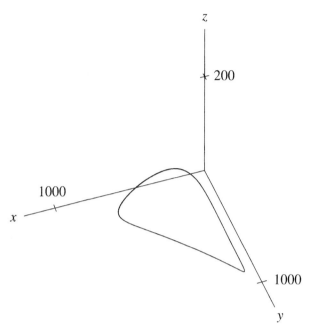

Fig. 12.14. Stable limit cycle for $\alpha_{31} = 0.0033$.

studying a Lotka–Volterra system,

$$\frac{dx}{dt} = x(r_1 + \alpha_{11} x + \alpha_{12} y + \alpha_{13} z), \qquad (12.50a)$$

$$\frac{dy}{dt} = y(r_2 + \alpha_{21} x + \alpha_{22} y + \alpha_{23} z), \qquad (12.50b)$$

$$\frac{dz}{dt} = z(r_3 + \alpha_{31} x + \alpha_{32} y + \alpha_{33} z), \qquad (12.50c)$$

for two competitors, x and y, and a keystone predator, z. The parameters for this system were chosen in such a way that x was the superior competitor, y was an inferior competitor, and z did better on x than on y,

$$r = \begin{pmatrix} 1 \\ 1 \\ -1 \end{pmatrix}, \quad A = \begin{pmatrix} -0.0010 & -0.0010 & -0.0150 \\ -0.0015 & -0.0010 & -0.0010 \\ \alpha_{31} & 0.0005 & 0.0000 \end{pmatrix}, \qquad (12.51)$$

with $0.002 \le \alpha_{31} \le 0.0055$ (Schaffer and Kot, 1986b; Schaffer *et al.*, 1986). For $\alpha_{31} = 0.002$ (see Figure 12.13), x may start to outcompete y. However, dominance by x inevitably leads to an irruption of the predator z. The system eventually settles down to a stable focus. As we increase the predator's predilection for the superior competitor, this equilibrium loses its stability in a Hopf bifurcation. The system begins to cycle between dominance by

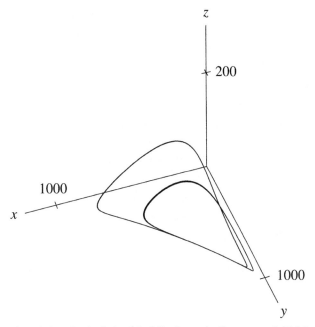

Fig. 12.15. Period-doubled limit cycle for $\alpha_{31} = 0.0036$.

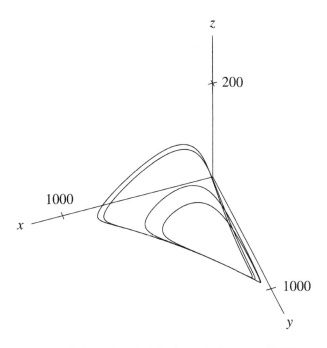

Fig. 12.16. Period-quadrupled limit cycle for $\alpha_{31} = 0.00372$.

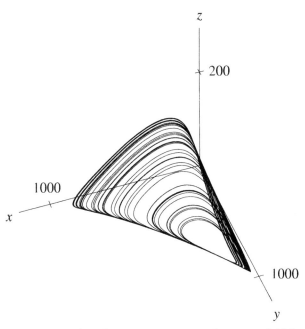

Fig. 12.17. Chaotic strange attractor for $\alpha_{31} = 0.004$.

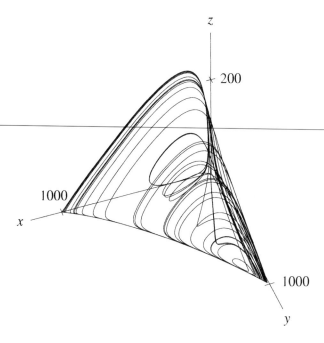

Fig. 12.18. Complicated strange attractor for $\alpha_{31} = 0.0055$.

the superior competitor, an outbreak of the predator, and an increase in the inferior competitor in the near absence of the superior competitor. Figure 12.14 shows a stable limit cycle for $\alpha_{31} = 0.0033$.

Additional increases in α_{31} cause the simple limit cycle to lose its stability in a series of period-doubling bifurcations (Figures 12.15 and 12.16). For sufficiently high α_{31}, the system possesses a strange attractor with chaotic dynamics (see Figures 12.17 and 12.18). The entire transition from equilibrium to chaotic strange attractor occurs over a remarkably narrow range of the parameter α_{31}.

Much of our fascination with competition clearly comes from the challenge of trying to understand how organisms avoid competition's dire consequences.

Recommended readings

Waltman's (1983) monograph on competition models in population biology is required reading for all mathematical ecologists. Grover (1997) provides an up-to-date and useful overview of recent research on resource competition. Adler (1990) develops the theory for discrete-time resource-competition models.

13 Mutualism models

Mutualism is an interaction in which species help one another. Mutualism has seldom received the attention of predation and competition. This neglect is quite surprising, given the ubiquity of mutualisms.

Janzen (1985), in a thorough review of the natural history of mutualisms, has argued that most mutualisms can be classified into one of four types: seed-dispersal mutualisms, pollination mutualisms, digestive mutualisms, and protection mutualisms. Let me give you a few examples of each type of mutualism.

Seed-dispersal mutualisms

A great many plants rely on animals to carry their seeds to favorable sites. Plants frequently produce fruits and nuts to attract and reward dispersal agents. Squirrels are undoubtedly the most familiar mammalian dispersal agents. Tree squirrels collect and eat acorns, walnuts, beechnuts and other nuts (Smith and Follmer, 1972; Stapanian and Smith, 1978). At times they will bury or cache seeds and then lose or forget them. It is this last bit of forgetfulness that makes the squirrel a mutualist of the adult tree.

I am an avid birdwatcher, so most of my favorite examples of seed-dispersal mutualism involve birds. In western North America, there is an extremely tight correspondence between the ranges of wingless-seeded soft pines such as white-bark pine (*Pinus albicaulis*), limber pine (*Pinus flexilis*), southwestern white pine (*Pinus strobiformis*), Colorado pinyon (*Pinus edulis*), and singleleaf pinyon (*Pinus monophylla*) and the ranges of corvids such as Clark's Nutcracker (*Nucifraga columbiana*), Pinyon Jay (*Gymnorhinus cyanocephalus*), Western Scrub Jay (*Aphelocoma californica*), and Stellar's Jay (*Cyanocitta stelleri*) (Lanner, 1996). The aforementioned trees have wingless seeds that sit within a cone. The corvids are adept at harvesting these seeds. They then carry the seeds long distances and cache the seeds in small clumps, where they may germinate. Clark's Nutcracker has a particularly well-developed sublingual pouch (Bock *et al.*, 1973) that enables it to carry on the order of 50 seeds on its caching trips of up to 22 km (Vander Wall and Balda, 1977).

There is a similar correspondence between the ranges of mistletoes and of the Phainopepla (*Phainopepla nitens*), a silky flycatcher, in the Southwest. Mistletoes are parasitic plants that parasitize tree branches. Mistletoes are frequently dispersed by birds – in this case Phainopeplas – that eat the berries and void the seeds some distance away on other tree branches.

Elephant-like gomphotheres are the most notorious seed dispersal agents. Starting some 20 million years ago, North America supported a large collection of large mammals. This mammalian megafauna included antelopes, camels, giant ground sloths, gomphotheres, mastodonts, rhinoceroses, and the like. This large megafauna went extinct, rather suddenly, 8 to 15 thousand years ago. Janzen and Martin (1982) have claimed that the loss of this megafauna had a profound effect on many plants and that many modern fruits are anachronisms, adapted for dispersal by members of the extinct megafauna. This megafaunal fruit hypothesis has been used to explain seemingly maladaptive tropical fruits that rot directly beneath the trees that produce them. (Some of these trees appear to have undergone range expansions since the reintroduction of horses – new dispersal agents – into the New World.) This hypothesis has been criticized by Howe (1985). Janzen (1986) has since argued that the megafauna also played a major role in shaping the ecology and evolution of cacti.

A similar avian example involves the putative mutualism between the extinct Dodo (*Raphus cucullatus*) and the tambalocoque tree (*Calvaria major*) of Mauritius in the Indian Ocean (Temple, 1977). The Dodo, a giant flightless dove, was an omnivore that crushed fruits and seeds with its powerful gizzard. It was driven extinct by 1690 as the result of hunting pressure from early European explorers and travelers and because of egg predation from introduced monkeys, hogs, and cats (Campbell and Lack, 1985; Fuller, 1987). Temple has suggested that the Dodo was the principle dispersal agent for the tambalocoque tree and that the seeds of this tree needed to pass through the gut of Dodo to germinate. The fruits of this tree have a pit that is too tough for most frugivores and by 1973 the tambalocoque was quite rare; there were only thirteen extremely old (over 300 years in age) individuals known to exist. Temple overcame the seed-coat dormancy by force-feeding pits to introduced turkeys.

Pollination mutualisms

Pollination is the transfer of a plant's pollen grain before fertilization. In gymnosperms, the transfer is from a pollen-producing cone directly to an ovule. In angiosperms, the transfer is from an anther to a stigma. Most gymnosperms are *anemophilous* or wind pollinated. In contrast, many angiosperms are *zoophilous* or animal pollinated. The obvious advantage of animal pollination is that pollen may be transferred far from the host anther in a way that promotes outbreeding and genetic variability. Angiosperm flowers often reward pollinators with nectar and pollen to promote this process.

When we think of animal pollinators, we usually think of butterflies, hive bees, and bumblebees. However, the earliest pollinators were probably winged insects from early in the fossil record, such as beetles (Coleoptera) and flies with biting mouthparts (Diptera) (Richards, 1997). Beetles and flies ate pollen and accidentally carried pollen to other flowers. Some modern-day plants with primitive flowers are still pollinated by these vectors. For example, magnolias produce little or no nectar, but large amounts of protein- and fat-rich pollen; they are visited, almost exclusive, by pollen-eating beetles and flies. Beetle-pollinated flowers are typically white (or dull-colored), strongly scented, and bowl shaped.

Insects that have tubular mouthparts for sucking nectar (bees, butterflies, moths, etc.) appeared much later in the fossil record, in the mid-Cretaceous. Their appearance was coeval with the main diversification of angiosperms and there has been much coevolution between flowers and nectar-sucking insects. These insects have a visual spectral range different from that of humans and this is reflected in the color of their flowers. Bees cannot distinguish between shades of red, but their vision does extend into the ultraviolet. Bee-pollinated flowers are typically bright yellow to blue; they frequently have nectar guide marks that may be seen on ultraviolet-sensitive film.

Modern plants use a variety of pollinators. These include birds and bats. Avian pollinators such as the hummingbirds of North and South America, the sunbirds of Africa, the honeyeaters of Australia, and the honey-creepers of Hawaii are responsive to, and distinguish well between, shades of orange and red, but do not distinguish well between yellows, blues, and purples. Most birds also have a bad sense of smell. It should come as no surprise, therefore, that most bird flowers have no scent. In contrast, flowers that are pollinated at night by moths and bats are often strongly scented.

Most pollinators are *oligolectic* or *polylectic* in that they visit several to many different taxa of plants. Fig wasps are a notable exception to this generalization. It is generally believed that a different species of fig wasp pollinates each of the 900-odd species of fig. Janzen (1979) provides an interesting introduction to the biology of this unusual system.

Digestive mutualisms

The guts of many animals are filled with mutualists (bacteria, yeast, protozoa) that help to break down food. Often, the host is unable to digest the food on its own. Ruminants, such as cattle, deer, and sheep, rely on bacteria to break down plant cellulose and hemicelluloses into digestible subunits (Ahmadjian and Paracer, 1986). The microbial flora of the rumen is typically quite diverse and may contain over 200 species of bacteria (Lengeler *et al.*, 1999). These bacteria are joined by various ciliates (holotrichs and entodiniomorphs) and fungi. Ruminants meet their nutritional needs by consuming some portion of their rumen residents.

Many termites survive by consuming wood. Although some higher termites produce their own cellulases, many lower termites depend on the microbial mutualists in their hindgut to break down cellulose (Lengeler *et al.*, 1999). Flagellated protozoa are responsible for 30% to 50% of the weight of some wood-eating termites (Ahmadjian and Paracer, 1986). These protozoa produce acetic acid and acetate that are absorbed by termites. Termite hindguts also contain a wide variety of bacteria. Some of these bacteria produce acetate, others fix nitrogen. The metabolic function of some bacteria (most notably spirochetes) is unknown.

Many plants rely on bacteria to turn atmospheric nitrogen into usable ammonia. Plants do not store their microsymbionts in a gut. Rather, they store them in root nodules. Legumes rely on bacteria of the genus *Rhizobium* to fix nitrogen. Some 200 species of woody dicotyledons dicots rely on actinomycetes to fix nitrogen (Ahmadjian and Paracer, 1986; Lengeler *et al.*, 1999).

Many of the most remarkable harvest and digestive mutualisms occur within animal and plant cells. Late in the 19th century, the German botanist A. F. W. Schimper (best known to ecologists for the neologism 'tropical rainforest') suggested that chloroplasts were cyanobacteria that lived as symbionts within plant cells. Ivan Wallin (1923, 1927) then proposed that mitochondria are bacterial cells that had formed a symbiotic relationship with a nucleus-containing host cell. These ideas were largely ignored until the 1960s when they were revitalized, expanded, and championed by Lynn Margulis (Sagan, 1967; Margulis, 1970, 1981). According to Margulis's endosymbiotic hypothesis, the eukaryotic cell is a chimera. This hypothesis was quite controversial, but is now supported by strong structural, molecular, and biochemical evidence.

Protection mutualisms

In 1874, the famous naturalist, Thomas Belt, described a remarkable mutualism between ants and acacias. The genus *Acacia* contains a large number of leguminous trees and shrubs native to the warm parts of both hemispheres. Many of the plants in this genus house, support, and employ ants (Janzen, 1966; Beattie, 1985; Keeler, 1989). In Central America, ants use hollowed-out, swollen acacia thorns as nest sites. In Africa, ants occupy stem galls near thorns. Both the ants and the acacias benefit from this arrangement. The ants guard the acacias against herbivorous predators and encroaching vines. Since the acacia contains their nest, the ants will often swarm at the merest approach of a predator. The ants are rewarded with a nest site and, in many instances, food. Many acacias have extrafloral nectories that provide water, sugars, and amino acids and leaf tip Beltian bodies that contain proteins and lipids. Acacias are examples of myrmecophytes or ant plants. Myrmecophytes are thought to occur in over 100 genera of 36 plant families.

Many other organisms also form protective mutualisms. A particularly well-known example involves tropical anemone fishes, or clown fishes, and their anemones. Clown fish are immune to the stinging nematocysts of giant sea anemones and will feed and nest amongst their tentacles. Horse mackerels appear to have a similar relationship with Portuguese man-of-war jellyfish.

So far, I have classified mutualisms according to the nature of the benefit. Several other characters are also used to classify mutualisms. One such character is the specificity of the mutualism. Are we dealing with a *diffuse* mutualism, involving many species, or a *specialized* or one-to-one mutualism? Most mutualisms are diffuse, but it is easier to start by modeling specialized mutualisms. Also, how dependent are the species on the interaction? Are we dealing with an *obligate* mutualism where neither mutualist can survive without the other, or a *facultative* mutualism in which the interaction is helpful, but not essential?

I will begin by looking at a simple model for a one-to-one facultative mutualism. I will follow this with an equally simple model for an obligate mutualism. These two models are Lotka–Volterra (competition) equations in which the negative competitive interactions have been turned into positive mutualistic interactions. Lotka–Volterra equations make reasonable predictions in the case of competition. This is not always the case for the corresponding mutualism models !

Let us begin as we did at the start of Chapter 12. Let us assume that there are two species, with population sizes N_1 and N_2, and that each species grows logistically in the absence of the other. Each species thus has a per capita growth rate that decreases linearly with density,

$$\frac{1}{N_1}\frac{dN_1}{dt} = r_1 \left(1 - \frac{N_1}{K_1}\right), \tag{13.1a}$$

$$\frac{1}{N_2}\frac{dN_2}{dt} = r_2 \left(1 - \frac{N_2}{K_2}\right). \tag{13.1b}$$

The mutualism ameliorates intraspecific competition. We may thus write

$$\frac{1}{N_1}\frac{dN_1}{dt} = r_1 \left[1 - \frac{(N_1 - \alpha_{12} N_2)}{K_1}\right], \tag{13.2a}$$

$$\frac{1}{N_2}\frac{dN_2}{dt} = r_2 \left[1 - \frac{(N_2 - \alpha_{21} N_1)}{K_2}\right], \tag{13.2b}$$

where the parameters α_{12} and α_{21} measure the strength of the positive effect of species 2 on species 1 and of species 1 on species 2.

Equations (13.2a) and (13.2b) may be rewritten, more succinctly, as

$$\frac{dN_1}{dt} = \frac{r_1}{K_1} N_1 (K_1 - N_1 + \alpha_{12} N_2), \tag{13.3a}$$

$$\frac{dN_2}{dt} = \frac{r_2}{K_2} N_2 (K_2 - N_2 + \alpha_{21} N_1). \tag{13.3b}$$

This is a model for facultative mutualism so far as

$$r_1 > 0, \ r_2 > 0, \ K_1 > 0, \ K_2 > 0. \tag{13.4}$$

Each species can, in other words, survive without its mutualist.

Let us look at the zero-growth isoclines for this system. The N_2 zero-growth isoclines are simply

$$N_2 = 0 \tag{13.5a}$$

and

$$N_2 = K_2 + \alpha_{21} N_1 \tag{13.5b}$$

(see Figure 13.1). Below line (13.5b), N_2 increases; above this line, N_2 decreases. The N_1 zero-growth isoclines, in turn, are given by

$$N_1 = 0 \tag{13.6a}$$

and

$$N_1 = K_1 + \alpha_{12} N_2 \tag{13.6b}$$

(see Figure 13.2). To the left of line (13.6b), N_1 increases; to the right of this line, N_1 decreases.

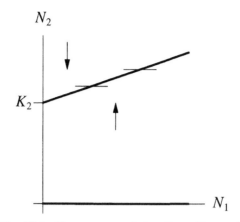

Fig. 13.1. N_2 zero-growth isoclines, $K_2 > 0$.

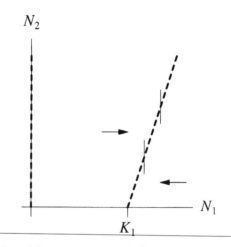

Fig. 13.2. N_1 zero-growth isoclines, $K_1 > 0$.

Zero-growth isoclines (13.5b) and (13.6b) may either converge or diverge. They converge if

$$\frac{1}{\alpha_{12}} > \alpha_{21} \tag{13.7}$$

or

$$\alpha_{12}\alpha_{21} < 1. \tag{13.8}$$

In this case, the two isoclines cross and orbits approach a stable node in the interior of the first quadrant (see Figure 13.3). Since the slopes of the two zero-growth isoclines are positive, the coordinates of this equilibrium

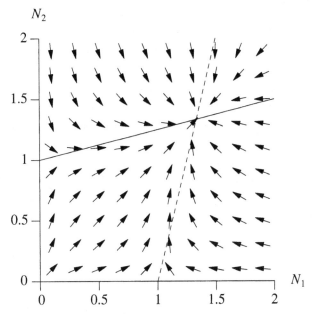

Fig. 13.3. Facultative mutualism phase portrait for
$\alpha_{12} \alpha_{21} < 1$.

are greater than the carrying capacities K_1 and K_2: each species surpasses its carrying capacity because of its mutualist.

If

$$\alpha_{12}\alpha_{21} > 1, \tag{13.9}$$

zero-growth isoclines (13.5b) and (13.6b) diverge. Now, the zero-growth isoclines do not cross and there is no nontrivial equilibrium in the first quadrant. The populations undergo unlimited growth (see Figure 13.4) in what has been called 'an orgy of mutual benefaction' (May, 1981).

Equations (13.3a) and (13.3b) can also be used as a model for obligate mutualism if we assume that

$$r_1 < 0, \; r_2 < 0, \; K_1 < 0, \; K_2 < 0. \tag{13.10}$$

Neither species can now survive on its own; each species is banking on the other to save it.

Equations (13.5a) and (13.5b) and equations (13.6a) and (13.6b) are still the correct equations for the N_2 and the N_1 zero-growth isoclines. However, since K_1 and K_2 are now negative, the lines (13.5b) and (13.6b) look rather different (see Figures 13.5 and 13.6).

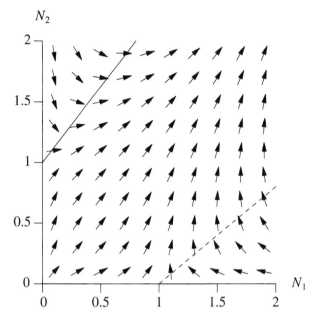

Fig. 13.4. Facultative mutualism phase portrait for $\alpha_{12}\,\alpha_{21} > 1$.

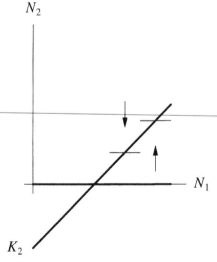

Fig. 13.5. N_2 zero-growth isoclines, $K_2 < 0$.

We are again confronted by two cases. In the first case,

$$\alpha_{12}\alpha_{21} < 1 \tag{13.11}$$

and the two isoclines (13.5b) and (13.6b) do not intersect in the first quadrant.

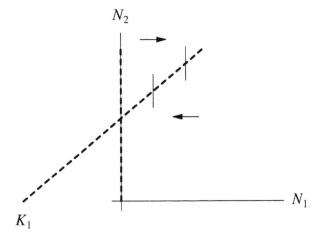

Fig. 13.6. N_1 zero-growth isoclines, $K_1 < 0$.

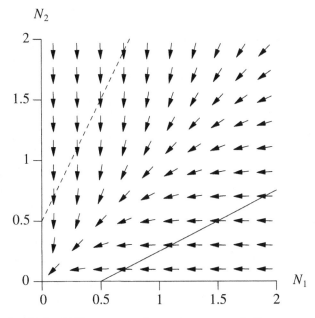

Fig. 13.7. Obligate mutualism phase portrait for
$\alpha_{12}\alpha_{21} < 1$.

The stable node at the origin is now the only equilibrium. Both populations decay to zero (see Figure 13.7). The two species rely on one another, but the interaction is too weak to rescue either species.

If the interaction is strong,

$$\alpha_{12}\alpha_{21} > 1, \tag{13.12}$$

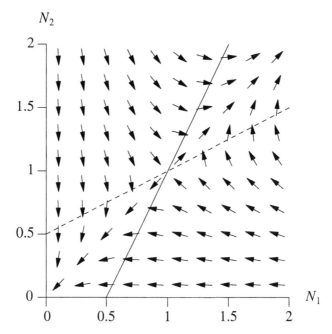

N_2

Fig. 13.8. Obligate mutualism phase portrait for $\alpha_{12}\,\alpha_{21} > 1$.

the two isoclines (13.5b) and (13.6b) do intersect in the first quadrant. There is now a saddle point in the first quadrant. If mutualist densities are low, both populations go extinct: the interaction is strong, but there are too few mutualists to rescue either population. If mutualist densities are high, both species increase in another orgy of mutual benefaction. Orbits now diverge to infinity.

All orbits of the two-species mutualism models that I have considered appear to tend to an equilibrium or to diverge to infinity. Can I be sure that there are not limit cycles that I have missed? Previously, I used the Bendixson–Dulac negative criterion to prove that systems do not possess limit cycles. Now I can use the fact that our models are *cooperative*.

Definition   The system

$$\frac{dN_1}{dt} = f(N_1, N_2), \tag{13.13a}$$

$$\frac{dN_2}{dt} = g(N_1, N_2), \tag{13.13b}$$

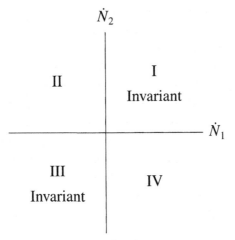

Fig. 13.9. $(\dot{N}_1, \dot{N}_2)$ plane.

defined on $D \subseteq \mathfrak{R}^2$, is *cooperative* if

$$\frac{\partial f}{\partial N_2} \geq 0, \ \frac{\partial g}{\partial N_1} \geq 0 \tag{13.14}$$

for all $(N_1, N_2) \in D$.

EXAMPLE
Let

$$f(N_1, N_2) = \frac{r_1}{K_1} N_1 (K_1 - N_1 + \alpha_{12} N_2), \tag{13.15a}$$

$$g(N_1, N_2) = \frac{r_2}{K_2} N_2 (K_2 - N_2 + \alpha_{21} N_1), \tag{13.15b}$$

on the (invariant) first quadrant of the (N_1, N_2) plane. Since

$$\frac{\partial f}{\partial N_2} = \alpha_{12} \frac{r_1}{K_1} N_1, \tag{13.16a}$$

$$\frac{\partial g}{\partial N_1} = \alpha_{21} \frac{r_2}{K_2} N_2, \tag{13.16b}$$

our mutualism models are cooperative on this quadrant. ◇

Theorem The orbits of a system that is cooperative either converge to equilibria or diverge to infinity.

Proof Let us look at the $(\dot{N}_1, \dot{N}_2)$ plane (see Figure 13.9). Each trajectory of a planar system generates an orbit in this $(\dot{N}_1, \dot{N}_2)$ plane.

If the planar system that we are looking at is everywhere cooperative, the first quadrant of the $(\dot{N}_1, \dot{N}_2)$ plane is invariant. To see this, consider an orbit that attempts to leave the first quadrant by crossing the positive $\dot{N}_2$-axis. In light of equation (13.13a) and our definition of cooperativity,

$$\frac{d^2 N_1}{dt^2} = \frac{\partial f}{\partial N_1}\frac{dN_1}{dt} + \frac{\partial f}{\partial N_2}\frac{dN_2}{dt} = \frac{\partial f}{\partial N_2}\frac{dN_2}{dt} > 0 \qquad (13.17)$$

on the positive $\dot{N}_2$-axis. The orbit can thus not cross the positive $\dot{N}_2$-axis. By a similar argument, the orbit cannot cross the positive $\dot{N}_1$-axis. Finally, the orbit cannot cross through the origin, since this would imply that the original trajectory passes through a rest point. Similar arguments show that the third quadrant is also invariant.

Ultimately (as $t \to \infty$), $\dot{N}_1$ and $\dot{N}_2$ are of constant sign. If you start in the first or third quadrant, you stay there. If you start in the second or fourth quadrant, you either stay in those quadrants or you enter into the first or third quadrants, which are invariant. Either way, N_1 and N_2 are ultimately monotonic functions of time. This precludes limit cycles and implies that trajectories either approach equilibria or diverge to infinity. □

The study of cooperative systems was initiated by Hirsch (1982, 1985, 1988, 1990) and has been extended to higher-dimensional and to competitive systems. Two-dimensional cooperative systems do not possess attractors more complicated than equilibria. Three-dimensional cooperative systems do not possess attractors more complicated than those of general two-dimensional autonomous systems.

We have looked at mutualism models that allow interacting populations to reap unbounded benefits and to grow in an unbounded manner. These models are unrealistic for most conditions. To reintroduce a semblance of reality, we must limit the benefits of mutualism. Some modelers have prevented unlimited growth by limiting per capita birth and death rates. Other modelers have tried to put mutualism on a sound footing by insisting on a resource-based formalism.

An example of the first approach can be found in the work of Wolin and Lawlor (1984). Wolin and Lawlor begin by considering a single-species population model in which the per capita birth rate decreases with density,

$$\mathscr{E} = b_0 - b\,N_1, \qquad (13.18)$$

and the per capita death rate increases with density,

$$d = d_0 + d\,N_1. \qquad (13.19)$$

The growth rate of this population is

$$\frac{dN_1}{dt} = (\ell - d) N_1 \tag{13.20}$$

or

$$\frac{dN_1}{dt} = (b_0 - d_0) N_1 - (b + d) N_1^2. \tag{13.21}$$

This is simply the logistic differential equation,

$$\frac{dN_1}{dt} = r N_1 - \frac{r}{K} N_1^2, \tag{13.22}$$

with

$$r \equiv b_0 - d_0, \quad K \equiv \frac{b_0 - d_0}{b + d}. \tag{13.23}$$

Now, assume (naively) that a facultative mutualist (N_2) increases the per capita birth rate of N_1,

$$\ell = b_0 - b N_1 + m N_2, \tag{13.24}$$

but has no effect on the per capita death rate,

$$d = d_0 + d N_1. \tag{13.25}$$

This quickly leads to equation (13.3a). However, equation (13.24) is suspect. The equation implies that the presence of many mutualists will raise the per capita birth rate above b_0. Note, however, that b_0 is the maximum per capita birth rate.

Let us imagine, instead, that the mutualist decreases the density dependence in the per capita birth rate of N_1,

$$\ell = b_0 - \frac{b}{1 + \alpha_{12} N_2} N_1, \tag{13.26}$$

and that it has no effect on the per capita death rate. If we make a similar assumption with regard to the second species (and add a subscript to each unsubscripted b, d, and r), we quickly derive the system

$$\frac{dN_1}{dt} = \left(r_1 - \frac{b_1 N_1}{1 + \alpha_{12} N_2} - d_1 N_1 \right) N_1, \tag{13.27a}$$

$$\frac{dN_2}{dt} = \left(r_2 - \frac{b_2 N_2}{1 + \alpha_{21} N_1} - d_2 N_2 \right) N_2, \tag{13.27b}$$

where

$$K_1 \equiv \frac{r_1}{b_1 + d_1}, \quad K_2 \equiv \frac{r_2}{b_2 + d_2}. \tag{13.28}$$

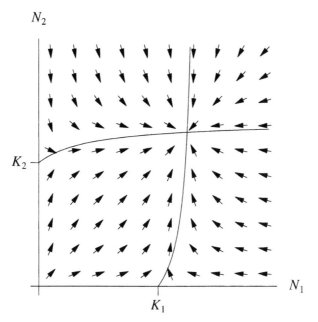

Fig. 13.10. Bent zero-growth isoclines.

The N_2 zero-growth isoclines are now

$$N_2 = 0 \tag{13.29}$$

and

$$N_2 = \frac{r_2(1 + \alpha_{21} N_1)}{b_2 + d_2(1 + \alpha_{21} N_1)}. \tag{13.30}$$

The N_1 zero-growth isoclines, in turn, are

$$N_1 = 0 \tag{13.31}$$

and

$$N_1 = \frac{r_1(1 + \alpha_{12} N_2)}{b_1 + d_1(1 + \alpha_{12} N_2)}. \tag{13.32}$$

The nontrivial zero-growth isoclines no longer diverge, as in Figure 13.4. Rather, they cross (Figure 13.10), producing an equilibrium in which both species are above their carrying capacities. There is no orgy of mutual benefaction.

An alternative approach to preventing unlimited growth is to construct a resource-based model. A simple example can be found in the work of Lee *et al.* (1976) on the interaction between a lactic acid bacterium, *Lactobacillus plantarum*, and a propionic acid bacterium, *Propionibacterium shermanii*.

Lactobacillus plantarum grows on glucose and produces lactic acid. *Propionibacterium shermanii* takes lactate (a salt of lactic acid) and metabolizes it to produce propionic acid and carbon dioxide. The interaction is of great interest to gourmands: the propionic acid gives Emmenthal cheese its distinct flavor. The carbon dioxide produces the holes or 'eyes' of the Swiss cheese. (If there is too little carbon dioxide to produce holes, the cheese is 'blind'.) Although this is a model for commensalism, a $(+, 0)$ interaction, the resulting framework can easily be extended to mutualism (Meyer *et al.*, 1975; Miura *et al.*, 1980; Dean, 1985).

Lee *et al.*'s model can be written

$$\frac{dS}{dT} = D\,(S_i - S) - \frac{1}{Y_1}\frac{\mu_1\,S\,N_1}{K_1 + S} - c\,N_1, \tag{13.33a}$$

$$\frac{dN_1}{dT} = \frac{\mu_1\,S\,N_1}{K_1 + S} - D\,N_1, \tag{13.33b}$$

$$\frac{dP}{dT} = a\,\frac{\mu_1\,S\,N_1}{K_1 + S} + b\,N_1 - \frac{1}{Y_2}\frac{\mu_1\,P\,N_2}{K_2 + P} - D\,P, \tag{13.33c}$$

$$\frac{dN_2}{dT} = \frac{\mu_1\,P\,N_2}{K_2 + P} - D\,N_2, \tag{13.33d}$$

where S is the substrate (glucose), P is the product (lactate), N_1 is the density of *L. plantarum*, and N_2 is the density of *P. shermanii*. The model contains the usual dilution rate, inflowing substrate concentration, functional responses, half-saturation constants, and yield coefficients that we first met in Chapter 10. Two added terms, $c\,N_1$ and $b\,N_1$, account for *L. plantarum's* large maintenance cost. (*Lactobacillus plantarum* must consume glucose without growing in order to stay viable.)

Problem 13.1 *Commensal growth*
Nondimensionalize equations (13.33a), (13.33b), (13.33c), and (13.33d). Find the equilibria. Determine their stability. Do any of the equations decouple from the others?

This model has three nonnegative equilibria, corresponding to (a) washout of both species, (b) washout of the propionic acid bacterium but survival of the lactic acid bacterium, or (c) survival of both bacteria. For each choice of the inflowing glucose concentration and dilution rate, only one equilibrium is stable. For each glucose concentration, there is a threshold dilution rate, below which *L. plantarum* first survives. There is also a second, lower dilution rate below which *P. shermanii* survives. At no point, do we find runaway growth. Both populations are ultimately limited by the concentration of glucose.

I have described the interaction between *L. plantarum* and *P. shermanii* as a commensalism. The interaction is, in fact, more complicated. Lee *et al.* (1974) showed that *P. shermanii* can use either glucose or lactate as a carbon and energy source. *Propionibacterium shermanii* actually experiences higher growth rates and higher densities while growing on glucose (with no *L. plantarum*). Even so, it invariably switches to lactate when growing in a medium that contains both glucose and lactate. If it failed to do so, it would lose in competition to *L. plantarum*. Does *P. shermanii* in fact 'benefit' from *L. plantarum?*

The challenge, in the future, will be to construct mechanistic or resource-based models that describe mutualistic interactions such as pollination and seed dispersal.

Recommended readings

The book by Boucher (1985) remains the best edited volume on mutualisms. The articles by Wolin (1985) and Dean (1985) in that volume are especially relevant to this chapter.

Section C
DYNAMICS OF EXPLOITED POPULATIONS

14 Harvest models and optimal control theory

I would like to turn our attention to some simple models in which humans interact with animal or plant populations by harvesting those populations. This is a problem of great practical importance and I am sure that we can all think of examples of successful and sustainable harvesting and of other examples of extreme overexploitation. What leads to one or the other of these outcomes?

We can use two rather different approaches to model harvesting. One approach emphasizes dynamics; the other relies on optimization theory. Let us look at one example of the dynamics approach and two examples of the optimization approach.

EXAMPLE Open-access fishery

An open-access fishery is an unregulated fishery in which fishermen are free to come and go as they please. One simple model for such a fishery (Schaefer, 1954) takes the form

$$\frac{dN}{dt} = r N \left(1 - \frac{N}{K} \right) - q E N, \tag{14.1a}$$

$$\frac{dE}{dt} = k (p q E N - c E), \tag{14.1b}$$

where N is the stock level, E is the level of fishing effort, r is the intrinsic rate of growth of the stock, K is the carrying capacity of the stock, q is the catchability, p is the price per fish, c is the (opportunity) cost per unit effort, and k is a proportionality constant. The first equation should look quite familiar. It contains logistic growth of the stock and a mass-action harvesting term proportional to both the stock level and the level of fishing effort. The second equation should also look familiar, though it now has an economic rather than an ecological interpretation. This second equation states that the rate of change of fishing effort is proportional to profit or economic rent, with $pqEN$ the revenues and cE the costs.

Despite the economic interpretation, this model is just the Lotka–Volterra model with prey-density dependence. Fishermen now play the predator. For r positive, there are two possible outcomes. The fishery may not be economically viable. Fishing effort will then decrease to zero and the fish stock will return to its carrying capacity. Alternatively, fish and fishing effort may coexist at an equilibrium in the interior of the (N, E) phase plane. The stock level at this equilibrium, $N^* = c/pq$, is determined by price, cost, and catchability. Since this equilibrium level is chiefly determined by economic rather than biological factors, it is not surprising that an open-access fishery may exhibit biological overfishing. The stock level may be below the level that maximizes the sustainable yield of fish.

Allee growth, different functional responses, and biological and economic delays may all lead to the same sorts of oscillations that one observes with predator-prey models. An example of an open-access fishery that has exhibited oscillations is the Lake Whitefish (*Coregonus clupeaformis*) fishery in Lesser Slave Lake, Canada (Bell *et al.*, 1977). ◇

EXAMPLE   Sole-owner fishery
At the opposite extreme, a fishery may be owned and regulated by a single individual. We will assume (pessimistically?) that this person's sole motive is to maximize his or her (long-term) profit. The owner's goal is to obtain

$$\max_{0 \le E(t) \le E_{\max}} \int_0^T e^{-\delta t} \left[p\, q\, E(t)\, N(t) - c\, E(t) \right] dt, \tag{14.2a}$$

subject to

$$\frac{dN}{dt} = r\, N \left(1 - \frac{N}{K} \right) - q\, E\, N, \tag{14.2b}$$

$$N(0) = N_0. \tag{14.2c}$$

The owner wishes to control the fishing effort, $E(t)$, so as to maximize the discounted net economic rent over the period T of ownership. The economic rent is discounted with rate δ for the simple reason that a dollar today is worth more than a dollar tomorrow, if only because the owner can invest today's dollar. There will typically be some maximum fishing effort that the owner can apply, limited, say, by the number of fishing boats. Any change in the fishing effort has an immediate effect on the state variable and stock level $N(t)$ through differential equation (14.2b). As stated, this is a problem in *optimal control theory*. ◇

EXAMPLE   Oligopoly
In an oligopoly, a small number of competitive firms control the market.
Each firm strives to obtain

$$\max_{0 \le E_i(t) \le E_{max}} \int_0^T e^{-\delta t} \left[p \, q \, E_i(t) \, N(t) - c \, E_i(t) \right] dt, \qquad (14.3a)$$

subject to

$$\frac{dN}{dt} = r N \left(1 - \frac{N}{K} \right) - q \left(\sum_i E_i \right) N, \qquad (14.3b)$$

$$N(0) = N_0. \qquad (14.3c)$$

That is, each firm tries to maximize its own discounted net economic rent.
Since there are several 'players', this is a problem in *differential game theory*
(which we won't really get into). Although this scenario appears intermediate
between the open-access fishery and the sole-owner fishery, it is more like
the open-access fishery. $\diamond$

Our goal is to analyze the sole-owner fishery. Since most of you have not
had a course in optimal control theory, let me start with some mathematical
background. The typical optimal control problem contains:

(1)   state variables $x(t)$,
(2)   control variables $u(t) \in U$,
(3)   a set U of admissible controls, and
(4)   an objective functional, performance index, or payoff, $J[x(t), u(t)]$.

Three well-known problems or formulations contain these ingredients.

Lagrange problem
The objective functional for the Lagrange problem is an integral: the payoff
accumulates through time. For a single state variable and a single control
variable, the goal is to obtain

$$\max_{u \in U} J[x, u], \qquad (14.4a)$$

where

$$J[x, u] = \int_{t_0}^{t_1} F[x(t), u(t), t] \, dt, \qquad (14.4b)$$

subject to

$$\frac{dx}{dt} = f(x, u, t). \qquad (14.4c)$$

We may also have initial and/or terminal conditions:

$$x(t_0) = x_0, \quad x(t_1) = x_1. \tag{14.4d}$$

EXAMPLE   Sole-owner fishery   ◇

EXAMPLE   Exhaustible resources (Hotelling, 1931)
The problem is much like the sole-owner fishery, except that the resource does not renew itself. The goal is to find

$$\max_{u(t) \in U} \int_0^T e^{-\delta t} p[u(t)]\, u(t)\, dt, \tag{14.5a}$$

subject to

$$\frac{dx}{dt} = -u \tag{14.5b}$$

with the initial condition

$$x(0) = x_0. \tag{14.5c}$$

The control variable u is the rate at which we mine or exploit the resource. The price $p(u)$ for the product depends on the extraction rate (i.e., on how badly we flood the market). The problem is to schedule the depletion of the resource so as to maximize discounted revenues.   ◇

Mayer problem
In a Mayer problem, all payoff occurs at the initial or terminal time. Thus, for the typical Mayer problem with one state variable and one control variable, the goal is to obtain

$$\max_{u(t) \in U} J[x, u], \tag{14.6a}$$

where

$$J[x, u] = G[t_0, x(t_0), t_1, x(t_1)], \tag{14.6b}$$

subject to

$$\frac{dx}{dt} = f(x, u, t). \tag{14.6c}$$

We may also have initial and/or terminal conditions:

$$x(t_0) = x_0, \quad x(t_1) = x_1. \tag{14.6d}$$

The function G is sometimes referred to as the *salvage function*.

EXAMPLE   Getting to class
Let x represent your position on the highway relative to the classroom and

suppose that you can accelerate or decelerate,

$$\frac{d^2x}{dt^2} = u, \tag{14.7}$$

up to some maximum rate,

$$|u| \leq u_{max}. \tag{14.8}$$

Imagine that you wish to go from

$$\left(x, \frac{dx}{dt}\right) = (a, b) \tag{14.9}$$

to

$$\left(x, \frac{dx}{dt}\right) = (0, 0) \tag{14.10}$$

in minimum time.

This problem is best thought of has having two state variables and one control variable. The state equations may be written

$$\frac{dx}{dt} = y, \quad \frac{dy}{dt} = u, \tag{14.11}$$

the initial conditions are

$$x(0) = a, \quad y(0) = b, \tag{14.12}$$

the terminal conditions are

$$x(T) = 0, \quad y(T) = 0, \tag{14.13}$$

and the goal is to find

$$\min_{|u(t)| \leq u_{max}} T \tag{14.14}$$

or, equivalently,

$$\max_{|u(t)| \leq u_{max}} -T. \tag{14.15}$$

$$\diamond$$

Bolza problem

The Bolza problem combines all the features of both the Lagrange problem and the Mayer problem. Part of the payoff is in an integral; part of it is in an initial or terminal function. The goal is thus to find

$$\max_{u \in U} J[x, u], \tag{14.16a}$$

where

$$J[x, u] = \int_{t_0}^{t_1} F[x(t), u(t), t] \, dt + G[t_0, x(t_0), t_1, x(t_1)], \tag{14.16b}$$

subject to

$$\frac{dx}{dt} = f(x, u, t).$$ (14.16c)

We may also have initial and/or terminal conditions:

$$x(t_0) = x_0, \quad x(t_1) = x_1.$$ (14.16d)

Different books state results using one or another of these formulations. The three formulations are equivalent: a problem of one type can always be reformulated as a problem of another type. I will state the main result of optimal control theory, the Pontryagin Maximum Principle, for the Bolza problem, since this version of the principle can also be applied, in a straightforward manner, to Lagrange and Mayer problems.

It is just as easy to state the Pontryagin Maximum Principle for the multidimensional Bolza problem as for the problem in one state and one control variable. The multidimensional problem is as above except that we may have a vector of n state variables and a vector of m control variables. In particular, we wish to find

$$\max_{u \in U} J[x, u],$$ (14.17a)

where

$$J[x, u] = \int_{t_0}^{t_1} F[x(t), u(t), t] \, dt + G[t_0, x(t_0), t_1, x(t_1)],$$ (14.17b)

subject to

$$\frac{dx_i}{dt} = f_i(x, u, t)$$ (14.17c)

for each of the $i = 1, \ldots, n$ state variables. In addition, any or all of the state variables may have initial or terminal conditions,

$$x_i(t_0) = x_{i0}, \quad x_i(t_1) = x_{i1}.$$ (14.17d)

We will also assume that each of the control variables is piecewise continuous and that F, G, and the f_i are suitably well behaved.

Finally, let me introduce the following:

(1) adjoint variables:

$$\lambda_0 \equiv \text{a constant},$$ (14.18a)

$$\lambda_i \equiv \lambda_i(t) \quad i = 1, \ldots, n,$$ (14.18b)

(2)   a Hamiltonian:

$$H(\mathbf{x}, \mathbf{u}, \boldsymbol{\lambda}, t) \equiv \lambda_0 F + \sum_{i=1}^{n} \lambda_i(t) f_i(\mathbf{x}, \mathbf{u}, t), \qquad (14.18c)$$

(3)   a maximized Hamiltonian:

$$M[\mathbf{x}(t), \boldsymbol{\lambda}(t), t] \equiv \sup_{\mathbf{u} \in U} H[\mathbf{x}(t), \mathbf{u}(t), \boldsymbol{\lambda}(t), t]. \qquad (14.18d)$$

We are now ready to state Pontryagin's Maximum Principle.

Pontryagin's Maximum Principle

If $\mathbf{u}(t)$ is an optimal control and if $\mathbf{x}(t)$ is the corresponding response, then:

(1)   There exist

$$\boldsymbol{\lambda}(t) = [\lambda_0, \lambda_1(t), \ldots, \lambda_n(t)], \qquad (14.19a)$$

with

$$\boldsymbol{\lambda}(t) \neq 0, \quad t_0 \leq t \leq t_1 \qquad (14.19b)$$

such that the *canonical equations*,

$$\frac{dx_i}{dt} = \frac{\partial H}{\partial \lambda_i}, \qquad (14.19c)$$

$$\frac{d\lambda_i}{dt} = -\frac{\partial H}{\partial x_i}, \qquad (14.19d)$$

are satisfied for each $i = 1, \ldots, n$.

(2) $\mathbf{u}(t)$ satisfies

$$H[\mathbf{x}(t), \mathbf{u}(t), \boldsymbol{\lambda}(t), t] = M[\mathbf{x}(t), \boldsymbol{\lambda}(t), t]. \qquad (14.19e)$$

(3)   The transversality condition

$$\lambda_0 \, dG + \left[M(t_1) \, dt_1 - \sum_{i=1}^{n} \lambda_i(t_1) \, dx_{i1} \right] - \left[M(t_0) \, dt_0 - \sum_{i=1}^{n} \lambda_i(t_0) \, dx_{i0} \right] = 0$$
$$(14.19f)$$

is satisfied.

Equation (14.19b) states that all of the adjoint variables cannot vanish simultaneously. λ_0 is always nonzero for $n = 1$; without loss of generality we may take $\lambda_0 = 1$ for $n = 1$. (For $n > 1$, the case $\lambda_0 = 0$ is somewhat pathological; it may arise when variables are overconstrained.) Equations (14.19c) and (14.19d) are a system of $2n$ ordinary differential equations for the state variables and the adjoint variables. Equation (14.19e) provides m additional conditions for the control variables. The transversality condition supplies missing initial or terminal conditions.

Yield

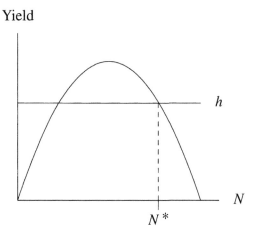

Fig. 14.1. Constant-rate harvesting.

PRELUDE TO EXAMPLE

Consider a stock of fish that is being harvested at a constant rate,

$$\frac{dN}{dt} = f(N) - h. \tag{14.20}$$

For sufficiently small harvest levels h, we may keep this model population at a constant equilibrium N^* and maintain a constant, sustainable yield,

$$\frac{dN}{dt} = 0 \;\Rightarrow\; h = f(N^*) \tag{14.21}$$

(see Figure 14.1). If we wish to maximize the sustainable yield, we require that

$$f'(N_{\text{MSY}}^*) = 0. \tag{14.22}$$

For example, for a logistically growing stock,

$$f(N) = r N \left(1 - \frac{N}{K}\right), \tag{14.23}$$

$$f'(N) = r \left(1 - \frac{2N}{K}\right) = 0 \;\Rightarrow\; N_{\text{MSY}}^* = \frac{K}{2}. \tag{14.24}$$

The maximum sustainable yield is thus given by

$$h = f(N_{\text{MSY}}^*) = r \frac{K}{2} \left(1 - \frac{K/2}{K}\right) = \frac{rK}{4}. \tag{14.25}$$

◇

EXAMPLE   Sole-owner fishery (no costs or discount rate)
Consider the problem of finding

$$\max_{0 \le E \le E_{max}} \int_0^T p \, q \, E(t) \, N(t) \, dt, \tag{14.26a}$$

subject to

$$\frac{dN}{dt} = f(N) - q \, E \, N \tag{14.26b}$$

and

$$N(0) = K. \tag{14.26c}$$

The stock level, $N(t)$, is the state variable while the fishing effort, $E(t)$, is the control variable. The initial condition tells us that the fishery is initially in pristine condition.

The first step in solving this problem is to form the Hamiltonian:

$$H = p \, q \, E \, N + \lambda [f(N) - q \, E \, N] \tag{14.27}$$

or

$$H = q \, (p - \lambda) \, E \, N + \lambda \, f(N). \tag{14.28}$$

The corresponding canonical equations are simply

$$\frac{dN}{dt} = \frac{\partial H}{\partial \lambda} = f(N) - q \, E \, N, \tag{14.29a}$$

$$\frac{d\lambda}{dt} = -\frac{\partial H}{\partial N} = -q \, (p - \lambda) \, E - \lambda \, f'(N). \tag{14.29b}$$

The Hamiltonian is linear in the control variable E. To maximize the Hamiltonian, we must thus choose

$$E(t) = \begin{cases} E_{max}, & \lambda(t) < p, \\ 0, & \lambda(t) > p. \end{cases} \tag{14.30}$$

However, it is also possible that $\lambda(t) = p$ along an entire interval. If this happens, it follows, by equation (14.29b), that

$$f'(N) = 0. \tag{14.31}$$

We must then keep the stock at a level that maximizes the sustainable yield,

$$N(t) = N_{MSY}^*. \tag{14.32}$$

The control variable, in turn, should now be chosen so as to keep the stock at that level,

$$E^*(t) = \frac{f(N_{MSY}^*)}{q \, N_{MSY}^*}. \tag{14.33}$$

Having applied part two of Pontryagin's Maximum Principle, we conclude that we must harvest at the maximum possible rate, leave the stock alone, or apply singular control (14.33), depending on the magnitude of $\lambda(t)$.

If we knew the adjoint variable, $\lambda(t)$, for all time, we would be finished. We do have a differential equation for the adjoint variable, but we do not have an initial condition for this variable. Transversality condition (14.19f) is supposed to provide us with missing initial or terminal conditions. For this problem, we do not have an initial or terminal payoff ($dG = 0$), the initial and terminal times are fixed ($dt_0 = 0$, $dt_1 = 0$), and the state variable is a fixed constant at $t_0 = 0$ ($dN_0 = 0$). The state variable is unconstrained at $t_1 = T$ so that dN_1 is arbitrary. Thus the transversality condition reduces to

$$\lambda(T) = 0. \tag{14.34}$$

The transversality condition gives us a terminal condition on the adjoint variable rather than an initial condition. The fact that we have an initial condition on the state variable and a terminal condition on the adjoint variable means that equations (14.29a) and (14.29b) constitute a boundary value problem rather than an initial value problem. This is typical of optimal control problems. It is what makes these problems challenging.

We have no way of computing the exact value of $\lambda(0)$ short of (numerically) solving the two-point boundary value problem. However, we may easily determine a range of possible values for $\lambda(0)$. First, could it be that $\lambda(0) > p$? If that were the case, the control variable would, by equation (14.30), initially satisfy

$$E(t) = 0. \tag{14.35}$$

Adjoint equation (14.29b) would, in turn, reduce to

$$\frac{d\lambda}{dt} = -\lambda f'(N) \tag{14.36}$$

and since $f'(K) < 0$, the adjoining variable would increase. Indeed, there would never be any harvesting and λ would continue to increase for all time, preventing us from satisfying terminal condition (14.34). Thus we must have that $\lambda(0) < p$.

Now, let us think through what happens if $0 < \lambda(0) < p$. Following equation (14.30), we begin by applying

$$E(t) = E_{\max}. \tag{14.37}$$

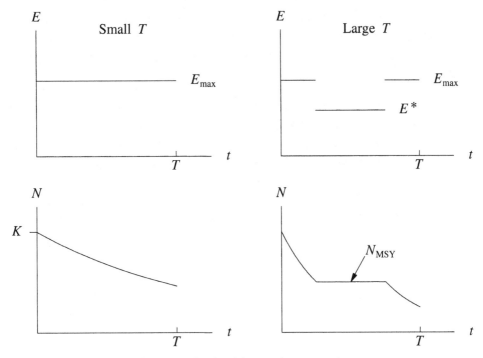

Fig. 14.2. Optimal harvesting scenarios.

What happens to the adjoint variable? Well, it depends. The first term on the right-hand side of canonical equation (14.29b) will be negative, but the second term will be positive. If $\lambda(0)$ is close to zero, the first term dominates. If, however, $\lambda(0)$ is close to p, the second term dominates. So, how should we choose $\lambda(0)$? Well, it all depends on how much time T you have. If you are going to own the fishery for a short time (T small), choose $\lambda(0)$ close to zero so that the adjoint variable decays to zero, in accordance with transversality condition (14.34), in the allotted time. Your strategy is to fish as hard as you can for as long as you can (and to leave it to the next owner to pick up the pieces).

If you will own the fishery longer, choose $\lambda(0)$ closer to p. The adjoint variable λ will then increase to p, at which point you can apply the singular control (14.33), keep the adjoint variable at p, and enjoy the benefits of maximum sustainable yield. However, sometime before turning over ownership of the fishery, you too will turn greedy, increase the harvest rate (to E_{max}) and bring the adjoint variable back down to $\lambda(T) = 0$. The two scenarios are illustrated in Figure 14.2.

I will leave it to you to work through the implications of $\lambda(0) < 0$. ◇

Problem 14.1 *A linear problem*
Suppose a fish population grows exponentially. The owner of the fishery wishes to maximize the catch,

$$\max_{0 \le E(t) \le h} \int_0^T E(t) N(t) \, dt, \tag{14.38a}$$

over a fixed interval T, subject to

$$\frac{dN}{dt} = N - E N \tag{14.38b}$$

and

$$N(0) = 1. \tag{14.38c}$$

$N(t)$ and $E(t)$ have their usual meanings. T is assumed large. Determine the optimal harvest strategy. (Everything is linear here; you should be able to solve for the state, control, and adjoint variables in closed form.)

EXAMPLE Discounting
Now consider finding

$$\max_{0 \le E \le E_{max}} \int_0^T e^{-\delta t} p q E(t) N(t) \, dt, \tag{14.39a}$$

subject to

$$\frac{dN}{dt} = f(N) - q E N \tag{14.39b}$$

and

$$N(0) = K. \tag{14.39c}$$

Note that we have added a discount rate that discounts the value of future revenues relative to current revenues. You may think of this as the 'Whimpy factor' (as in 'I will gladly pay you Tuesday for a hamburger today'): current revenues are more valuable than future revenues, if only because current revenues can be invested with some rate of return.
   The Hamiltonian now takes the form

$$H = e^{-\delta t} p q E N + \lambda [f(N) - q E N] \tag{14.40}$$

or

$$H = q (p e^{-\delta t} - \lambda) N E + \lambda f(N). \tag{14.41}$$

Note that this expression is linear in the control variable E. The part of the coefficient of E that may change sign,

$$\sigma \equiv p e^{-\delta t} - \lambda, \qquad (14.42)$$

will play a particularly important role; I will refer to it as the *switching function*. The canonical equations may, in turn, be written as

$$\frac{dN}{dt} = \frac{\partial H}{\partial \lambda} = f(N) - q E N, \qquad (14.43a)$$

$$\frac{d\lambda}{dt} = -\frac{\partial H}{\partial N} = -q(p e^{-\delta t} - \lambda) E - \lambda f'(N). \qquad (14.43b)$$

In order to maximize the Hamiltonian with respect to the control variable E, we must look at the slope as indicated by the switching function and choose

$$E(t) = \begin{cases} E_{\max}, & \lambda < p e^{-\delta t}, \\ 0, & \lambda > p e^{-\delta t}. \end{cases} \qquad (14.44)$$

However, it is also possible that the switching function is zero on an entire interval so that

$$\lambda = p e^{-\delta t} \qquad (14.45)$$

and

$$\frac{d\lambda}{dt} = -\delta p e^{-\delta t} = -\delta \lambda. \qquad (14.46)$$

Moreover, the canonical equation (14.43b) tells that

$$\frac{d\lambda}{dt} = -\lambda f'(N) \qquad (14.47)$$

along the singular path. If we compare equations (14.46) and (14.47), we conclude that

$$f'(N) = \delta \qquad (14.48)$$

along the singular path.

At what level should the stock be kept at for the owner of the fishery to comply with equation (14.48)? Let us look at this graphically (see Figure 14.3). Maximum sustainable yield is attained when $f'(N) = 0$. For simple compensatory growth (i.e., for density-dependent models such as the logistic equation) the slope of f is a monotonically decreasing function; this slope is positive only if $N < N_{\mathrm{MSY}}$. The higher the value of δ, the lower N^*.

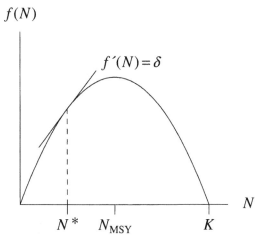

Fig. 14.3. Biological overfishing.

Therefore:

The Fundamental Principle of Renewable Resources

LARGER DISCOUNT   **LESS BIOLOGICAL**

RATES $\Rightarrow$ **CONSERVATION**

Indeed, if the discount rate is larger than the intrinsic rate of growth, $\delta > r$, we predict that the owner will, all other things being equal (i.e., ignoring costs, etc.) drive the stock to extinction.

Why should this be? In effect, if the stock grows too slowly, the owner is (economically) better off if he or she 'cashes in' the resource and invests the money in something that has a higher rate of return. Baleen whale populations, for example, have small growth rates of 2% to 5% per year and have suffered extreme overexploitation. Mind you, I am not saying that this is in any sense good (quite the contrary) or 'optimal'. What I am saying is that anyone who believes that there have been strong economic incentives for maintaining whale stocks is being naive. $\diamond$

Problem 14.2 *Timber harvesting*

Consider a stand of trees that grows according to

$$\frac{dV}{dt} = \frac{r}{1 + at} V \left(1 - \frac{V}{K} \right).$$

(14.49)

(1) Find $V(t)$ and show that $\lim_{t \to \infty} V(t) = K$.

(2) Determine the effort $E(t)$ that will yield

$$\max_{0 \le E(t) \le E_{\max}} \int_0^T e^{-\delta t} p E(t) V(t)\, dt + p e^{-\delta T} V(T), \qquad (14.50\text{a})$$

subject to

$$\frac{dV}{dt} = \frac{r}{1 + at} V \left(1 - \frac{V}{K}\right) - E V, \qquad (14.50\text{b})$$

$$V(0) = V_0, \qquad (14.50\text{c})$$

where

$$r = 0.2\,\text{yr}^{-1}, \quad \delta = 0.1\,\text{yr}^{-1}, \quad a = 0.01\,\text{yr}^{-1}, \qquad (14.50\text{d})$$

$$T = 70\,\text{yrs}, \quad K = 500\,\text{TCF}, \quad V_0 = 10\,\text{TCF}, \qquad (14.50\text{e})$$

and TCF stands for thousand cubic feet (1 TCF = 28.3 m^3).

EXAMPLE   Nonlinear revenue

So far, we have considered only examples that are linear in the control variable. This has lead to 'bang-bang' controls. Let us assume that the harvest rate depends on the fishing effort through, for example,

$$h = q E N, \qquad (14.51)$$

but that revenues are now a nonlinear, increasing, but concave-down function of harvest,

$$R'(h) > 0, \qquad (14.52\text{a})$$

$$R''(h) < 0 \qquad (14.52\text{b})$$

(see Figure 14.4). We are assuming that revenues increase with catch, but at decreasing rates. One can flood the market and reduce the price per fish.

The sole owner who wishes to maximize discounted revenues must now find

$$\max_{0 \le h \le h_{\max}} \int_0^T e^{-\delta t} R(h)\, dt, \qquad (14.53\text{a})$$

subject to

$$\frac{dN}{dt} = f(N) - h \qquad (14.53\text{b})$$

and

$$N(0) = K. \qquad (14.53\text{c})$$

Having written the problem this way, we will view the harvest rate h as our

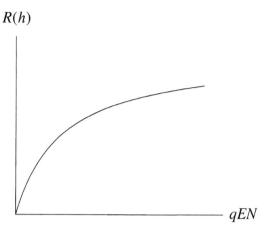

Fig. 14.4. Diminishing revenue returns.

control variable and avoid any more discussion of the fishing effort or of the dependence of the harvest rate on stock levels.

The Hamiltonian for this problem is

$$H = e^{-\delta t} R(h) + \lambda [f(N) - h]. \qquad (14.54)$$

Since this Hamiltonian is a nonlinear function of harvest rate, we will look for a local (interior) maximum:

$$\frac{\partial H}{\partial h} = e^{-\delta t} R'(h) - \lambda = 0 \qquad (14.55)$$

or

$$\lambda = e^{-\delta t} R'(h). \qquad (14.56)$$

You might note, in passing, that

$$\frac{\partial^2 H}{\partial h^2} = e^{-\delta t} R''(h) < 0, \qquad (14.57)$$

so that we really are talking about a local maximum (and not a minimum).

It is somewhat painful to think about, but equation (14.56) for λ also implies that

$$\frac{d\lambda}{dt} = -\delta e^{-\delta t} R'(h) + e^{-\delta t} R''(h) \frac{dh}{dt}. \qquad (14.58)$$

From the adjoint equation, we also have that

$$\frac{d\lambda}{dt} = -\frac{\partial H}{\partial N} = -\lambda f'(N), \qquad (14.59)$$

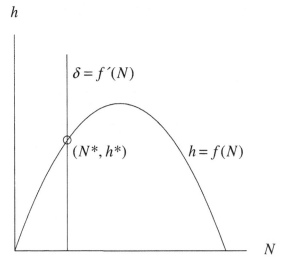

Fig. 14.5. (N, h) phase plane.

or, substituting for λ,

$$\frac{d\lambda}{dt} = -e^{-\delta t} R'(h) f'(N). \tag{14.60}$$

If we compare equations (14.58) and (14.60), it follows that

$$\frac{dh}{dt} = [\delta - f'(N)] \frac{R'(h)}{R''(h)}. \tag{14.61}$$

The adjoint variable is simply a means to an end. We are ultimately interested in the stock level and the harvest rate and, most conveniently, equations (14.53b) and (14.61) give us an autonomous system of two first-order differential equations:

$$\frac{dN}{dt} = f(N) - h, \tag{14.62a}$$

$$\frac{dh}{dt} = [\delta - f'(N)] \frac{R'(h)}{R''(h)}. \tag{14.62b}$$

This system can be analyzed using traditional phase-plane methods. There are two zero-growth isoclines (see Figure 14.5) and an equilibrium at the intersection of these zero-growth isoclines.

The nature of the equilibrium can be discerned by looking at the Jacobian

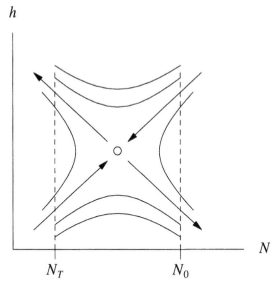

Fig. 14.6. Fishery saddle point.

of equations (14.62a) and (14.62b) evaluated at the equilibrium:

$$J = \left[\begin{array}{cc} f'(N) & -1 \\ -f''(N)[R'(h)/R''(h)] & [\delta - f'(N)]\,[R'(h)/R''(h)]' \end{array} \right]_{(N^*, h^*)} \tag{14.63}$$

$$J = \left[\begin{array}{cc} \delta & -1 \\ -f''(N^*)\,[R'(h^*)/R''(h^*)] & 0 \end{array} \right]. \tag{14.64}$$

The corresponding characteristic equation is

$$\lambda^2 - \delta\lambda - f''(N^*)\,[R'(h^*)/R''(h^*)] = 0 \tag{14.65}$$

and if we note the sign changes in the coefficients of this equation $(+, -, -)$ and apply Descartes's rule of signs, we discover that the equilibrium point is a saddle point.

Typically, the flow near the equilibrium will look like Figure 14.6. We may determine the directions of the eigenvectors from Jacobian (14.64). If we are given the initial and terminal conditions for the state variable, we must find an orbit that moves between these values in the specified time T. As T increases, we find ourselves choosing orbits that come closer and closer (and that linger longer and longer) near the singular point (N^*, h^*). However, rather than having bang-bang controls and a most rapid approach to this singular point, we will have graded controls. ◇

Until now, we have ignored costs. Let us correct this.

EXAMPLE   Including costs

Consider a sole owner whose goal is to find

$$\max_{0 \le E \le E_{\max}} \int_0^T e^{-\delta t} [p q N(t) - c] E(t)\, dt, \tag{14.66a}$$

subject to

$$\frac{dN}{dt} = f(N) - q E N \tag{14.66b}$$

and

$$N(0) = K. \tag{14.66c}$$

The Hamiltonian for this problem is given by

$$H = e^{-\delta t}(pqN - c) E + \lambda[f(N) - q E N] \tag{14.67}$$

or

$$H = [e^{-\delta t}(pqN - c) - \lambda qN] E + \lambda f(N). \tag{14.68}$$

The coefficient of E,

$$\sigma = [e^{-\delta t}(pqN - c) - \lambda qN], \tag{14.69}$$

is the switching function and once again plays a pivotal role. We are interested in intervals where the switching function is identically zero. On such intervals, we have singular controls that dominate the problem. There will also be a most rapid approach to the singular state at the beginning of the problem and some profit-taking at the end of the problem. Setting the switching function equal to zero gives us

$$\lambda = e^{-\delta t}\left(p - \frac{c}{qN}\right). \tag{14.70}$$

We can, as usual, compute the time derivative of λ from this expression (in conjunction with equation (14.66b)). In addition, we can compute a second expression for this time derivative from the second canonical equation,

$$\frac{d\lambda}{dt} = -\frac{\partial H}{\partial N} = -e^{-\delta t}pqE + \lambda qE - \lambda f'(N). \tag{14.71}$$

By equating these two expressions, we get (after much algebra)

$$f'(N) = \delta - \frac{c f(N)}{N(pq N - c)}. \tag{14.72}$$

This is an implicit equation for the unknown N. Any N^* that satisfies this equation is a singular solution of our original control problem. And, even though this equation appears quite messy, it reduces to a quadratic equation for logistic growth.

It is convenient to rewrite equation (14.72) as

$$f'(N) = \delta - \frac{\frac{c}{qN^2} f(N)}{p - \frac{c}{qN}}. \tag{14.73}$$

If we multiply through by the denominator on the right-hand side and rearrange terms, we obtain

$$\left[\left(p - \frac{c}{qN}\right) f'(N) + \frac{c}{qN^2} f(N)\right] = \delta \left(p - \frac{c}{qN}\right) \tag{14.74}$$

or

$$\frac{dS}{dN} = \delta \left(p - \frac{c}{qN}\right), \tag{14.75}$$

where

$$S \equiv \left(p - \frac{c}{qN}\right) f(N). \tag{14.76}$$

I will call S the sustainable rent.

Let us see if we can make some sense of equation (14.75). It is easiest to begin by looking at this equation in the case $\delta = 0$,

$$\frac{dS}{dN} = 0. \tag{14.77}$$

This corresponds to choosing a stock level that maximizes the sustainable rent. What stock level are we talking about?

For logistic growth, we have that

$$\frac{dN}{dt} = rN\left(1 - \frac{N}{K}\right) - qEN. \tag{14.78}$$

If the fishery is kept at an equilibrium level ($dN/dt = 0$), we require that

$$qE^* = r\left(1 - \frac{N^*}{K}\right). \tag{14.79}$$

Thus, the equilibrium stock level is a decreasing function of the corresponding effort level,

$$N^* = K\left(1 - \frac{qE^*}{r}\right) \tag{14.80}$$

and the equilibrium yield is a quadratic or parabolic function of effort,

$$qE^* N^* = qKE^*\left(1 - \frac{qE^*}{r}\right). \tag{14.81}$$

$/time

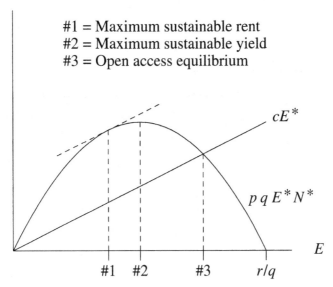

#1 = Maximum sustainable rent
#2 = Maximum sustainable yield
#3 = Open access equilibrium

cE^*

pqE^*N^*

E

#1 #2   #3 r/q

Fig. 14.7. Maximum sustainable rent.

Okay, but where does the sustainable rent enter into all this? Well, the sustainable rent can be written

$$S = pqE^*N^* - cE^*. \tag{14.82}$$

The first term on the right-hand side is proportional to the yield and is a quadratic function of effort (see Figure 14.7). The second term is the cost of fishing (per unit time) and is a linear function of effort. The sustainable rent is simply the difference between this revenue and cost for a given level of effort. To maximize the rent, we need to maximize the difference between the revenue and cost functions. We can do this by choosing the effort level so that the slope on revenue curve is identical to the slope along the cost function (see Figure 14.7). Typically, this effort will be less than that needed for maximum sustainable yield. Costs leads to reduced fishing and greater amounts of biological conservation.

Let us now look at the special case $\delta = \infty$. Although the discount rate is infinite, there is no reason to suspect that dS/dN is anything but bounded. By equation (14.75), we require that

$$pqN^* = c. \tag{14.83}$$

The sustainable rent is thus zero. If there are costs, increases in the discount rate need not lead to the extinction of the stock. Rather, the sole-owner fishery begins to look more and more like an open-access fishery: all economic rent is squandered. ◇

Problem 14.3 *Whales versus tuna*

Contrast the bioeconomics of whaling and tuna fishing, comparing such quantities as the maximum sustainable yield, the bionomic (open-access) equilibrium, maximum sustainable rent, optimal (singular) population and effort levels, etc. Be explicit about your assumptions and any formulae that you use.

	Antarctic baleen whales	Pacific yellowfin tuna
r	0.05 per year	2.6 per year
K	400 000 BWU	250 000 metric tons
q	0.0016 per WCY	0.0000385 per SFD
p	$7 000 per BWU	$600 per metric ton
c	$600 000 per WCY	$2 500 per SFD
δ	0 to 10%	0 to 10%

r, intrinsic rate of growth; K, carrying capacity; q, catchability; p, price; c, cost; δ, discount rate; BWU, blue whale units; WCY, whale-catcher year; SFD, standard fishing day.

Mathematical meanderings

Proofs of the Maximum Principle

In a sense, all of the varied proofs of the Maximum Principle are beyond the scope of this course. Published proofs typically come in one of three different flavors:

(1) Rigorous proofs. The proof in Pontryagin *et al.* (1962) runs to about 33 pages! Beavis and Dobbs (1990) have an outline of the proof that is only about 11 pages long.

(2) Heuristic proofs that build upon the calculus of variations. See, for example, Tu (1991). Unfortunately, the calculus of variations cannot handle linear problems.

(3) Heuristic proofs that use dynamic programming. See, for example, Clark (1990). These proofs typically assume a form of strong differentiability far too stringent for real problems.

Economic meaning of λ and H

I will, once again, write $x(t)$ for the state variable and $u(t)$ for the control variable. Let $w(t, x)$ represent the present value of the stock, assuming optimal exploitation that starts at $x(t) = x$ and that ends at $x(t_1) = x_1$,

$$w(t, x) = \max_{u \in U} J[x(t), u(t)]. \tag{14.84}$$

(The endpoints are fixed but the initial time and state are variable.) The function w is called the *current value return function*, the *payoff function* or the *optimal-return function*. It can be shown that

$$\lambda = \frac{\partial w}{\partial x}. \tag{14.85}$$

The adjoint variable may thus be thought of as the marginal value of the stock or as the *shadow price of capital*. In effect, λ answers the question 'If I add one more fish to the stock, how much will I affect the value of the fishery?'

EXAMPLE   To find

$$\max_{0 \leq E \leq E_{\max}} \int_0^T p \, q \, E(t) \, N(t) \, dt, \tag{14.86a}$$

subject to

$$\frac{dN}{dt} = f(N) - q \, E \, N, \tag{14.86b}$$

$$N(0) = K, \tag{14.86c}$$

we concluded that the control variable E must satisfy

$$E(t) = \begin{cases} E_{\max}, & \lambda(t) < p, \\ 0, & \lambda(t) > p. \end{cases} \tag{14.87}$$

Namely, if the contribution of a fish to the present value is greater than the price you will receive for that fish, don't fish.

The Hamiltonian for this problem,

$$H = p \, q \, E \, N + \lambda \, [f(N) - q \, E \, N], \tag{14.88}$$

also has a nice economic meaning. The first term can be interpreted as a *flow of dividends* whereas the second term may be interpreted as a *rate of change of capital*. Maximizing this Hamiltonian amounts to maximizing the rate of change of assets. ◇

Transversality conditions

For a problem with a single state variable and a single control variable, the transversality condition may be written as

$$dG + [M(t_1) \, dt_1 - \lambda(t_1) \, dx_1] - [M(t_0) \, dt_0 - \lambda(t_0) \, dx_0] = 0. \tag{14.89}$$

A number of interesting special cases fall out of this condition. If the initial and terminal times and the initial value of the state variable are specified, but the terminal value of the state variable is not, we have a problem with a *free terminal value*. It then follows that we must satisfy

$$\lambda(t_1) = \frac{\partial G}{\partial x_1}. \tag{14.90}$$

In particular, if there is no salvage function, the strategy is to drive the shadow price of capital to zero at the terminal time. The economic argument that is usually made is that capital is valuable to the sole owner only for its ability to generate profits. With a rigid terminal time t_1, it is only the profits made within the period $[t_0, t_1]$ that matter. Whatever marginal value there is at t_1 occurs too late to be put to any use – unless there is a terminal payoff.

Another interesting case arises if the initial time and the initial and terminal values of the state variable are specified, but the *terminal time* is *unspecified*. In this instance the transversality condition dictates that

$$\frac{\partial G}{\partial t_1} + M(t_1) = 0. \tag{14.91}$$

If there is no salvage function, this reduces to

$$M(t_1) = 0. \tag{14.92}$$

In other words, stop when you are no longer adding to your assets.

Several state variables

All the ideas that I have introduced apply for several state or control variables. The only difference is that problems with several state and control variables are typically harder.

EXAMPLE   Irreversible investment (Clark *et al.*, 1979)
Consider an example where the fishing effort,

$$E(t) = a(t) S(t), \tag{14.93}$$

is limited by the number of ships, with $S(t)$, the number of ships, a state variable, and $a(t)$, the percentage of ships that are active, a control variable. An interesting two-state-variable, two-control-variable problem is to find

$$\max_{\substack{0 \le a \le 1 \\ 0 \le b \le \infty}} \int_0^T e^{-\delta t} \left[p\, q\, a(t)\, S(t)\, N(t) - c_v\, a(t)\, S(t) - c_f\, b(t) \right] dt, \tag{14.94a}$$

subject to

$$\frac{dN}{dt} = f(N) - q\, a\, S\, N, \tag{14.94b}$$

$$\frac{dS}{dt} = b - \gamma\, S, \tag{14.94c}$$

and

$$N(0) = N_0, \tag{14.94d}$$

$$S(0) = S_0. \tag{14.94e}$$

The control variable $b(t)$ describes the rate at which boats are being built. The parameters c_v and c_f are the variable cost of running the ships and the fixed cost of building the ships. The parameter γ controls the rate at which ships deteriorate. This problem was analyzed by Clark *et al.* (1979). The authors derived the singular solution in one afternoon. However, it took them 11 months to determine the optimal approach path. $\diamondsuit$

Discrete-time harvesting problems

Throughout this book, I have tried to stress the importance of both differential and difference equations. Organisms with discrete, nonoverlapping generations are often modeled using difference equations. How do we handle the sole-owner fishery if the underlying stock is described by difference equations?

Let's consider a discrete-time sole-owner fishery without costs in which the owner is trying to determine

$$\max_{\substack{0 \le u_t \le u_{max} \\ 0 \le u_{max} \le 1}} \sum_{t=0}^{T-1} p\, u_t\, f(x_t), \tag{14.95a}$$

subject to

$$x_{t+1} = (1 - u_t)\, f(x_t), \tag{14.95b}$$

$$x_0 = K. \tag{14.95c}$$

The state variable u_t is the fraction of each year's recruitment that is harvested. To make life easier, I will assume that f is a monotonically increasing function (i.e., no chaotic dynamics).

As written, this is a problem in nonlinear programming and can be handled using Lagrange multipliers. Let

$$L \equiv \sum_{t=0}^{T-1} \{p\, u_t\, f(x_t) - \lambda_{t+1}\, [x_{t+1} - (1 - u_t)\, f(x_t)]\}. \tag{14.96}$$

I will rearrange this a bit, so that the terms involving the control variable u_t are together:

$$L \equiv \sum_{t=0}^{T-1} \{(p - \lambda_{t+1})\, u_t f(x_t) + \lambda_{t+1}\, [f(x_t) - x_{t+1}]\}. \tag{14.97}$$

We now need to maximize this Lagrangian L with respect to both u_t and x_t.

Our Lagrangian is linear in u_t. There are three possibilities. It could be the Lagrangian is independent of u_t,

$$(p - \lambda_{t+1})\, f(x_t) = 0, \tag{14.98}$$

with

$$\lambda_{t+1} = p. \tag{14.99}$$

Or, the maximum could be at the left endpoint, $u_t = 0$, with

$$\frac{\partial L}{\partial u_t} < 0, \tag{14.100}$$

so that

$$\lambda_{t+1} > p. \tag{14.101}$$

Finally, the maximum could be at the right endpoint, with

$$\frac{\partial L}{\partial u_t} > 0. \tag{14.102}$$

In this case,

$$\lambda_{t+1} < p. \tag{14.103}$$

To maximize with respect to x_t, we set

$$\frac{\partial L}{\partial x_k} = 0, \quad k = 1,\dots, T - 1, \tag{14.104}$$

so that

$$f'(x_k)(p - \lambda_{k+1}) u_k + \lambda_{k+1} f'(x_k) - \lambda_k = 0. \tag{14.105}$$

It then follows that

$$\lambda_k = f'(x_k) [p u_k + \lambda_{k+1}(1 - u_k)], \tag{14.106}$$

which can be thought of as a backwards difference equation. For the singular control

$$\lambda_{k+1} = \lambda_k = p, \tag{14.107}$$

this difference equation yields

$$f'(x^*) = 1. \tag{14.108}$$

Finally,

$$\frac{\partial L}{\partial x_T} = 0 \tag{14.109}$$

implies

$$\lambda_T = 0. \tag{14.110}$$

These conditions were derived without much rigor. Even so, they appear to be similar to the conditions that we obtained using optimal control theory.

Infinite horizon problems

The problems that we've looked at contain finite time horizons. There are a number of technical difficulties in moving to infinite horizons. The integrals at the heart of the objective functionals may no longer converge. In addition, there are subtleties involving the transversality conditions. Chiang (1992) provides a thorough discussion of these topics.

Recommended readings

Mathematical bioeconomics originated with the publication, in 1976, of Colin Clark's book of the same title. This chapter borrows heavily from the second edition of Clark's book (Clark, 1990) and from Conrad and Clark (1987). There are a number of good introductory and advanced treatments of optimal control theory including those by Chiang (1992), Intrilligator (1971), Boltyanskii (1971), and Cesari (1983).

Part II Structured population models

Section D
SPATIALLY STRUCTURED MODELS

15 Formulating spatially structured models

Why should we study spatially structured models? There are several reasons:

(1) The environment may be heterogeneous. In Part I we studied the logistic differential equation,

$$\frac{dN}{dt} = r N \left(1 - \frac{N}{K} \right). \tag{15.1}$$

We took the intrinsic rate of growth, r, and the carrying capacity, K, to be constants. These parameters may, in fact, vary with geographical location. There may be good locations or bad locations. Resources may vary smoothly, along clines or gradients, or they may be distributed in a patchy manner.

(2) Even if the environment is homogeneous, a population may have a heterogeneous initial distribution. If organisms are vagile, they may spread from regions of high density to unoccupied habitat. In many instances, this leads to traveling waves of invasion.

During the winter of 1997, Cattle Egrets (*Bubulcus ibis*) were spotted in the state of Washington. This species is a natural transoceanic invader from Africa. It first appeared in South America in substantial numbers in the 1930s. It spread rapidly, at over 100 km/year (van den Bosch *et al.*, 1992), and is now found throughout much of North America. Egrets are harmless; other invaders are less endearing. At the start of World War II, an epizootic of rabies broke out along the Russian–Polish border. This epizootic has spread westward, across Europe, at 30–60 km/year (Murray, 1989). It reached Belgium and France in 1966, crossed the Alps into northern Italy in 1982, and is of great concern to the British.

(3) Different species may have different rates of spread. The predator–prey models in Part I treated the predator and the prey as 'well mixed.' If a predator and its prey disperse at different rates, patterns may arise, even in the absence of environmental heterogeneity. Many scientists are interested in understanding the mechanisms that lead to spatial pattern (Ball, 1999).

Table 15.1. *Spatially structured models*

	Continuous time	Discrete time
Continuous space	Reaction-diffusion equations	Integrodifference equations
Discrete space	Coupled-patch models Metapopulation models	Coupled lattice maps Cellular automata

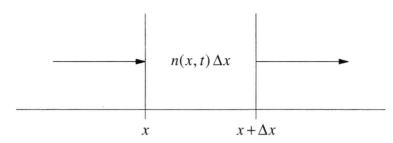

Fig. 15.1. A small interval of space.

You will need to make some important decisions in modeling spatially structured populations. You can consider equations (see Table 15.1) that are either continuous or discrete in time. In addition, you can treat the other independent variable, space, as either continuous or discrete. In the latter instance, the environment is thought of as patchy. You can even choose between models that treat the dependent variable as continuous (as a density) and those that treat it as discrete (corresponding to the presence or absence of a species). In Part II, I will concentrate on those continuous-time, continuous-space models that are known as reaction-diffusion equations.

A classical approach

I will begin with a single-species reaction-diffusion model in one spatial dimension. In particular, let

$$n(x, t)\, dx \equiv \quad \begin{array}{l} \text{number of individuals located in the} \\ \text{interval } (x, x + dx) \text{ at time } t. \end{array} \tag{15.2}$$

If we consider the rate of change of the number of individuals in a given interval of space (see Figure 15.1) we may write

$$\frac{\partial}{\partial t}\, [n(x, t)\, \Delta x] = \left[\begin{array}{l} + \text{ growth rate in } (x, x + \Delta x) \\ + \text{ rate of entry at } x \\ - \text{ rate of departure at } x + \Delta x \end{array} \right] \tag{15.3}$$

or

$$\frac{\partial n}{\partial t} \Delta x = f(x, t) \Delta x + J(x, t) - J(x + \Delta x, t), \qquad (15.4)$$

where $f(x, t)$ is the growth rate of the population per unit length and $J(x, t)$ is the positive (left to right) 'flux' of individuals at position x and time t. Dividing by Δx gives us

$$\frac{\partial n}{\partial t} = f(x, t) - \left[\frac{J(x + \Delta x, t) - J(x, t)}{\Delta x} \right]. \qquad (15.5)$$

Taking the limit as Δx approaches zero produces a conservation law for the density of individuals:

$$\frac{\partial n}{\partial t} = f(x, t) - \frac{\partial J}{\partial x}. \qquad (15.6)$$

This general balance equation can be generalized to higher spatial dimensions. Let us take a quick look at the three-dimensional case. Let Ω represent some, more or less, arbitrary region in space with boundary $\delta\Omega$; organisms move into and out of this region in any direction. Now,

$$\frac{d}{dt} \int_{\Omega} n \, dV = \int_{\Omega} f \, dV - \int_{\delta\Omega} \mathbf{J} \cdot d\mathbf{S}, \qquad (15.7)$$

where $d\mathbf{S}$ is the vector element of surface and $\mathbf{J}$ is a flux vector. In order to simplify this equation, we use Gauss's divergence theorem to convert the third integral into a volume integral,

$$\int_{\delta\Omega} \mathbf{J} \cdot d\mathbf{S} = \int_{\Omega} \nabla \cdot \mathbf{J} \, dV. \qquad (15.8)$$

This causes equation (15.7) to reduce to

$$\int_{\Omega} \left(\frac{\partial n}{\partial t} - f + \nabla \cdot \mathbf{J} \right) dV = 0. \qquad (15.9)$$

Since this integral is zero for arbitrary Ω, it must be that

$$\frac{\partial n}{\partial t} = f - \nabla \cdot \mathbf{J}. \qquad (15.10)$$

We thus have two conservation or general balance equations:
one-dimensional:

$$\frac{\partial n}{\partial t} = f(x, t) - \frac{\partial J}{\partial x}, \qquad (15.11)$$

three-dimensional:

$$\frac{\partial n}{\partial t} = f(x, y, z, t) - \nabla \cdot \mathbf{J}. \qquad (15.12)$$

What are reasonable choices for the growth rate f? Well, ultimately f

depends on the independent variables, space and time. However, it often does so through the dependent variable, the population density. Indeed, all of the simple growth equations that we have studied can be applied in this setting. Here are some examples:

(1)   No births/deaths, only flux:

$$f = 0, \tag{15.13}$$

(2)   Exponential growth:

$$f = r\,n, \tag{15.14}$$

(3)   Logistic growth:

$$f = r\,n \left(1 - \frac{n}{K}\right), \tag{15.15}$$

(4)   Logistic growth (with spatial variation in the parameters):

$$f = r(x)\,n(x,\,t)\left[1 - \frac{n(x,\,t)}{K(x)}\right], \tag{15.16}$$

(5)   Hutchinson–Wright equation:

$$f = r\,n(x,\,t)\left[1 - \frac{n(x,\,t-\tau)}{K}\right], \tag{15.17}$$

(6)   Allee effect:

$$f = r\,n \left(\frac{n}{K_o} - 1\right)\left(1 - \frac{n}{K}\right). \tag{15.18}$$

Modeling the flux J correctly is equally important. What are reasonable choices for the flux? Again, there are a number of possibilities:

(1)   *Advection/convection*: One can imagine organisms that move horizontally (advection) or vertically (convection) with velocity $v = v(x)$ so that

$$J = J_A = v(x)\,n(x,\,t). \tag{15.19}$$

Equation (15.11) then reduces to

$$\frac{\partial n}{\partial t} + \frac{\partial}{\partial x}\,[v(x)\,n] = f(n). \tag{15.20}$$

In three dimensions, we could have an entire vector field of velocities. For the flux, we would then write

$$\boldsymbol{J}_A = \boldsymbol{v}(x,\,y,\,z)\,n. \tag{15.21}$$

The general balance equation would, in turn, reduce to

$$\frac{\partial n}{\partial t} + \nabla \cdot (n\,\boldsymbol{v}) = f(n). \tag{15.22}$$

Equations (15.20) and (15.22) simplify considerably for organisms that move with constant velocity. We would then write

$$\frac{\partial n}{\partial t} + v\frac{\partial n}{\partial x} = f(n) \tag{15.23}$$

in one dimension, or

$$\frac{\partial n}{\partial t} + \mathbf{v} \cdot \nabla n = f(n) \tag{15.24}$$

in three dimensions. Either way, we have a simple scalar reaction-advection equation.

(2) *Random motion/diffusion*: Biological species are often reminiscent of chemical species. Many chemical species move down a density gradient, from regions of high concentration to regions of low concentration. For these species, the flux is proportional to the negative concentration gradient of density. (This is called Fick's law.) In one dimension, we may thus write

$$J_D = -D\frac{\partial n}{\partial x} \tag{15.25}$$

so that general balance equation (15.11) reduces to

$$\frac{\partial n}{\partial t} = f(n) - \frac{\partial}{\partial x}\left(-D\frac{\partial n}{\partial x}\right) \tag{15.26}$$

or

$$\frac{\partial n}{\partial t} = f(n) + D\frac{\partial^2 n}{\partial x^2}. \tag{15.27}$$

The proportionality constant D is known as the *diffusivity*.

In three dimensions, the gradient is

$$\nabla n = \left(\frac{\partial n}{\partial x}, \frac{\partial n}{\partial y}, \frac{\partial n}{\partial z}\right). \tag{15.28}$$

We may thus write the diffusive flux as

$$\mathbf{J}_D = -D\nabla n.$$

General balance equation (15.12) now reduces to

$$\frac{\partial n}{\partial t} = f(n) - \nabla \cdot (-D\nabla n) \tag{15.29}$$

or

$$\frac{\partial n}{\partial t} = f(n) + D\nabla^2 n, \tag{15.30}$$

where $\nabla^2 n$, the *Laplacian*, is defined as

$$\nabla^2 n \equiv \frac{\partial^2 n}{\partial x^2} + \frac{\partial^2 n}{\partial y^2} + \frac{\partial^2 n}{\partial z^2}. \tag{15.31}$$

Equations (15.27) and (15.30) are both scalar reaction-diffusion equations.

EXAMPLE

$$\frac{\partial n}{\partial t} = D \frac{\partial^2 n}{\partial x^2} \tag{15.32}$$

and

$$\frac{\partial n}{\partial t} = D \nabla^2 n \tag{15.33}$$

have no growth (reaction) terms and simple Fickian diffusion and are known as the *one-dimensional heat equation* and the *three-dimensional heat equation*. ◇

EXAMPLE
The simple linear model

$$\frac{\partial n}{\partial t} = r n + D \frac{\partial^2 n}{\partial x^2}$$

is frequently referred to, by ecologists, as the KISS (Kierstead and Slobodkin and Skellam) model. It has been used to describe red tide outbreaks (Kierstead and Slobodkin, 1953). ◇

EXAMPLE
The simple nonlinear model with logistic growth and simple Fickian diffusion,

$$\frac{\partial n}{\partial t} = r n \left(1 - \frac{n}{K}\right) + D \frac{\partial^2 n}{\partial x^2}, \tag{15.34}$$

is known as the *Fisher equation* (Fisher, 1937). This is the paradigmatic example of a scalar reaction-diffusion equation with a simple traveling wave solution. ◇

EXAMPLE

$$\frac{\partial n}{\partial t} = n(n - a)(1 - n) + D \frac{\partial^2 n}{\partial x^2} \tag{15.35}$$

is frequently referred to as the *Nagumo* (Nagumo *et al.*, 1962) or *bistable equation*. Note the Allee effect. ◇

(3) *Density-dependent diffusion*: There are instances in which the diffusivity D is not constant but depends instead on the density of the organism (e.g., $D = D_0 n^m$). It is then a natural extension of our reaction-diffusion framework to write the diffusive flux as

$$J_{DD} = -D_0 n^m \frac{\partial n}{\partial x} \tag{15.36}$$

in one dimension, or as

$$\boldsymbol{J}_{DD} = -D_0 n^m \nabla n \tag{15.37}$$

in three dimensions. The one-dimensional general balance equation, equation (15.11), then takes the form

$$\frac{\partial n}{\partial t} = f(n) + D_0 \frac{\partial}{\partial x}\left(n^m \frac{\partial n}{\partial x}\right), \tag{15.38}$$

while the three-dimensional general balance equation, equation (15.12) reduces to

$$\frac{\partial n}{\partial t} = f(n) + D_0 \nabla \cdot (n^m \nabla n). \tag{15.39}$$

These equations may give more realistic predictions than equations (15.27) and (15.30).

(4) *Chemotaxis*: The last two choices of flux may have given the (false) impression that all organisms move either randomly or to get away from one another. Many organisms move to pursue food or because of a chemical signal in the environment. If we go back to our advective formulation, we can now imagine that the velocity of motion is proportional to the gradient of some attractant a with the proportionality constant also a function of a,

$$\boldsymbol{J}_C = \chi(a) \nabla a\, n. \tag{15.40}$$

Our general balance equation now reduces to

$$\frac{\partial n}{\partial t} = f(n) - \nabla \cdot [n\,\chi(a)\,\nabla a].$$

Real organisms often experience more than one flux. The total flux may be some combination of advective, convective, diffusive, and chemotactic flux. In addition, there are movement patterns that do not fit into any of these categories.

Let us start with simplest diffusion model that we have considered, the one-dimensional heat equation,

$$\frac{\partial n}{\partial t} = D \frac{\partial^2 n}{\partial x^2}. \tag{15.41}$$

The heat equation is the canonical example of a parabolic evolution equation. As equations go, it is quite nice. In particular, it is *linear*. We will find that we can use the linear superposition principle to construct general solutions from simpler solutions.

However, this equation does not live in a vacuum. Since it is an evolution equation, it needs *initial conditions*. If the spatial domain is finite, it also needs boundary conditions. And, not all conditions will do. Ideally, we want our problems to be *well posed*. This means that (a) a solution should exist, (b) the solution should be unique, and (c) the solution should be stable, i.e., the solution should depend continuously on initial and boundary data.

Which ancillary data will lead to well-posed problems for the heat equation? Here are some examples:

I. *Cauchy problem (initial value problem)*: These are problems with an infinite domain. All one needs is an initial condition (and possibly some boundedness condition at infinity):

$$n(x, 0) = n_0(x), \quad x \in \mathfrak{R}. \tag{15.42}$$

II. *Initial-boundary value problems*: The spatial domain is now bounded. For the sake of specificity, let us imagine that we are interested in solving a problem on the finite domain $0 \le x \le L$. We must then specify an initial condition,

$$n(x, 0) = n_0(x), \quad 0 \le x \le L, \tag{15.43}$$

and boundary conditions (at $x = 0$ and $x = L$). We will restrict our attention to three major classes of boundary conditions:

A. Dirichlet conditions (or boundary conditions of the first kind):

$$n(0, t) = b_1(t), \tag{15.44a}$$
$$n(L, t) = b_2(t). \tag{15.44b}$$

With Dirichlet boundary conditions, we specify the density on the boundary. If the region outside the domain is uninhabitable, so that $b_1(t)$ and $b_2(t)$ are both zero, we have homogeneous Dirichlet boundary conditions.

B. Neumann conditions (or boundary conditions of the second kind):

$$\frac{\partial n}{\partial x}(0, t) = b_1(t), \tag{15.45a}$$
$$\frac{\partial n}{\partial x}(L, t) = b_2(t). \tag{15.45b}$$

In our derivation of reaction-diffusion equation (15.27), we assumed that the flux was proportional to the negative gradient of density. With Neumann conditions, we are thus making a direct statement about the flux at the boundary. In particular, with homogeneous Neumann conditions, we have no flux through the boundaries.

C. Robin conditions (or boundary conditions of the third kind):

$$-D\frac{\partial n}{\partial x}(0, t) = h[b_1(t) - n(0, t)], \tag{15.46a}$$
$$-D\frac{\partial n}{\partial x}(L, t) = h[n(L, t) - b_2(t)]. \tag{15.46b}$$

In this instance, the flux at each boundary is proportional to the difference between the density at the boundary and a specified function. The parameter h is sometimes referred to as an exchange coefficient. Dirichlet conditions and Neumann conditions may be thought of as limiting cases of Robin conditions.

We are now ready to look at problems with heterogeneous environments, heterogeneous initial conditions, or differences in diffusivity.

Recommended readings

Logan (1994) gives a modern introduction to partial differential equations. Britton (1986), Edelstein-Keshet (1988), Grindrod (1996), Murray (1989), and Okubo (1980) focus on reaction-diffusion equations and provide examples from biology and ecology.

16 Spatial steady states: linear problems

Let us begin with a straightforward example. Let us suppose that we have a population that inhabits a patch of length L, $0 \le x \le L$, that does not grow, and that is subject to simple Fickian diffusion,

$$\frac{\partial n}{\partial t} = D \frac{\partial^2 n}{\partial x^2}. \tag{16.1}$$

We must, of course, have an initial condition,

$$n(x, 0) = n_0(x), \tag{16.2}$$

and boundary conditions. Let us assume that the region outside the patch is so inhospitable that we have homogeneous Dirichlet boundary conditions:

$$n(0, t) = 0, \tag{16.3a}$$
$$n(L, t) = 0. \tag{16.3b}$$

This population is doomed. There is no growth and once organisms diffuse out of the patch they die. We would like to describe the time course of this population's collapse.

To solve this problem, we will apply a classical technique introduced by Jean Le Rond d'Alembert (1752) and used extensively, and effectively, by Joseph Fourier (1822). This technique, separation of variables, is the method of choice in dealing with linear homogeneous partial differential equations on a finite domain with homogeneous boundary conditions. The strategy (as with virtually all techniques for solving partial differential equations) is to convert the partial differential equation into one or more ordinary differential equations. In this instance, we make the assumption that we can separate the solution into a product of spatial and temporal terms,

$$n(x, t) = S(x) T(t). \tag{16.4}$$

Since this product is a solution, it must satisfy our partial differential equation. Plugging equation (16.4) into equation (16.1) gives us

$$\dot{T} S = D T S'', \tag{16.5}$$

where the dot corresponds to a time derivative and the primes correspond to space derivatives. If we rearrange terms,

$$\frac{1}{D} \frac{\dot{T}}{T} = \frac{S''}{S}, \tag{16.6}$$

we have a function of time on one side and a function of space on the other. Space and time are independent variables. The only way that a function of time can equal a function of space, for all t and x, is if both functions equal a constant. This constant could be positive, negative, or zero. With some foresight, I will take it to be negative and write it as $-\lambda$,

$$\frac{1}{D} \frac{\dot{T}}{T} = \frac{S''}{S} = -\lambda. \tag{16.7}$$

We can now separate equation (16.7) into two ordinary differential equations,

$$\frac{1}{D} \frac{\dot{T}}{T} = -\lambda, \quad \frac{S''}{S} = -\lambda. \tag{16.8}$$

These two equations are conveniently written as

$$\dot{T} = -\lambda D T, \tag{16.9a}$$
$$S'' + \lambda S = 0. \tag{16.9b}$$

I will leave it to you to show that we do not get any nontrivial solutions that agree with our boundary conditions for λ zero or negative. However, for λ positive, we can write the solutions to equations (16.9a) and (16.9b) as

$$T(t) = c e^{-\lambda D t}, \tag{16.10a}$$
$$S(x) = a \sin \sqrt{\lambda} x + b \cos \sqrt{\lambda} x. \tag{16.10b}$$

By combining these two functions (see equation (16.4)), we obtain

$$n(x, t) = e^{-\lambda D t} (A \sin \sqrt{\lambda} x + B \cos \sqrt{\lambda} x) \tag{16.11}$$

as a potential solution.

Does this solution satisfy our boundary conditions? It follows from equation (16.3a) that

$$B = 0. \tag{16.12}$$

Boundary condition (16.3b), in turn, implies that

$$A \sin \sqrt{\lambda} L = 0. \tag{16.13}$$

Since we are looking for nontrivial solutions, we will assume that $A \neq 0$. It then follows that

$$\sqrt{\lambda} = \frac{k\pi}{L} \tag{16.14}$$

for k an integer. We thus have a countable infinity of solutions of the form

$$n_k(x, t) = A_k e^{-D(k\pi/L)^2 t} \sin\left(\frac{k\pi x}{L}\right). \tag{16.15}$$

Better yet, the superposition principle ensures that a linear combination of these solutions forms a more general solution,

$$n(x, t) = \sum_{k=1}^{\infty} A_k e^{-D(k\pi/L)^2 t} \sin\left(\frac{k\pi x}{L}\right). \tag{16.16}$$

Equation (16.16) implies that our solution is made up of independent spatial modes, that each spatial mode decays with its own characteristic decay constant, and that the high frequency (short wavelength) modes decay more rapidly than the low frequency modes.

But how do we determine the A_k? We need to use initial condition (16.2). Our initial condition implies that

$$n_0(x) = \sum_{k=1}^{\infty} A_k \sin\left(\frac{k\pi x}{L}\right). \tag{16.17}$$

Our initial function can thus be written as a Fourier sine series; the coefficients in this sine series will be the coefficients in our solution. These coefficients can also be thought of as coordinates with the various sine functions as basis vectors. We can take advantage of the orthogonality of this set of basis vectors,

$$\int_0^L \sin\left(\frac{j\pi x}{L}\right) \sin\left(\frac{k\pi x}{L}\right) dx = \begin{cases} 0, & j \neq k, \\ L/2, & j = k, \end{cases} \tag{16.18}$$

by multiplying equation (16.17) through by a sine and integrating from 0 to L,

$$\int_0^L n_0(x) \sin\left(\frac{j\pi x}{L}\right) dx = \sum_{k=1}^{\infty} \int_0^L A_k \sin\left(\frac{j\pi x}{L}\right) \sin\left(\frac{k\pi x}{L}\right) dx. \tag{16.19}$$

The orthogonality of the sine functions causes this last equation to reduce to

$$\int_0^L n_0(x) \sin\left(\frac{k\pi x}{L}\right) dx = \frac{L}{2} A_k \tag{16.20}$$

or

$$A_k = \frac{2}{L} \int_0^L n_0(x) \sin\left(\frac{k\pi x}{L}\right) dx. \tag{16.21}$$

This is an explicit formula for our coefficients.

EXAMPLE

$$\frac{\partial n}{\partial t} = D \frac{\partial^2 n}{\partial x^2}, \quad 0 < x < 1, \tag{16.22a}$$

$$n(0, t) = 0, \tag{16.22b}$$

$$n(1, t) = 0, \tag{16.22c}$$

$$n(x, 0) = 100. \tag{16.22d}$$

The solution may be written

$$n(x, t) = \sum_{k=1}^{\infty} A_k e^{-(k\pi)^2 Dt} \sin k\pi x, \tag{16.23}$$

with

$$A_k = 2 \int_0^1 100 \sin k\pi x \, dx = \begin{cases} 0, & k \text{ even,} \\ \dfrac{400}{k\pi}, & k \text{ odd.} \end{cases} \tag{16.24}$$

For large times,

$$n(x, t) \approx \frac{400}{\pi} e^{-\pi^2 Dt} \sin \pi x. \tag{16.25}$$

The relative size of the first neglected term is given by

$$\epsilon = \frac{\left(\frac{400}{3\pi}\right) e^{-9\pi^2 Dt}}{\left(\frac{400}{\pi}\right) e^{-\pi^2 Dt}} = \frac{1}{3} e^{-8\pi^2 Dt}. \tag{16.26}$$

Thus, for $\pi^2 Dt = 0.5$, $\epsilon \approx e^{-4}/3 \approx 0.006$. For $D \approx 1$, this relative error is less than 1% after one-twentieth of a time unit. ◇

Problem 16.1 *Neumann boundary conditions*
Determine the solution of equation (16.1) in the presence of homogeneous Neumann boundary conditions.

Hint Be sure to consider all possible signs for λ.

Notice that for solutions (16.16) and (16.23)

$$\lim_{t \to \infty} n(x, t) = 0. \tag{16.27}$$

The system approaches the steady-state solution $n^*(x) = 0$. We could have found this steady state directly by requiring

$$\frac{\partial n}{\partial t} = 0 \tag{16.28}$$

so that

$$D \frac{\partial^2 n^*}{\partial x^2} = 0. \tag{16.29}$$

This has the solution

$$n^*(x) = A x + B \tag{16.30}$$

and as soon as we impose boundary conditions (16.3a) and (16.3b) we observe that

$$n^*(x) = 0. \tag{16.31}$$

In following this direct approach, we would, however, have missed the exact solution and the exact approach to the steady state.

The heat equation is only as good as its assumptions. In the 1860s, William Thomson, Lord Kelvin, used the heat equation to estimate the age of the earth (Thomson, 1863, 1864). Thomson's estimate of a mere 100 million years was a direct challenge to the geological uniformitarianism of James Hutton and Sir Charles Lyell and to Charles Darwin's theory of evolution. Although Lord Kelvin's calculations of the cooling of the earth's interior were impeccable, he knew nothing of radioactivity (an important heat source). It was not until 1896 that Henri Becquerel discovered radioactivity. In 1906, R. J. Strutt (the fourth Lord Rayleigh) demonstrated that the heat generated by radioactivity allowed for a much older earth (Strutt, 1905).

From a biological viewpoint, there is much that is missing from equation (16.1). We have, for example, ignored the tremendous potential of populations to grow. Fortunately, Skellam (1951) and Kierstead and Slobodkin (1953) realized, some time ago, that the addition of growth makes equation (16.1) ecologically interesting. One then has a race between growth in the interior of the patch and diffusion and death through the boundary of the patch. Kierstead and Slobodkin's paper makes for especially interesting reading. Kierstead and Slobodkin were interested in phytoplankton blooms (e.g., red tide outbreaks). They assumed simple exponential growth, diffusion, and homogeneous Dirichlet boundary conditions:

$$\frac{\partial n}{\partial t} = rn + D\frac{\partial^2 n}{\partial x^2}, \tag{16.32a}$$

$$n(0, t) = 0, \tag{16.32b}$$

$$n(L, t) = 0, \tag{16.32c}$$

$$n(x, 0) = n_0(x). \tag{16.32d}$$

We could, of course, use separation of variables. However, it is easier to reduce this new problem to a previously solved problem. Since the population grows exponentially with no diffusion, it is reasonable to look for an exponential somewhere in the solution. So, let

$$n(x, t) \equiv e^{rt} u(x, t). \tag{16.33}$$

Substituting (16.33) into (16.32a) produces

$$e^{rt}\frac{\partial u}{\partial t} + r e^{rt} u = r e^{rt} u + D e^{rt}\frac{\partial^2 u}{\partial x^2}. \tag{16.34}$$

It follows that

$$\frac{\partial u}{\partial t} = D\frac{\partial^2 u}{\partial x^2}, \tag{16.35a}$$

$$u(0, t) = 0, \tag{16.35b}$$

$$u(L, t) = 0, \tag{16.35c}$$

$$u(x, 0) = n_0(x). \tag{16.35d}$$

This should look familiar. It is simply the heat equation with homogeneous Dirichlet boundary conditions (and u instead of n). We know that the solution is

$$u(x, t) = \sum_{k=1}^{\infty} A_k e^{-D(k\pi/L)^2 t} \sin\left(\frac{k\pi x}{L}\right) \tag{16.36}$$

and so, by definition (16.33),

$$n(x, t) = \sum_{k=1}^{\infty} A_k e^{[r - D(k\pi/L)^2]t} \sin\left(\frac{k\pi x}{L}\right). \tag{16.37}$$

Table 16.1. *Diffusion coefficients*

Example	Diffusivity
Sucrose in water	4.6×10^{-6} cm^2/s
Oxygen in air	2.0×10^{-1} cm^2/s
Swarming midges	1.0×10^{1} cm^2/s
Oaks	1.8×10^{-1} km^2/generation
Black plague	1.0×10^{4} mi^2/year

We would like to know if this population grows or collapses. It will grow if

$$r - D\,(k\pi/L)^2 > 0 \tag{16.38}$$

for some k (so that the coefficient of time in the exponent is positive). We will focus on the dominant or slowest decaying spatial mode. For $k = 1$,

$$L < \pi\sqrt{\frac{D}{r}} \tag{16.39}$$

ensures the death of the population.

The critical length on the right-hand side of inequality (16.39) is commonly referred to as the KISS size. If the patch is shorter than this length, the phytoplankton population collapses. If it is longer than this length, a bloom occurs. Increasing the size of the patch leads to a bifurcation. The zero steady state loses its stability at the bifurcation point. Estimated values for the KISS size for dinoflagellates range from 1 to 50 km (Okubo, 1978).

The above results can also be applied to many other problems. The square root of the diffusivity plays an important role in all of these problems. Table 16.1 shows that the diffusivity can vary dramatically from problem to problem.

The analysis of the KISS problem was facilitated by the existence of a complete and countable set of mutually orthogonal eigenfunctions that satisfy equation (16.9b) with homogeneous Dirichlet boundary conditions. How typical is this?

Definition The differential equation

$$\frac{d}{dx}\left[p(x)\frac{dy}{dx}\right] + [\lambda w(x) - q(x)]\,y = 0, \quad a < x < b, \tag{16.40}$$

with *unmixed* boundary conditions,

$$\alpha_0\, y(a) + \alpha_1\, y'(a) = 0, \tag{16.41a}$$

$$\beta_0\, y(b) + \beta_1\, y'(b) = 0, \tag{16.41b}$$

is said to be a *regular Sturm–Liouville problem* if

(1) (a, b) is finite,
(2) p, p', q, and w are real and continuous on $[a, b]$, and
(3) $p > 0$ and $w > 0$ on $[a, b]$.

Theorem   For a regular Sturm–Liouville problem:

(1) All eigenvalues λ are real.
(2) The eigenvalues form an infinite ordered sequence

$$\lambda_1 < \lambda_2 < \lambda_3 < \cdots$$

with $\lambda_i \to \infty$ as $i \to \infty$.
(3) Corresponding to each eigenvalue λ_i there is an eigenfunction $y_i(x)$. This eigen-vector is unique to within an arbitrary multiplicative constant. Eigenfunction y_i has exactly $i - 1$ zeros for $a < x < b$.
(4) Eigenfunctions corresponding to distinct eigenvalues are orthogonal with weighting function $w(x)$. In other words,

$$\int_a^b y_i(x)\, y_j(x)\, w(x)\, dx \; = \; 0, \quad \text{if } \lambda_i \neq \lambda_j. \tag{16.42}$$

(5) The eigenfunctions form a complete set: there is an orthonormal basis of $L_w^2\,(a, b)$ consisting of eigenfunctions. Any piecewise smooth function can be represented by a generalized Fourier series of the eigenfunctions,

$$f(x) \; \sim \; \sum_{i=1}^{\infty} a_i\, y_i(x). \tag{16.43}$$

This infinite series converges to $[f(x+) + f(x-)]/2$ for $a < x < b$. If the function $f \in C^2$ on $[a, b]$ and satisfies the boundary conditions, the generalized Fourier series converges uniformly to f.

I will not prove this theorem. I will, however, point out that the theorem relies on the fact that the operator

$$L \equiv \frac{d}{dx}\left[p(x)\,\frac{d}{dx}\right] - q(x) \tag{16.44}$$

is self-adjoint,

$$\int_a^b u\,Lv \; - \; v\,Lu\, dx \; = \; 0, \tag{16.45}$$

for boundary conditions (16.41a) and (16.41b). Imagine two eigenvectors, $y_i(x)$ and $y_j(x)$, that satisfy equation (16.40):

$$Ly_i(x) + \lambda_i w(x) y_i(x) = 0, \tag{16.46a}$$

$$Ly_j(x) + \lambda_j w(x) y_j(x) = 0. \tag{16.46b}$$

If we multiply (16.46a) by $y_j(x)$ and (16.46b) by $y_i(x)$, we may take the difference between these two equations, and integrate, to obtain

$$(\lambda_i - \lambda_j) \int_a^b y_i(x) y_j(x) w(x) \, dx \;=\; \int_a^b (y_j L y_i - y_i L y_j) \, dx \;=\; 0. \tag{16.47}$$

The orthogonality of distinct eigenfunctions is an immediate consequence of equation (16.47).

A great many ordinary and partial differential equations classes cover Sturm–Liouville theory. The standard references include Birkhoff and Rota (1978), Coddington and Levinson (1955), and Stakgold (1979). The existence of a complete set of eigenfunctions, as in the KISS problem, is typical of regular Sturm–Liouville problems.

The examples get tougher very soon. Now is a good time for you to do a homework problem. Try your hand at the following:

Problem 16.2 *Critical patch size with Robin boundary conditions*
Determine the critical patch for a growing and diffusing population with homogeneous Robin boundary conditions:

$$\frac{\partial n}{\partial t} = r n + D \frac{\partial^2 n}{\partial x^2}, \tag{16.48a}$$

$$\frac{\partial n}{\partial x}(0, t) = +\alpha n(0, t), \tag{16.48b}$$

$$\frac{\partial n}{\partial x}(L, t) = -\alpha n(L, t), \tag{16.48c}$$

$$n(x, 0) = n_0(x). \tag{16.48d}$$

Discuss the limiting cases $\alpha = 0$ and $\alpha \to \infty$.

A Sturm–Liouville problem is *singular* if $p(x)$ or $w(x)$ vanish or become infinite at an endpoint or if the interval itself is infinite. The general theory of singular Sturm–Liouville problems is complicated; it is not unusual for solutions to become unbounded or for discrete spectra to give way to continuous spectra. Fortunately, interesting ecological scenarios can generate singular Sturm–Liouville problems that are tractable. Two examples follow.

EXAMPLE   The radially symmetric two-dimensional KISS problem
The two-dimensional Laplacian may be written in polar coordinates as

$$\nabla^2 n = \frac{\partial^2 n}{\partial x^2} + \frac{\partial^2 n}{\partial y^2} = \frac{1}{r}\frac{\partial}{\partial r}\left(r\frac{\partial n}{\partial r}\right) + \frac{1}{r^2}\frac{\partial^2 n}{\partial \theta^2}. \tag{16.49}$$

If we assume simple exponential growth and radial symmetry, we must thus solve

$$\frac{\partial n}{\partial t} = \alpha n + D\frac{1}{r}\frac{\partial}{\partial r}\left(r\frac{\partial n}{\partial r}\right), \tag{16.50a}$$

$$n(R, t) = 0, \tag{16.50b}$$

$$n(r, 0) = n_0(r). \tag{16.50c}$$

Note that we are using α rather than r as the intrinsic rate of growth. There is also only one boundary condition: the habitat is assumed to be inhospitable beyond $r = R$.
   The substitution

$$n(r, t) = e^{\alpha t} u(r, t) \tag{16.51}$$

reduces our system to the more familiar

$$\frac{\partial u}{\partial t} = D\frac{1}{r}\frac{\partial}{\partial r}\left(r\frac{\partial u}{\partial r}\right), \tag{16.52a}$$

$$u(R, t) = 0, \tag{16.52b}$$

$$u(r, 0) = n_0(r). \tag{16.52c}$$

We can again separate our solution into spatial and temporal components,

$$u(r, t) = T(t)S(r). \tag{16.53}$$

Substituting (16.53) into equation (16.52a), separating the function of time from that of space, and setting both sides equal to a negative constant gives us

$$\frac{1}{D}\frac{\dot{T}}{T} = \frac{1}{S}\frac{1}{r}\frac{d}{dr}\left(r\frac{dS}{dr}\right) = -\lambda. \tag{16.54}$$

We thus have

$$\dot{T} = -\lambda D T, \tag{16.55}$$

as the temporal equation and

$$\frac{d}{dr}\left(r\frac{dS}{dr}\right) + \lambda r S = 0 \tag{16.56}$$

as the spatial equation. Equation (16.56) also inherits the boundary condition

$$S(R) = 0. \tag{16.57}$$

Equation (16.56) with boundary condition (16.57) is *not* a regular Sturm–Liouville problem. In particular, $p(r) = r$ vanishes at $r = 0$. Nevertheless, equation (16.56) is a well-studied equation. It may be rewritten

$$r^2 \frac{d^2 S}{dr^2} + r \frac{dS}{dr} + \lambda r^2 S = 0. \tag{16.58}$$

Equation (16.58) is just a minor variant of Bessel's differential equation,

$$r^2 \frac{d^2 S}{dr^2} + r \frac{dS}{dr} + (r^2 - v^2) S = 0. \tag{16.59}$$

Indeed, the solution of (16.58) can be written as

$$S(r) = c_1 J_0(\sqrt{\lambda} r) + c_2 Y_0(\sqrt{\lambda} r), \tag{16.60}$$

where $J_0(r)$ is a Bessel function (of the first kind) of order zero and $Y_0(r)$ is a Bessel function of the second kind of order zero. (Bessel functions of the second kind are also referred to as Neumann functions or, more rarely, as Weber functions.)

The first of these functions is typically expressed as a power series. Since equations (16.58) and (16.59) both have regular singular points at the origin, the preferred method for obtaining this solution is to let

$$S(r) = r^m (a_0 + a_1 r + a_2 r^2 + \cdots). \tag{16.61}$$

This is the method of Frobenius. The exponent m may be a negative integer, a fraction, or even an irrational number. The exponent is chosen to make a_0 nonzero. It can then be shown that

$$J_0(r) = \sum_{k=0}^{\infty} \frac{(-1)^k}{(k!)^2} \left(\frac{r}{2}\right)^{2k}. \tag{16.62}$$

This is a special case of the more general

$$J_v(r) = \sum_{k=0}^{\infty} \frac{(-1)^k}{k!\, \Gamma(k + v + 1)} \left(\frac{r}{2}\right)^{2k+v}. \tag{16.63}$$

At this point, matters become more complicated. So much so, that I will give only the briefest sketch. Standard power series methods fail to give $Y_0(r)$. The usual solution is to define

$$Y_0(r) = \lim_{v \to 0} \frac{\cos(v\pi) J_v(r) - J_{-v}(r)}{\sin(v\pi)}. \tag{16.64}$$

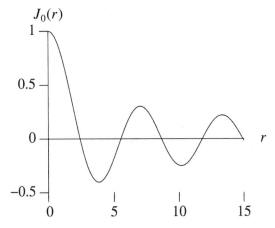

Fig. 16.1. Bessel function of the first kind.

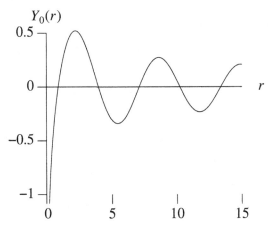

Fig. 16.2. Bessel function of the second kind.

This clearly involves the use of l'Hôpital's rule. The most important property of $Y_0(r)$ for our purposes is its asymptotic behavior as $r \rightarrow 0$:

$$Y_0(r) \sim \frac{2}{\pi} \ln \left(\frac{r}{2} \right) \quad \text{as } r \rightarrow 0. \tag{16.65}$$

The function $Y_0(r)$ blows up as $r \rightarrow 0$, whereas $J_0(r)$ remains bounded (see Figures 16.1 and 16.2). You can solve many problems by remembering that $Y_0(r)$ is unbounded at the origin. Since we expect boundedness for finite times (this is our second boundary condition) we need

$$c_2 = 0. \tag{16.66}$$

Since we also require that $S(R) = 0$, we have that

$$J_0(\sqrt{\lambda}\, R) = 0 \tag{16.67}$$

or

$$\lambda_k = \left(\frac{\beta_k}{R}\right)^2, \quad k = 1, 2,\dots, \tag{16.68}$$

where β_k is a zero of our zero-order Bessel function, $J_0(\beta_k) = 0$.

A little effort now shows that the solution of equations (16.50a), (16.50b), and (16.50c) must now be of the form

$$n(r, t) = \sum_{k=1}^{\infty} c_k e^{[\alpha - (\beta_k/R)^2 D]t} J_0\left(\frac{\beta_k}{R}r\right). \tag{16.69}$$

The coefficients c_k in this solution are determined by the initial condition. In particular, the initial distribution can be represented by a Fourier–Bessel series,

$$n_0(r) = \sum_{k=1}^{\infty} c_k J_0\left(\frac{\beta_k}{R}r\right). \tag{16.70}$$

The coefficients in this Fourier–Bessel series also show up in our solution. Orthogonality of the Bessel functions (with respect to the weighting function r),

$$\int_0^R J_0\left(\frac{\beta_j}{R}r\right) J_0\left(\frac{\beta_k}{R}r\right) r\, dr = 0, \quad \text{if } j \neq k, \tag{16.71}$$

can be used to show that the coefficients are just

$$c_k = \frac{\int_0^R n_0(r) J_0\left(\frac{\beta_k}{R}r\right) r\, dr}{\int_0^R J_0^2\left(\frac{\beta_k}{R}r\right) r\, dr}, \quad k = 1, 2, \dots, \tag{16.72}$$

or

$$c_k = \frac{2}{R^2 J_1^2(\beta_k)} \int_0^R n_0(r) J_0\left(\frac{\beta_k}{R}r\right) r\, dr, \quad k = 1, 2, \dots. \tag{16.73}$$

For the radially symmetric, two-dimensional KISS problem, the population grows if

$$\alpha - (\beta_k/R)^2 D > 0, \tag{16.74}$$

for some k. It collapses if

$$R < \beta_1 \sqrt{\frac{D}{\alpha}}. \tag{16.75}$$

β_1 is the first zero of the zero-order Bessel function of the first kind. $\beta_1 \approx 2.4$.

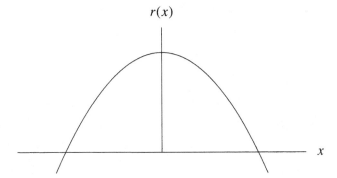

Fig. 16.3. Spatial variation in growth rate.

The resulting critical patch size is similar to that for the one-dimensional problem. ◇

EXAMPLE The Schrödinger equation in ecology?

Homogeneous Dirichlet boundary conditions are an extremely crude way of capturing spatial heterogeneity. An alternative approach is that of Gurney and Nisbet (1975). They considered a population on an infinite domain in which the intrinsic rate of growth decreases with the square of the distance from the center of the range (see Figure 16.3). They took the equation

$$\frac{\partial n}{\partial t} = r_0 \left[1 - \left(\frac{x}{x_0} \right)^2 \right] n + D \frac{\partial^2 n}{\partial x^2}, \quad -\infty < x < \infty, \tag{16.76}$$

with the boundedness condition

$$\lim_{|x| \to \infty} n(x, t) = 0. \tag{16.77}$$

The change of variables

$$n(x, t) \equiv e^{r_0 t} u(x, t) \tag{16.78}$$

produces

$$\frac{\partial u}{\partial t} = -r_0 \left(\frac{x}{x_0} \right)^2 u + D \frac{\partial^2 u}{\partial x^2}. \tag{16.79}$$

Separating out the temporal and spatial parts of the solution,

$$u(x, t) = T(t) S(x), \tag{16.80}$$

now leaves us with two ordinary differential equations:

$$\dot{T} = -\lambda D\, T, \tag{16.81a}$$

$$\frac{d^2 S}{dx^2} + \left[\lambda - \frac{r_0}{D} \left(\frac{x}{x_0} \right)^2 \right] S = 0. \tag{16.81b}$$

Equation (16.81b) is of Sturm–Liouville form. Since the domain is infinite, this is a singular Sturm–Liouville problem. It is common for singular Sturm–Liouville problems on infinite domains to have continuous spectra. Fortunately, the fact that $q(x)$ approaches infinity as x approaches infinity is enough to guarantee us that we have a point spectrum (Titchmarsh, 1946). Indeed, equation (16.81b) is a well-known and well-studied differential equation. If we let

$$a^2 \equiv \frac{r_0}{D\,x_0^2}, \quad \hat{x} \equiv \sqrt{a}\,x, \quad 2p + 1 = \frac{\lambda}{a}, \tag{16.82}$$

equation (16.81b) may be rewritten as

$$\frac{d^2 S}{d\hat{x}^2} + (2p + 1 - \hat{x}^2) S = 0. \tag{16.83}$$

This is 'simply' the quantum mechanical Schrödinger equation for a one-dimensional harmonic oscillator.

Equation (16.83) is in Sturm–Liouville form. However, the interval is infinite and so this is a singular Sturm–Liouville problem. If we try to solve this equation directly by power series, we get a three-term recursion formula for the coefficients that is too inconvenient for words. However, when $\hat{x}$ is large,

$$\frac{d^2 S}{d\hat{x}^2} \approx \hat{x}^2 S. \tag{16.84}$$

If we factor this,

$$\left(\frac{d}{d\hat{x}} - \hat{x} \right) \left(\frac{d}{d\hat{x}} + \hat{x} \right) S \approx 0, \tag{16.85}$$

we see that

$$S \approx e^{\pm \hat{x}^2/2} \tag{16.86}$$

are approximate solutions for large $\hat{x}$. We may throw the first approximate solution out, since it does not approach zero as $|\hat{x}| \to \infty$, but we may use the second solution as the basis of a change of variables,

$$S \equiv y e^{-\hat{x}^2/2}, \tag{16.87}$$

with the hope that it will simplify equation (16.83)†.

Whatever you may think of this strategy, it works. In particular, it transforms equation (16.83) into

$$\frac{d^2 y}{d\hat{x}^2} - 2\hat{x}\frac{dy}{d\hat{x}} + 2py = 0. \tag{16.88}$$

Equation (16.88) is Hermite's equation. It is sometimes written in the Sturm–Liouville form

$$\frac{d}{d\hat{x}}\left(e^{-\hat{x}^2}\frac{dy}{d\hat{x}}\right) + 2pe^{-\hat{x}^2}y = 0. \tag{16.89}$$

It is easy to solve Hermite's equation using power series methods. If you try a solution of the form

$$y(\hat{x}) = \sum_{k=0}^{\infty} a_k \hat{x}^k, \tag{16.90}$$

you can show that

$$a_{k+2} = -\frac{2(p-k)}{(k+1)(k+2)} a_k. \tag{16.91}$$

This is a two-term recursion formula for the coefficients of two independent series solutions,

$$y_1(\hat{x}) = 1 - \frac{2p}{2!}\hat{x}^2 + \frac{2^2 p(p-2)}{4!}\hat{x}^4 - \frac{2^3 p(p-2)(p-4)}{6!}\hat{x}^6 + \cdots \tag{16.92}$$

and

$$y_2(\hat{x}) = x - \frac{2(p-1)}{3!}\hat{x}^3 + \frac{2^2(p-1)(p-3)}{5!}\hat{x}^5$$
$$- \frac{2^3(p-1)(p-3)(p-5)}{7!}\hat{x}^7 + \cdots. \tag{16.93}$$

The first series arises for $a_0 = 1$. The second arises for $a_1 = 1$. Both series converge for all $\hat{x}$.

We want something more than just convergence. We want $S(\hat{x})$, the product of $y(\hat{x})$ and $\exp(-\hat{x}^2/2)$, to tend to zero as the absolute value of $\hat{x}$ tends to infinity. This will happen if $y_1(\hat{x})$ and $y_2(\hat{x})$ are finite-degree polynomials.

---

† This approximation also suggests that organisms should be normally distributed about the center of their range.

Thus we require $p = 0, 2, 4,\ldots$ for $y_1(\hat{x})$, or $p = 1, 3, 5,\ldots$ for $y_2(\hat{x})$. For each eigenvalue p, we get a corresponding polynomial eigenfunction. It is traditional to normalize these polynomials so that the terms containing the highest power of $\hat{x}$ are of the form $2^p \hat{x}^p$, in which case

$$H_0(\hat{x}) = 1, \tag{16.94a}$$
$$H_1(\hat{x}) = 2\,\hat{x}, \tag{16.94b}$$
$$H_2(\hat{x}) = 4\,\hat{x}^2 - 2, \tag{16.94c}$$
$$H_3(\hat{x}) = 8\,\hat{x}^3 - 12\,\hat{x}, \tag{16.94d}$$

and so on.

Backtracking through the various transformations that I have introduced, the eigenvalues and eigenfunctions of equation (16.81b) are simply

$$\lambda_p = a(2p + 1) \tag{16.95}$$

and

$$S_p(x) = H_p(\sqrt{a}\,x)\,e^{-a x^2/2}, \tag{16.96}$$

where $p = 0, 1, 2,\ldots$, $H_p(x)$ are the Hermite polynomials, and a, as before, satisfies

$$a^2 \equiv \frac{r_0}{D x_0^2}. \tag{16.97}$$

For large times, the first eigenvalue and the first eigenfunction dominate the solution. We may then approximate the solution to our original problem, equation (16.76), by

$$n(x, t) \approx c_0\,e^{(r_0 - D\lambda_0)t}\,e^{-a x^2/2}. \tag{16.98}$$

The population will either grow or decay depending on whether r_0 or $D\lambda_0$ is larger. The population will grow if

$$2x_0 > 2\sqrt{\frac{D}{r_0}}. \tag{16.99}$$

$\diamond$

Table 16.2 summarizes the previous three examples. All three problems give similar estimates of the critical patch size. This suggests that the essential feature that determines critical size, the balance between diffusion and growth, is preserved in each formulation.

Table 16.2. *Critical patch size*

Example	Critical patch size
1D KISS	$L = \pi \sqrt{\frac{D}{r}}$
2D radial KISS	$R = 2.4 \sqrt{\frac{D}{r}}$
Parabolic decline	$2x_0 = 2 \sqrt{\frac{D}{r}}$

Problem 16.3 *Parabolic fall-off in polar coordinates*
Determine the critical patch size for a population that satisfies

$$\frac{\partial n}{\partial t} = \alpha_0 \left[1 - \left(\frac{r}{R} \right)^2 \right] n + D \frac{1}{r} \frac{\partial}{\partial r} \left(r \frac{\partial n}{\partial r} \right) \qquad (16.100)$$

for $0 < r < \infty$.

Recommended readings

Farlow (1982) and Haberman (1983) provide straightforward introductions to separation of variables.

17 Spatial steady states: nonlinear problems

All of the models in Chapter 16 were linear. The populations that we considered tended to grow or decay exponentially. However, we have already seen that there is more to population dynamics than booms and busts. Many populations are limited by density-dependent factors. Let us turn then to the nonlinear model

$$\frac{\partial n}{\partial t} = r n \left(1 - \frac{n}{K} \right) + D \frac{\partial^2 n}{\partial x^2}, \quad 0 < x < L, \tag{17.1a}$$

$$n(0, t) = 0, \tag{17.1b}$$

$$n(L, t) = 0, \tag{17.1c}$$

$$n(x, 0) = n_0(x). \tag{17.1d}$$

Partial differential equation (17.1a) contains logistic growth and simple Fickian diffusion. It is often called the Fisher equation after the geneticist R. A. Fisher. He was interested in the traveling waves of advance of an advantageous allele. We will not study traveling waves in this chapter. Rather, we will key in on some of the steady states that arise in this and other nonlinear models.

Equation (17.1a) contains three parameters: r, K, and D. These parameters can be eliminated by rescaling the dependent and the independent variables: time can be rescaled to remove the intrinsic rate of growth r, space can be rescaled to remove the diffusivity D, and the density can be rescaled to eliminate the carrying capacity K. All of this is mathematically appealing, but it is easy to lose sight of the biology by going too far in this direction. Indeed, we have already seen that r and D play an important role in critical patch-size problems. I will therefore rescale density,

$$u(x, t) \equiv \frac{n(x, t)}{K}, \tag{17.2}$$

but leave space and time alone. Our system now takes the form

$$\frac{\partial u}{\partial t} = r u (1 - u) + D \frac{\partial^2 u}{\partial x^2}, \quad 0 < x < L, \tag{17.3a}$$

$$u(0, t) = 0, \tag{17.3b}$$

$$u(L, t) = 0, \tag{17.3c}$$

$$u(x, 0) = u_0(x). \tag{17.3d}$$

The steady-state solutions of this system do not depend on time and satisfy

$$r u (1 - u) + D u'' = 0, \tag{17.4a}$$

$$u(0) = 0, \quad u(L) = 0. \tag{17.4b}$$

These are equilibria in time; they may still vary in space. We are interested only in those solutions of equation (17.4a) for which $u(x) \geq 0$. The trivial solution, $u = 0$, satisfies our system for all values of L. We would like to know whether there are other, nontrivial, steady states.

Equation (17.4a) is a nonlinear, second-order, ordinary differential equation. We will look at its behavior in the phase plane. This is an unusual phase plane with space, rather than time, as the independent variable. Nevertheless, we may rewrite equation (17.4a) as

$$u' = v, \quad v' = -\frac{r}{D} u (1 - u). \tag{17.5}$$

This system has two phase-plane equilibria, $(0, 0)$ and $(1, 0)$. The linearization of system (17.5) about $(0, 0)$ is

$$\begin{pmatrix} u' \\ v' \end{pmatrix} = \begin{pmatrix} 0 & 1 \\ -r/D & 0 \end{pmatrix} \begin{pmatrix} u \\ v \end{pmatrix}. \tag{17.6}$$

This linear system has purely imaginary eigenvalues,

$$\lambda = \pm i \sqrt{r/D}. \tag{17.7}$$

The linearization of system (17.5) about $(1, 0)$ is

$$\begin{pmatrix} 0 & 1 \\ r/D & 0 \end{pmatrix} \tag{17.8}$$

and has

$$\lambda = \pm \sqrt{r/D} \tag{17.9}$$

as its eigenvalues. The second equilibrium is clearly a saddle point, both for the linearized system and for the original nonlinear system. The first equilibrium is a center – for the linearized system. However, linearization is

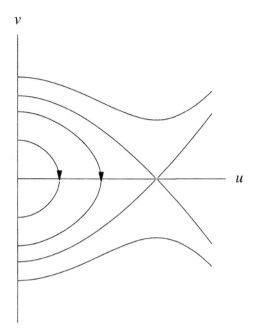

Fig. 17.1. Phase plane for the steady states of the Fisher equation.

unreliable for nonhyperbolic fixed points; we cannot, as yet, be sure of the true nature of the first equilibrium point.

Fortunately, equation (17.4a) has a first integral. Multiplying equation (17.4a) through by u',

$$D u'' u' + r u(1 - u) u' = 0, \qquad (17.10)$$

and integrating (with respect to x) produces

$$\frac{D}{2} (u')^2 + r \left(\frac{1}{2} u^2 - \frac{1}{3} u^3 \right) = c. \qquad (17.11)$$

This equation may be rewritten as

$$\frac{1}{2} v^2 + \frac{r}{D} \left(\frac{1}{2} u^2 - \frac{1}{3} u^3 \right) = c. \qquad (17.12)$$

The level curves of this equation are orbits in the phase plane (see Figure 17.1). The phase portrait is symmetric in $v = u'$. This makes the origin a center.

Each orbit in this phase portrait satisfies equation (17.4a). However, we are interested only in those orbits that satisfy boundary conditions (17.4b). The origin of the phase plane, $u(x) = 0$, $u'(x) = 0$, is one such solution.

There may or may not be other solutions, depending on whether we can find an orbit that starts on the positive v axis (corresponding to $u = 0$) at $x = 0$ and that ends on the negative v-axis at $x = L$. The parameter L, the length of the patch, plays a pivotal role, just as it did for the KISS problem. We expect orbits to 'slow down' close to the saddle point at $(1, 0)$. Orbits that approach the saddle point should correspond to large values of L. At the other extreme, orbits close to the origin should have a (spatial) half-period close to $\pi \sqrt{D/r}$, since equation (17.12) is then similar to the first integral for a harmonic oscillator. There is one orbit for each patch size between $\pi \sqrt{D/r}$ and infinity.

Let us see whether we can be more precise. Let us assume that we have a solution to equation (17.4a) that satisfies the boundary conditions. Equation (17.12) may now be rewritten

$$\frac{1}{2} v^2 + \frac{r}{D} F(u) = \frac{r}{D} F(\mu),$$

(17.13)

where $u = \mu$ when $v = 0$ at $x = L/2$. Thus

$$v = \frac{du}{dx} = \begin{cases} +\sqrt{\dfrac{2r}{D} [F(\mu) - F(u)]}, & 0 < x < L/2, \\[2ex] -\sqrt{\dfrac{2r}{D} [F(\mu) - F(u)]}, & L/2 < x < L. \end{cases}$$

(17.14)

If we separate equation (17.14) and integrate over the first half of the orbit, we get

$$\sqrt{\frac{D}{2r}} \int_0^\mu \frac{du}{\sqrt{F(\mu) - F(u)}} = \int_0^{L/2} dx.$$

(17.15)

Similarly, if we integrate over the second half of the orbit, we obtain

$$\sqrt{\frac{D}{2r}} \int_\mu^0 \frac{-du}{\sqrt{F(\mu) - F(u)}} = \int_{L/2}^L dx.$$

(17.16)

Either way,

$$L = \sqrt{\frac{2D}{r}} \int_0^\mu \frac{du}{\sqrt{F(\mu) - F(u)}}.$$

(17.17)

It is convenient to make the substitution

$$z \equiv \frac{u}{\mu}.$$

(17.18)

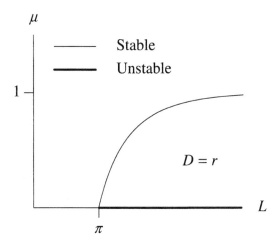

Fig. 17.2. Bifurcation diagram.

Now,

$$L = \sqrt{\frac{2D}{r}} \int_0^1 \frac{\mu}{\sqrt{F(\mu) - F(\mu z)}} \, dz. \tag{17.19}$$

This formula is referred to as a *time-map* since it maps orbits (parametrized by the maximum density μ) to the 'time' (space) it takes to traverse those orbits.

Equation (17.19) may be evaluated as an elliptic integral. However, it can also be evaluated numerically. One can plot L as a function of μ and reflect the graph about the 45° line in order to get a plot of the maximum density μ as a function of L (see Figure 17.2).

The following facts may also be derived from equation (17.19):

(1) L is an increasing function of μ for $0 \le \mu < 1$,
(2) L is concave up for $0 \le \mu < 1$,
(3) $\lim_{\mu \uparrow 1} L(\mu) \to \infty$,
(4) $\lim_{\mu \downarrow 0} L(\mu) = L_c = \pi \sqrt{D/r}$.

The last fact is important because it gives us the KISS size. Let us look at it a little more carefully:

$$L_c = \lim_{\mu \to 0} \sqrt{\frac{2D}{r}} \int_0^1 \frac{\mu}{\sqrt{F(\mu) - F(\mu z)}} \, dz, \tag{17.20}$$

$$L_c = \lim_{\mu \to 0} \sqrt{\frac{2D}{r}} \int_0^1 \frac{\mu}{\sqrt{\frac{1}{2}\mu^2(1-z)[1+z+\frac{2}{3}\mu(1+z+z^2)]}} \, dz, \tag{17.21}$$

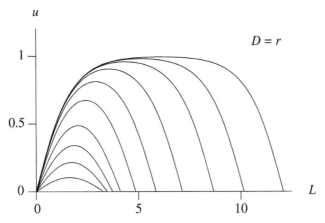

Fig. 17.3. Nonuniform steady states.

$$L_c = 2 \sqrt{\frac{D}{r}} \int_0^1 \frac{1}{\sqrt{1 - z^2}} \, dz, \tag{17.22}$$

$$L_c = \lim_{\mu \to 0} 2 \sqrt{\frac{D}{r}} \sin^{-1} z \Big]_0^1, \tag{17.23}$$

$$L_c = \pi \sqrt{\frac{D}{r}}. \tag{17.24}$$

All in all, we see that there is a unique, nonuniform, stationary solution for the Fisher equation with homogeneous Dirichlet boundary conditions if $L > \pi \sqrt{D/r}$. This solution coexists with the trivial solution $u(x) = 0$. Figure 17.3 depicts the shape of the nonuniform steady state for a variety of patch sizes L. This steady state starts out looking like a simple sine function; it flattens out as the domain increases and as the corresponding phase-plane orbit approaches the saddle point at $(1, 0)$.

What can we say about the stability of the trivial and nontrivial steady states? Analyzing the stability of the trivial solution is straightforward. If we linearize the Fisher equation, we recapture the KISS problem. The trivial steady state is thus stable for $L < \pi \sqrt{D/r}$ and unstable for $L > \pi \sqrt{D/r}$.

Analyzing the stability of the nonuniform steady state is harder. Many authors use subsolutions, supersolutions, and comparison theorems to study the stability of nonuniform steady states (Ludwig *et al.*, 1979). For the Fisher equation, a direct approach, due to Skellam (1951), also works.

The nonuniform steady state $u(x)$ satisfies equations (17.4a), and (17.4b). If we let

$$u(x, t) \equiv u(x) + \phi(x, t) \tag{17.25}$$

and consider the linearization of equation (17.3a) about this nonuniform steady state, we quickly obtain

$$\frac{\partial \phi}{\partial t} = r \left[1 - 2u(x)\right] \phi + D \frac{\partial^2 \phi}{\partial x^2}, \tag{17.26a}$$

$$\phi(0, t) = 0, \tag{17.26b}$$

$$\phi(L, t) = 0, \tag{17.26c}$$

with some appropriate initial condition. This system can, as usual, be attacked by separation of variables,

$$\phi(x, t) \equiv S(x) T(t). \tag{17.27}$$

This procedure quickly yields the temporal equation

$$\frac{dT}{dt} = -\lambda D T, \tag{17.28}$$

and the spatial equations

$$\frac{d^2 S}{dx^2} + \left\{\lambda + \frac{r}{D}\left[1 - 2u(x)\right]\right\} S = 0, \tag{17.29a}$$

$$S(0) = 0, \tag{17.29b}$$

$$S(L) = 0. \tag{17.29c}$$

(I have used $-\lambda$ as the separation constant.) By equation (17.28), all sufficiently small perturbations $\phi(x, t)$ will decay if all of the eigenvalues of the spatial system are positive.

The difficulty with equation (17.29a) is that the nonuniform steady state $u(x)$ is unknown. However, the equation and its boundary conditions do form a regular Sturm–Liouville problem. We can immediately put this fact to good use. Classical Sturm–Liouville theory implies that the spectrum of equations (17.29a), (17.29b), and (17.29c) consists of a discrete set of eigenvalues $\lambda_1 < \lambda_2 < \lambda_3 < \cdots$ and that the eigenfunction $S_1(x)$ corresponding to λ_1 does not change sign on $(0, L)$. If we now take equation (17.29a) (with $\lambda = \lambda_1$ and $S(x) = S_1(x)$), multiply by $u(x)$, and integrate by parts between 0 and L; take equation (17.4a), multiply by $S_1(x)$, and integrate by parts between 0 and L; and take the difference between these two equations, we obtain

$$\lambda_1 = \frac{\dfrac{r}{D} \displaystyle\int_0^L u^2(x) S_1(x)\, dx}{\displaystyle\int_0^L u(x) S_1(x)\, dx}. \tag{17.30}$$

Since $u(x)$ is positive away from the boundaries and since S_1 does not change sign on $(0, L)$, it follows that λ_1 (and all of the other eigenvalues) are positive. This guarantees the stability of the nonuniform steady state.

The picture that emerges then, is of a simple bifurcation at $L_c = \pi \sqrt{D/r}$, as illustrated in Figure 17.2. Below L_c the trivial steady state is stable. Above L_c, the trivial steady state is unstable and the nonuniform state is stable.

The phase-plane approach that we have considered can also be used for other boundary conditions. For example, if we modify equations (17.3a), (17.3b), (17.3c), and (17.3d) so as to incorporate Robin boundary conditions,

$$\frac{\partial u}{\partial t} = r u (1 - u) + D \frac{\partial^2 u}{\partial x^2}, \quad 0 < x < L, \tag{17.31a}$$

$$\frac{\partial u}{\partial x}(0, t) = + \alpha u(0, t), \tag{17.31b}$$

$$\frac{\partial u}{\partial x}(L, t) = - \alpha u(L, t), \tag{17.31c}$$

$$u(x, 0) = u_0(x). \tag{17.31d}$$

we are again led to equations (17.5) for the steady state, but with the ancillary boundary conditions

$$v(0) = + \alpha u(0), \tag{17.32a}$$

$$v(L) = - \alpha u(L). \tag{17.32b}$$

We must now look for orbits that start on one ray (at $x = 0$) and that reach the other ray at $x = L$ (see Figure 17.4). As α approaches zero, our boundary conditions turn into Neumann boundary conditions and our nonuniform solution approaches the homogeneous steady state $u \equiv 1$.

Other growth functions can lead to more interesting phase portraits, with more interesting bifurcation diagrams. The system

$$\frac{\partial u}{\partial t} = u (u - a) (1 - u) + \frac{\partial^2 u}{\partial x^2}, \tag{17.33a}$$

$$u(0, t) = 0, \tag{17.33b}$$

$$u(L, t) = 0, \tag{17.33c}$$

$$u(x, 0) = u_0(x). \tag{17.33d}$$

is sometimes referred to as the Nagumo equation. It may be thought of as the dimensionless version of a model with an Allee effect, simple Fickian diffusion, and homogeneous Dirichlet boundary conditions. The corresponding steady-state solutions satisfy

$$u'' + u (u - a) (1 - u) = 0. \tag{17.34}$$

This second-order ordinary differential equation may be rewritten as

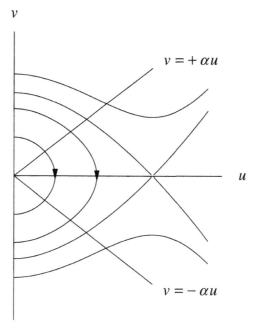

Fig. 17.4. Phase plane for Robin boundary conditions.

the system

$$u' = v, \quad v' = -u(u-a)(1-u). \tag{17.35}$$

Typical phase portraits for this system are shown in Figures 17.5, 17.6, and 17.7. The phase portraits include a saddle point at $u = 0$, a center at $u = a$, and another saddle at $u = 1$. The trivial steady state is always stable. Are there any other orbits that satisfy the boundary conditions? The answer depends on a. If, $a \geq \frac{1}{2}$, no orbit extends from the positive v-axis to the negative v-axis. In contrast, for $0 < a < \frac{1}{2}$, there are such orbits. By our previous reasoning, orbits that connect the positive and negative v-axes and that approach either saddle point correspond to large-patch solutions.

Equation (17.34) also has a first integral. One can multiply equation (17.34) by u' and integrate to produce

$$\frac{1}{2}(u')^2 + F(u) = F(\mu). \tag{17.36}$$

I have chosen $F(u)$ so that

$$\frac{dF}{du} = u(u-a)(1-u), \tag{17.37a}$$

$$F(0) = 0, \tag{17.37b}$$

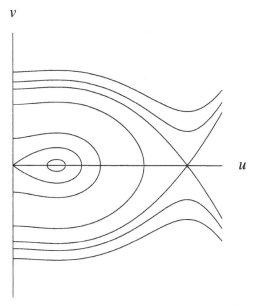

Fig. 17.5. Nagumo equation: $0 < a < \frac{1}{2}$.

and so that $u = \mu$ when $v = 0$. Let us assume that we have a solution of equation (17.34) that satisfies the boundary conditions. If we separate (17.36) and integrate over the first half of the orbit, we get

$$\frac{1}{\sqrt{2}} \int_0^\mu \frac{du}{\sqrt{F(\mu) - F(u)}} = \int_0^{L/2} dx \qquad (17.38)$$

or

$$L = \sqrt{2} \int_0^\mu \frac{du}{\sqrt{F(\mu) - F(u)}}. \qquad (17.39)$$

Smoller and Wasserman (1981) have proven that this time map has a single critical point, a minimum. One can numerically compute this integral for different values of μ and then plot μ as a function of L (see Figure 17.8).

Below the critical point, we have only the trivial steady state. For large enough L, two new nontrivial steady states emerge. I will not prove it here, but the small-amplitude nonuniform steady state is unstable and the large-amplitude steady state is stable. The outcome is similar to what we observed in Chapter 2 for harvesting with an Allee effect. Decreasing the patch size has the same effect as increasing the harvest rate in our earlier model: it leads to a saddle-node bifurcation and a population collapse.

I have concentrated on finite-domain problems. However, infinite-domain

v

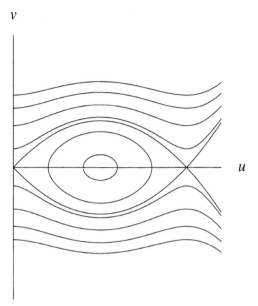

Fig. 17.6. Nagumo equation: $a = \frac{1}{2}$.

problems may also possess steady states. Consider

$$\frac{\partial u}{\partial t} = f(u) + \frac{\partial^2 u}{\partial x^2} \qquad (17.40)$$

over $-\infty < x < \infty$. The stationary solutions satisfy

$$u''(x) + f(u(x)) = 0. \qquad (17.41)$$

The corresponding first integral is

$$\frac{1}{2}(u')^2 + F(u) = c, \qquad (17.42)$$

with $F'(u) = f(u)$. Space (rather than time) is the independent variable. Even so, equation (17.41) is reminiscent of Newton's law for a conservative field. In the same way, equation (17.42) is similar to the law of conversation of energy, the fact that kinetic energy plus potential energy equals total energy. We can now fall back on our everyday intuition and imagine a marble rolling (in space rather than in time) on a hilly surface with no friction (see Figure 17.9). The 'potential energy' function, $F(u)$, determines the height of each hill. As the marble approaches the top of a hill, it loses 'kinetic energy' but gains 'potential energy'. The marble's eventual height is determined by the 'total energy' c.

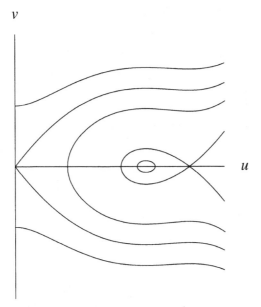

Fig. 17.7. Nagumo equation: $\frac{1}{2} < a < 1$.

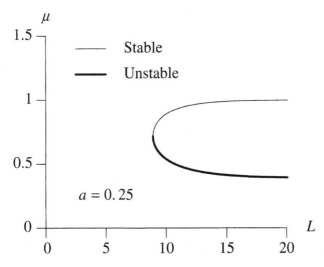

Fig. 17.8. Typical bifurcation diagram for Nagumo equation.

If you stare at this figure long enough, you will convince yourself that there are only a few types of steady-state solutions:

(A)   solutions that attain a single maximum value of $F(u)$ and for which $u \rightarrow \pm\infty$ (there is an 'or' in there) as $x \rightarrow \pm\infty$;

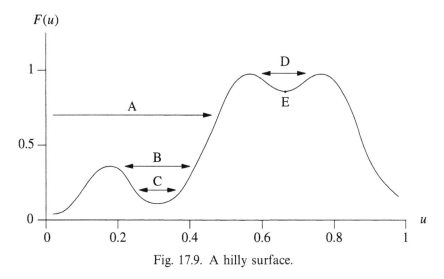

Fig. 17.9. A hilly surface.

(B) solutions that attain a single maximum value of $F(u)$ and for which $u \to K$, a constant, as $x \to \pm\infty$;

(C) solutions that are periodic in space;

(D) solutions that monotonically increase or decrease with x; and

(E) uniform solutions.

See if you can figure out what the corresponding phase-plane orbits look like. Fife (1979) discusses the stability of these different classes of steady-state solutions.

Problem 17.1 *Clinal solutions*

A cline is a gradual change in density across space. One of the simplest examples of a cline is that due to Haldane (1948). Haldane was interested in changes in gene frequency. It is easy enough, however, to give his example an ecological interpretation.

Consider a population that has a positive growth rate for positive x and a negative growth rate for negative x. In particular, let

$$\frac{\partial u}{\partial t} = +ru(1-u) + D\frac{\partial^2 u}{\partial x^2}, \quad x > 0, \tag{17.43a}$$

$$\frac{\partial u}{\partial t} = -ru(1-u) + D\frac{\partial^2 u}{\partial x^2}, \quad x < 0. \tag{17.43b}$$

(This is not quite right. Negative growth rate usually implies negative carrying capacity. However, we'll overlook this 'detail' for convenience.)

We're interested in densities that lie in the interval $0 < u < 1$. To make things easier, I'll rescale time and space,

$$\tilde{t} \equiv rt, \quad \tilde{x} \equiv \sqrt{\frac{r}{D}}\,x, \tag{17.44}$$

and drop the tildes so that equations (17.43a) and (17.43b) may be rewritten

$$\frac{\partial u}{\partial t} = +u(1-u) + \frac{\partial^2 u}{\partial x^2}, \quad x > 0, \tag{17.45a}$$

$$\frac{\partial u}{\partial t} = -u(1-u) + \frac{\partial^2 u}{\partial x^2}, \quad x < 0. \tag{17.45b}$$

This problem has boundary conditions. We expect the densities and fluxes to match at the origin,

$$u(0^-) = u(0^+), \tag{17.46a}$$

$$u'(0^-) = u'(0^+). \tag{17.46b}$$

The population should tend towards its carrying capacity for large positive x and towards 0 for large negative x,

$$\lim_{x \to +\infty} u \longrightarrow 1, \quad \lim_{x \to +\infty} u' \longrightarrow 0, \tag{17.46c}$$

$$\lim_{x \to -\infty} u \longrightarrow 0, \quad \lim_{x \to -\infty} u' \longrightarrow 0. \tag{17.46d}$$

Find and sketch the steady-state cline that satisfies the given boundary conditions.

In most of the examples that we have considered, steady states arose through a balance between growth and diffusion. However, steady states can also arise from a balance between diffusion and advection.

An advection–diffusion model is typically written

$$\frac{\partial n}{\partial t} = -\frac{\partial}{\partial x}(v\,n) + \frac{\partial}{\partial x}\left(D\,\frac{\partial n}{\partial x}\right). \tag{17.47}$$

The first term on the right-hand side corresponds to advection (to the right for positive v) while the last term accounts for diffusion. In this example, the domain is infinite.

Now, consider a population of insects of fixed size that is attracted towards the origin (Figure 17.10) (Okubo, 1980). In particular, let us assume that the attraction is advective and of constant velocity,

$$v = -v_0\,\mathrm{sgn}\,(x). \tag{17.48}$$

Fig. 17.10. Advection towards the origin.

To make life interesting, we will balance this advection with density-dependent diffusion. In particular, we will assume that the diffusivity increases with density,

$$D = D_0 n^m, \quad m > 0. \tag{17.49}$$

Substituting equations (17.48) and (17.49) into equation (17.47) produces

$$\frac{\partial n}{\partial t} = v_0 \frac{\partial}{\partial x} [\operatorname{sgn}(x) n] + D_0 \frac{\partial}{\partial x} \left(n^m \frac{\partial n}{\partial x} \right). \tag{17.50}$$

After some decent interval, a steady state may be reached. The steady state may be found by setting $\partial n/\partial t = 0$. We thus obtain

$$D_0 \frac{d}{dx} \left(n^m \frac{dn}{dx} \right) + v_0 \frac{d}{dx} [\operatorname{sgn}(x) n] = 0. \tag{17.51}$$

This steady-state equation has a simple first integral,

$$D_0 \left(n^m \frac{dn}{dx} \right) + v_0 [\operatorname{sgn}(x) n] = 0. \tag{17.52}$$

The constant of integration for this first integral has been set to zero because of our expectation that $n(\pm\infty) = 0$.

It is tempting to rewrite equation (17.52) as

$$\frac{dn}{dx} + \frac{v_0}{D_0} \operatorname{sgn}(x) n^{1-m} = 0. \tag{17.53}$$

Notice, however, that $n = 0$ is always a solution of equation (17.52), but that it is not necessarily a solution of equation (17.53).

Equation (17.52) is separable. For positive x,

$$\int_{n_0}^{n(x)} n^{m-1} dn = -\frac{v_0}{D_0} \int_0^x dx. \tag{17.54}$$

The constant n_0 is the density at the origin. (Unfortunately, we have, once again thrown out the $n = 0$ solution.) There are now two cases.

(1) *Case a*: For $m = 0$ (simple diffusion),

$$\ln \left(\frac{n}{n_0} \right) = -\frac{v_0}{D_0} x. \tag{17.55}$$

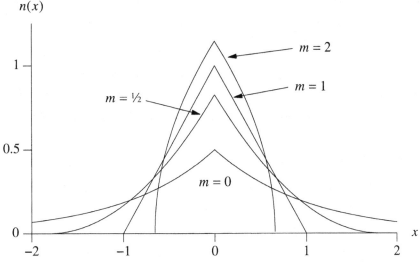

Fig. 17.11. Steady-state solutions.

Since we are looking for a solution that is symmetric about the origin, we may write

$$n(x) \;=\; n_0\, e^{-(v_0/D_0)|x|}. \tag{17.56}$$

There are no births or deaths in this model. The constant n_0 is chosen so that the total population size at time t is identical to the size at the start of the problem.

(2) *Case b*: For $m \neq 0$, quadrature produces

$$\frac{1}{m}\,[n^m(x) - n_0^m] \;=\; -\frac{v_0}{D_0}\,x \tag{17.57}$$

so that

$$n(x) \;=\; \left(n_0^m - \frac{m\,v_0}{D_0}\,x\right)^{1/m}. \tag{17.58}$$

This solution is fine to the extent that it produces positive densities. If the densities are negative, we must avail ourselves of the other solution to equation (17.54), $n = 0$. Since our solution is symmetric about the origin, we thus write

$$n(x) \;=\; \begin{cases} \left(n_0^m - \dfrac{m\,v_0}{D_0}\,|x|\right)^{1/m}, & |x| \le \dfrac{n_0^m\, D_0}{m\,v_0}, \\[2.5ex] 0, & |x| > \dfrac{n_0^m\, D_0}{m\,v_0}. \end{cases} \tag{17.59}$$

The density at the origin, n_0, is again chosen to make the population at the end of the problem the same as the population size at the start of the problem.

Our steady-state solution is a *weak* solution: the derivatives required by our partial differential equation may not exist on the boundary of the support of n. I have sketched the solution in Figure 17.11 for the special case $v_0 = D_0$ and for $m = 0, \frac{1}{2}, 1, 2$. For increasing m, the steady-state solution becomes progressively more clumped. This may seem counterintuitive: an increase in m should increase the diffusion at high densities. However, an increase in m also decreases the diffusion at low densities.

Recommended readings

Okubo's (1980) book is a classic of mathematical ecology and should be read by all aspiring mathematical ecologists. It contains many interesting examples of diffusion. Britton (1986), Grindrod (1991), and Smoller (1983) provide useful introductions to equilibria and to linear stability for scalar reaction-diffusion equations.

18 Models of spread

So far, I have emphasized steady states. Reaction-diffusion equations also allow us to model the spread of invading populations. I will begin with the analysis of two linear models for the spatial spread of a growing population. I'll then turn to some nonlinear models.

EXAMPLE   Exponential growth and spread on an infinite domain
Consider a simple model for a population with exponential growth and simple Fickian diffusion:

$$\frac{\partial n}{\partial t} = r n + D \frac{\partial^2 n}{\partial x^2}, \tag{18.1a}$$

$$n(x, 0) = f(x). \tag{18.1b}$$

The substitution

$$n(x, t) \equiv e^{rt} u(x, t) \tag{18.1c}$$

reduces this partial differential equation to the heat equation,

$$\frac{\partial u}{\partial t} = D \frac{\partial^2 u}{\partial x^2}, \tag{18.2a}$$

$$u(x, 0) = f(x). \tag{18.2b}$$

We will assume that the initial function $f(x)$ goes to zero and that the solution $u(x, t)$ is bounded in the limit of large positive or negative x. I am especially interested in the point release of organisms, where the first of these conditions is trivially satisfied.

Those of you in the know will realize that I will have to use a Fourier transform to solve this problem. Let us ease into this by separating variables in the usual manner,

$$u(x, t) = S(x) T(t), \tag{18.3}$$

so that

$$\frac{1}{D}\frac{\dot{T}}{T} = \frac{S''}{S} = -\lambda = -k^2. \tag{18.4}$$

I have set the constant in this last equation negative because we have already removed all growth from this process with our substitution. I do not expect $T(t)$ to grow with simple diffusion.

Equation (18.4) implies that

$$S''(x) + k^2 S(x) = 0. \tag{18.5}$$

The real solutions to this equation can be written in terms of sines and cosines. For convenience, I will instead use the complex exponentials e^{+ikx} and e^{-ikx}. For finite domains, we uncovered a countable infinity of eigenfunctions that satisfied the boundary conditions. Now, however, the only boundary condition is boundedness at infinity. Since all of the functions that we are considering satisfy this condition, there is no restriction on k. In addition, we do not need to distinguish between e^{+ikx} and e^{-ikx}. Since $-\infty < k < +\infty$, e^{+ikx} will suffice.

It is clear, in short, that there are countless functions of the form

$$u(x, t, k) = F(k) e^{ikx - k^2 D t} \tag{18.6}$$

that satisfy equation (18.2a). With this many functions, we need an integral,

$$u(x, t) = \int_{-\infty}^{+\infty} F(k) e^{ikx - k^2 D t} \, dk, \tag{18.7}$$

to write out the general solution. Initial condition (18.2b) must, in turn, satisfy

$$f(x) = \int_{-\infty}^{+\infty} F(k) e^{ikx} \, dk. \tag{18.8}$$

Equation (18.8) can be inverted using the Fourier inversion formula,

$$F(k) = \frac{1}{2\pi} \int_{-\infty}^{+\infty} f(x) e^{-ikx} \, dx. \tag{18.9}$$

Equations (18.8) and (18.9) together form a Fourier transform pair. Substituting equation (18.9) into solution (18.7) implies that

$$u(x, t) = \frac{1}{2\pi} \int_{-\infty}^{+\infty} e^{ikx - k^2 D t} \left(\int_{-\infty}^{+\infty} f(\xi) e^{-ik\xi} \, d\xi \right) dk \tag{18.10}$$

or that

$$u(x, t) = \frac{1}{2\pi} \int_{-\infty}^{+\infty} \int_{-\infty}^{+\infty} f(\xi) e^{-D t k^2 + ik(x - \xi)} \, d\xi \, dk. \tag{18.11}$$

Since the variable k occurs only in the exponent of this integral, it makes sense for us to reverse the order of integration and to integrate with respect to k first. To make the integral tractable, we need to complete the square in the exponent by writing

$$e^{-Dt\left[k^2 - \frac{i(x-\xi)}{Dt}k\right]} = e^{-Dt\left[k - \frac{i(x-\xi)}{2Dt}\right]^2} e^{-(x-\xi)^2/4Dt}. \qquad (18.12)$$

After this fancy piece of algebra, we get

$$u(x, t) = \frac{1}{2\pi} \int_{-\infty}^{+\infty} I\, e^{-(x-\xi)^2/4Dt} f(\xi)\, d\xi, \qquad (18.13)$$

where

$$I = \int_{-\infty}^{+\infty} e^{-Dt\left[k - \frac{i(x-\xi)}{2Dt}\right]^2} dk. \qquad (18.14)$$

Okay, what can we do with I? Well, the obvious substitution,

$$z \equiv k - \frac{i(x - \xi)}{2Dt}, \qquad (18.15)$$

gives us

$$I = \int_{-\infty + ic}^{+\infty + ic} e^{-Dtz^2} dz \qquad (18.16)$$

with

$$c = -\frac{x - \xi}{2Dt}. \qquad (18.17)$$

The integrand of I is analytical; it also tends to zero as $|Re\,(z)| \rightarrow \infty$. We can safely deform our contour back to the real axis:

$$I = \int_{-\infty}^{+\infty} e^{-Dtz^2} dz. \qquad (18.18)$$

The substitution

$$y \equiv \sqrt{Dt}\, z, \qquad (18.19)$$

now yields

$$I = \frac{1}{\sqrt{Dt}} \int_{-\infty}^{+\infty} e^{-y^2} dy = \sqrt{\frac{\pi}{Dt}}. \qquad (18.20)$$

We thus have that the solution to the heat equation is

$$u(x, t) = \frac{1}{2\sqrt{\pi Dt}} \int_{-\infty}^{+\infty} f(\xi)\, e^{-(x-\xi)^2/4Dt}\, d\xi. \qquad (18.21)$$

We may rewrite this solution

$$u(x, t) = \int_{-\infty}^{+\infty} f(\xi)\, g(x - \xi, t)\, d\xi \qquad (18.22)$$

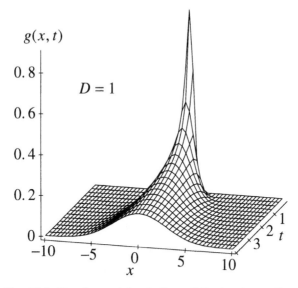

Fig. 18.1. Fundamental solution of the heat equation.

with

$$g(x - \xi, t) = \frac{1}{2\sqrt{\pi D t}} e^{-(x-\xi)^2/4D t}. \tag{18.23}$$

Equation (18.23) is known as the *fundamental solution* of the heat equation. It describes the evolution of a point release of heat. Figure 18.1 shows this function for various times. Equation (18.22), in turn, tells us that any solution can be thought of as a combination of evolving point releases.

If we return to equation (18.1a) and substitution (18.1c), we see that a point source of size n_0 will grow like

$$n(x, t) = \frac{n_0}{2\sqrt{\pi D t}} e^{rt - \frac{x^2}{4Dt}}. \tag{18.24}$$

How fast does this population spread? Let us imagine that the population is detectable when it reaches a certain threshold density n_c. If we set the density to this critical level, we can solve for

$$\frac{x}{t} = \pm\sqrt{4Dr - \frac{4D}{t} \ln\left(2\frac{n_c}{n_0}\sqrt{\pi D t}\right)}. \tag{18.25}$$

In the limit of large times,

$$\frac{x}{t} = \pm 2\sqrt{rD}. \tag{18.26}$$

The left-hand side of equations (18.25) and (18.26) can be interpreted as

Table 18.1. *Area occupied*
by muskrats

Year	Area (km²)
1905	0
1909	5 400
1911	14 400
1915	37 700
1920	79 300
1927	201 600

the average velocity of expansion. This average velocity of expansion tends towards a constant determined by the intrinsic rate of growth and the diffusivity. ◇

EXAMPLE Exponential growth and radially symmetric spread
Let us consider a population undergoing exponential growth and radial diffusion,

$$\frac{\partial n}{\partial t} = \alpha n + D \frac{1}{r} \frac{\partial}{\partial r} \left(r \frac{\partial n}{\partial r} \right), \tag{18.27}$$

where r is the radial coordinate and α is the intrinsic rate of growth.

Solving our one-dimensional growth and diffusion equation was nasty business. Let me go ahead and give you the solution to the radially symmetric, two-dimensional problem. You should, of course, verify this solution. The solution, corresponding to a point release of n_0 individuals concentrated at $r = 0$ at $t = 0$, is

$$n(r, t) = \frac{n_0}{4\pi D t} e^{\alpha t - \frac{r^2}{4Dt}}. \tag{18.28}$$

If n_c is the lowest detectable population density,

$$\frac{r}{t} = \sqrt{4\alpha D - \frac{4D}{t} \ln \left(4\pi \frac{n_c}{n_0} D t \right)}. \tag{18.29}$$

For large times, the average velocity of expansion now tends towards

$$\frac{r}{t} = 2 \sqrt{\alpha D}. \tag{18.30}$$

◇

If equations (18.26) and (18.30) are correct, invading organisms should have ranges that increase linearly in time. Skellam (1951) pointed out a historical example that obeys this law. The muskrat (*Ondatra zibethicus*) is quite common in Europe and Asia. Its range extends from Sweden and

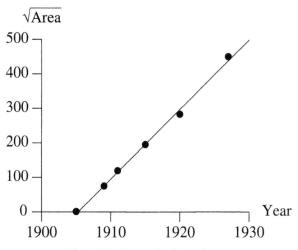

Fig. 18.2. Spread of muskrats.

France in the west to the major rivers systems of Siberia in the east. And yet, muskrats are a recent addition from the New World. The earliest introduction was in 1905 when a landowner in Bohemia accidentally released five muskrats. This initial cohort increased rapidly. (There were additional introductions later, when the value of muskrat fur had increased.) Table 18.1 shows the area (Banks, 1994) occupied by muskrats in Central Europe soon after their initial invasion.

If we plot the square root of the area occupied by muskrats as a function of time (see Figure 18.2), we see that the muskrats' range increased linearly with time. This agrees with the predictions of our first two models.

EXAMPLE   The Fisher equation
The equation

$$\frac{\partial n}{\partial t} = rn\left(1 - \frac{n}{K}\right) + D\frac{\partial^2 n}{\partial x^2} \tag{18.31}$$

was introduced by Fisher (1937) as a model for the spread of an advantageous allele. This equation has also been used as an ecological model to describe the spread of a population enjoying logistic growth and simple Fickian diffusion.

It is convenient to rescale the variables so that

$$u \equiv \frac{n}{K}, \ \tilde{t} \equiv rt, \ \tilde{x} \equiv \sqrt{\frac{r}{D}}\,x, \tag{18.32}$$

and to then drop the tildes so that

$$\frac{\partial u}{\partial t} = u(1 - u) + \frac{\partial^2 u}{\partial x^2}. \tag{18.33}$$

In earlier chapters, we looked for steady-state solutions. Since we are trying to model the spread of an organism, we will now look for a traveling wave solution. The easiest way to do so is to introduce the new variable

$$z \equiv x - ct \tag{18.34}$$

and to look for a rightward moving wave of the form

$$u(x, t) = u(x - ct) = u(z). \tag{18.35}$$

This allows us to transform our partial differential equation into an ordinary differential equation since

$$\frac{\partial u}{\partial t} = \frac{\partial u}{\partial z} \frac{\partial z}{\partial t} = -c u', \tag{18.36a}$$

$$\frac{\partial u}{\partial x} = \frac{\partial u}{\partial z} \frac{\partial z}{\partial x} = u', \tag{18.36b}$$

where prime means differentiation with respect to z. Equation (18.33) now reduces to

$$-c u' = u(1 - u) + u'' \tag{18.37}$$

or

$$u'' + c u' + u(1 - u) = 0. \tag{18.38}$$

Equation (18.38) may, in turn, be rewritten as the system

$$u' = v, \tag{18.39a}$$

$$v' = -c v - u(1 - u). \tag{18.39b}$$

This last system has two equilibria, at $(1, 0)$ and at $(0, 0)$. For a rightward moving traveling wave, we want

$$\lim_{z \to -\infty} (u, v) \to (1, 0), \tag{18.40a}$$

$$\lim_{z \to +\infty} (u, v) \to (0, 0). \tag{18.40b}$$

We must therefore find a heteroclinic connection between the two equilibria. We have every reason to believe that our solution is bounded below by zero: we are only interested in those heteroclinic orbits that remain nonnegative (in u).

Is such a nonnegative heteroclinic connection possible? Let us begin by determining the nature of the two phase-plane equilibria. For the equilibrium at $(1, 0)$, the Jacobian is

$$J = \begin{pmatrix} 0 & 1 \\ -1 + 2u & -c \end{pmatrix}_{(1, 0)} = \begin{pmatrix} 0 & 1 \\ 1 & -c \end{pmatrix} \tag{18.41}$$

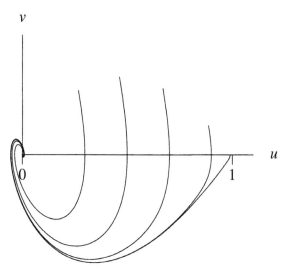

Fig. 18.3. Focus at the origin for $c < 2$.

with characteristic equation

$$\begin{vmatrix} -\lambda & 1 \\ 1 & -c-\lambda \end{vmatrix} = \lambda^2 + c\lambda - 1 = 0. \tag{18.42}$$

It is clear, using Descartes's rule of signs, that this equilibrium is a saddle point for positive c. We can also determine the stable and unstable eigenvectors for this saddle point. These are compatible with a heteroclinic connection.

For the equilibrium point at the origin, the Jacobian is

$$J = \begin{pmatrix} 0 & 1 \\ -1 & -c \end{pmatrix}. \tag{18.43}$$

The characteristic equation satisfies

$$\begin{vmatrix} -\lambda & 1 \\ -1 & -c-\lambda \end{vmatrix} = \lambda^2 + c\lambda + 1 = 0. \tag{18.44}$$

The Routh–Hurwitz criterion implies that the equilibrium at the origin is asymptotically stable. By solving the characteristic equation,

$$\lambda = \frac{-c \pm \sqrt{c^2 - 4}}{2}, \tag{18.45}$$

we see that the origin is a stable focus for $0 < c < 2$ (see Figure 18.3). For this range of c, all orbits close to the origin oscillate and we cannot find a *nonnegative* heteroclinic connection between $(1, 0)$ and $(0, 0)$.

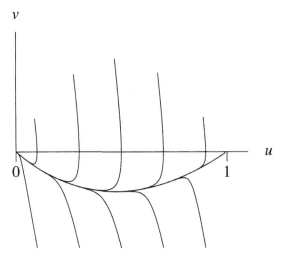

Fig. 18.4. Nonnegative heteroclinic orbit for $c > 2$.

For $c \geq 2$, the origin is no longer a stable focus. It is now a stable node and a nonnegative heteroclinic connection may now be possible (see Figure 18.4). We thus have that $c \geq 2$ is a necessary condition for a traveling wave. This provides us with a minimum wave speed for our traveling waves. In terms of our original coordinates,

$$c_{\min} = \frac{d\tilde{x}}{d\tilde{t}} = \frac{1}{\sqrt{Dr}} \frac{dx}{dt} = 2, \tag{18.46}$$

so that the speed is simply

$$\frac{dx}{dt} = 2\sqrt{Dr}. \tag{18.47}$$

We have just seen that $c \geq 2$ is a necessary condition for our heteroclinic connection. Is it also a sufficient condition? In particular, how do we know that there is an orbit that leaves $(1, 0)$ and that makes it to $(0, 0)$? Let us look at our phase portrait more carefully. Consider the triangular region in the (u, v) phase plane that I have drawn in Figure 18.5. I have labeled the three sides of the triangle I, II, and III. These three sides are defined as all or part of

(I) $\quad u = 1, v < 0$;
(II) $\quad v = 0, 0 < u < 1$;
(III) $v = -\alpha u, 0 < u < 1$.

For $c \geq 2$, there is an unstable manifold that emanates out of the equilibrium at $(1, 0)$ and into the triangular region. Let us now consider the

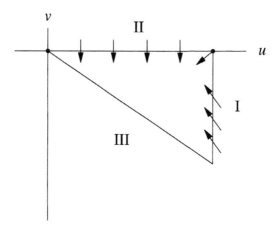

Fig. 18.5. Triangular region in the (u, v) phase plane.

behavior of the vector field on each of each edge of the triangle. On leg I,

$$u' = v < 0, \qquad (18.48a)$$
$$v' = -cv > 0, \qquad (18.48b)$$

so that the vector field points into the triangular region. On leg II, we have that

$$u' = 0, \qquad (18.49a)$$
$$v' = -u(1 - u) < 0, \qquad (18.49b)$$

so that the vector field again points into the triangle.

This leaves us with hypotenuse III. To make matters easy, I will consider the inner product between the normal to the hypotenuse (as given by the gradient of the level curve) and the vector field:

$$p = (\alpha, 1) \cdot [v, -cv - u(1 - u)] = \alpha v - cv - u(1 - u). \qquad (18.50)$$

Remember, however, that we are on the line $v = -\alpha u$. Thus

$$p = -\alpha^2 u + \alpha cu - u(1 - u) = -u[\alpha^2 - c\alpha + (1 - u)]. \qquad (18.51)$$

If this inner product is positive, the inward normal $(\alpha, 1)$ and the vector field make an acute angle and the vector field crosses into the triangular region across the hypotenuse. We are free, moreover, to choose α so that this is true. How should we choose α? Clearly, we want

$$\alpha^2 - c\alpha + 1 - u < 0. \qquad (18.52)$$

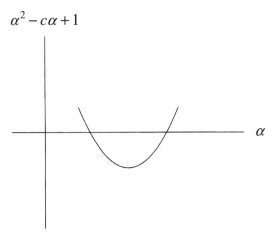

$\alpha^2 - c\alpha + 1$

α

Fig. 18.6. A desirable parabola.

Since $0 < u < 1$,

$$\alpha^2 - c\alpha + 1 - u < \alpha^2 - c\alpha + 1, \tag{18.53}$$

and we may simply choose α so that

$$\alpha^2 - c\alpha + 1 \leq 0. \tag{18.54}$$

The left-hand side of inequality (18.54) describes a parabola. We want this parabola to resemble Figure 18.6 so that we can choose a positive α that satisfies our inequality. This will occur if the parabola has two positive real roots. The roots of this parabola are

$$\alpha = \frac{c \pm \sqrt{c^2 - 4}}{2}. \tag{18.55}$$

For $c \geq 2$, the two roots are positive and real. We may thus choose an α that will force the vector field to flow into the triangle across the hypotenuse.

We now have a triangle that looks like Figure 18.7. So what? Well, let us think about that unstable separatrix that is leaving the saddle point at $(1, 0)$. Where can it go? It cannot cross any of the sides of the triangle. There is no equilibrium in the interior of the triangle. Could there be a closed orbit inside the triangle? Well, if we use Bendixon's negative criterion,

$$\frac{\partial}{\partial u}(v) + \frac{\partial}{\partial v}[-cv - u(1-u)] = -c \neq 0, \tag{18.56}$$

we see that there cannot be a closed orbit inside the triangle. The separatrix thus has no choice but to go to the origin and to make that heteroclinic connection. There must be a traveling wave solution.

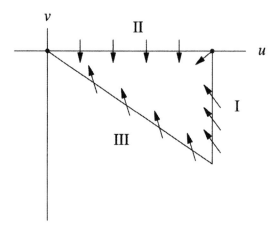

Fig. 18.7. Trapping an orbit.

We have shown that a necessary and sufficient condition for the Fisher equation to have a traveling wave solution is that $c \geq 2$. If we allow for the various changes in variables, $c = 2$ is also the speed of advance for our earlier example with exponential growth and simple Fickian diffusion. We certainly do not expect logistic growth to spread faster than exponential growth (this can be made rigorous) and so it is fair to expect that along with $c \geq 2$ we also have $c \leq 2$. In fact, one can show that, for the Fisher equation, initial conditions with compact support always lead to traveling waves with a speed of exactly $c = 2$. Initial conditions with tails that are not exponentially bounded can lead to faster waves, both for the Fisher equation and for exponential growth with simple Fickian diffusion (Murray, 1989).

It would be nice if we could also obtain an analytical representation of our traveling wave solution, both in the phase plane and in the traveling wave coordinate z. Let us begin with the heteroclinic connection in the phase plane (Canosa, 1973).

The heteroclinic orbit in Figure 18.4 satisfies

$$\frac{dv}{du} = \frac{-cv - u(1 - u)}{v}. \tag{18.57}$$

If we let

$$v \equiv \frac{1}{c} y, \tag{18.58}$$

equation (18.57) may be rewritten

$$\epsilon y \frac{dy}{du} = -y + u(u - 1), \tag{18.59}$$

where

$$\epsilon \equiv \frac{1}{c^2}. \tag{18.60}$$

Since c must be at least two for a traveling wave, we may think of ϵ as a small parameter and expand y as a power series in ϵ,

$$y(u, \epsilon) = y_0(u) + \epsilon y_1(u) + \epsilon^2 y_2(u) + \cdots. \tag{18.61}$$

If we plug this expression into equation (18.59) and match powers of ϵ, we get

ϵ^0:

$$y_0(u) = u(u - 1), \tag{18.62a}$$

ϵ^1:

$$y_1(u) = -y_0 y_0', \tag{18.62b}$$

ϵ^2:

$$y_2(u) = -y_0 y_1' - y_0' y_1, \tag{18.62c}$$

and so on, where prime now indicates differentiation with respect to u. We can easily chain our way up this sequence of equations. It follows that the heteroclinic connection takes the form

$$y(u) = u(u - 1) - \epsilon u(u - 1)(2u - 1) + \cdots. \tag{18.63}$$

This can also be written, in terms of the original u and v, as

$$v(u) = \epsilon^{1/2}(u^2 - u) - \epsilon^{3/2}(2u^3 - 3u^2 + u) + \cdots. \tag{18.64}$$

The heteroclinic connection is clearly a tangible object.

We are even more interested in the traveling wave solution in the physical plane. For this, we must return to equation (18.38). If we let

$$s \equiv \frac{z}{c}, \tag{18.65}$$

equation (18.38) takes the form

$$\epsilon u'' + u' + u(1 - u) = 0. \tag{18.66}$$

Equation (18.66) looks like a singular perturbation problem in that it has an epsilon in front of the highest-order derivative. It looks like we could have trouble satisfying boundary conditions (18.40a) and (18.40b) when we set $\epsilon = 0$, since we would be reducing the order of the equation. Luckily, the resulting first-order equation satisfies both boundary conditions; we are really dealing with a regular perturbation problem.

We may again think of ϵ as a small parameter. This time, we will expand u in powers of ϵ,

$$u(s, \epsilon) = u_0(s) + \epsilon u_1(s) + \cdots. \tag{18.67}$$

After substituting series (18.67) into equation (18.66) and matching powers of ϵ we obtain

ϵ^0:

$$u_0' = -u_0(1 - u_0), \tag{18.68a}$$

ϵ^1:

$$u_1' + (1 - 2u_0)u_1 = -u_0'', \tag{18.68b}$$

and so on. Equation (18.66) and our traveling wave solution are both invariant to translations in the independent variable s, so we may as well take $s = 0$ to be point where $u = 1/2$, for all ϵ. In other words, we will supplement our differential equations with the conditions

$$u_0 = 1/2, \tag{18.69a}$$

$$u_i = 0, \quad i = 1, 2, 3, \ldots. \tag{18.69b}$$

Equation (18.68a) with condition (18.69a) has the very simple solution

$$u_0(s) = \frac{1}{1 + e^s}. \tag{18.70}$$

This solution can now be inserted into equation (18.68b). Solving the resulting equation subject to condition (18.69b) gives us

$$u_1(s) = \frac{e^s}{(1 + e^s)^2} \ln\left[\frac{4e^s}{(1 + e^s)^2}\right]. \tag{18.71}$$

In terms of the original z variable,

$$u(z, \epsilon) = \frac{1}{1 + e^{z/c}} + \frac{1}{c^2} \frac{e^{z/c}}{(1 + e^{z/c})^2} \ln\left[\frac{4e^{z/c}}{(1 + e^{z/c})^2}\right] + O(c^{-4}). \tag{18.72}$$

This series is least accurate for $c = 2$, but numerically computed wavefronts show that the first term *alone* gives solutions that are accurate to within a few per cent even at $c = 2$.

Finally, it would be nice to know something about the stability of the traveling wave solution for the Fisher equation. I will not try to demonstrate stability for all perturbations. The traveling wave solution is, in fact, unstable to perturbations in the far field, i.e., in the limit as $|x| \to \infty$. Rather, I will follow Canosa (1973) and show that the traveling wave solution is asymptotically stable to small perturbations of compact support imposed in the moving coordinate frame of the wave.

Let me begin by rewriting the Fisher equation in a moving coordinate frame,

$$\frac{\partial u}{\partial t} = u(1 - u) + c\frac{\partial u}{\partial z} + \frac{\partial^2 u}{\partial z^2}, \tag{18.73}$$

with

$$z \equiv x - ct \tag{18.74}$$

and t as the two independent variables, and with $u(x, t) = u(z, t)$ as the (new) dependent variable. We are interested in the case $c \geq 2$. I will assume that we have found a traveling wave solution of the form

$$u(z, t) = u_c(z) \tag{18.75}$$

for some arbitrary choice of c in this range. To study the stability of this solution, I will consider a small perturbation $v(z, t)$ off $u_c(z)$,

$$u(z, t) = u_c(z) + v(z, t). \tag{18.76}$$

I will also impose the rather stringent condition that this perturbation must vanish outside some finite interval L (in the moving frame) for all time,

$$v(z, t) = 0, \quad |z| \geq L. \tag{18.77}$$

We may substitute (18.76) into equation (18.73) to obtain

$$\frac{\partial v}{\partial t} = [1 - 2u_c(z)]v - v^2 + c\frac{\partial v}{\partial z} + \frac{\partial^2 v}{\partial z^2}. \tag{18.78}$$

If we neglect higher-order terms, this simplifies to

$$\frac{\partial v}{\partial t} = [1 - 2u_c(z)]v + c\frac{\partial v}{\partial z} + \frac{\partial^2 v}{\partial z^2}. \tag{18.79}$$

I will look for a solution of the form

$$v(z, t) = g(z) e^{-\lambda t}. \tag{18.80}$$

The function $g(z)$ satisfies

$$g'' + cg' + [\lambda + 1 - 2u_c(z)]g = 0. \tag{18.81}$$

Finally, the 'well-known' Liouville–Green transformation,

$$g(z) = h(z) e^{-cz/2}, \tag{18.82}$$

allows us to eliminate the first derivative in our differential equation; $h(z)$ now satisfies

$$h'' + [\lambda - q(z)] h = 0, \tag{18.83a}$$

$$h(-L) = h(+L) = 0, \tag{18.83b}$$

with

$$q(z) = 2 u_c(z) + \frac{c^2}{4} - 1. \tag{18.84}$$

This is a regular Sturm–Liouville problem. If we can show that all of the eigenvalues of this problem are positive, we will have shown, by equation (18.80) that the traveling wave is asymptotically stable to perturbations of compact support in the traveling reference frame.

Let us imagine that $h(z)$ is an eigenfunction of equation (18.83a) for eigenvalue λ. If we multiply equation (18.83a) by $h(z)$ and integrate, we obtain

$$\int_{-L}^{+L} \left(h \frac{d^2 h}{dz^2} - q h^2 \right) dz + \lambda \int_{-L}^{+L} h^2 \, dz = 0. \tag{18.85}$$

If we now integrate the first term of the left-hand side by parts and solve for λ, we obtain the *Rayleigh quotient*,

$$\lambda = \frac{\int_{-L}^{+L} \left[(h')^2 + q h^2 \right] dz}{\int_{-L}^{+L} h^2 \, dz}. \tag{18.86}$$

Equation (18.86) is not particularly useful as a direct formula for eigenvalues, since it requires knowledge of the eigenfunctions, which in turn requires knowledge of the eigenvalues. However, for $q(z) > 0$, all eigenvalues are clearly positive. Fortunately,

$$q(z) = 2 u_c(z) + \frac{c^2}{4} - 1 \geq 2 u_c(z) > 0, \tag{18.87}$$

since $c \geq 2$. The traveling wave solution is thus stable to small perturbations of compact support in the moving coordinate frame. ◇

Problem 18.1 *An exact traveling wave*
Consider the reaction-diffusion model

$$\frac{\partial n}{\partial t} = f(n) + \frac{\partial^2 n}{\partial x^2} \tag{18.88}$$

with

$$f(n) = \begin{cases} n, & 0 \le n \le \frac{1}{2}, \\ 1 - n, & \frac{1}{2} \le n \le 1. \end{cases} \qquad (18.89)$$

Shift to traveling wave coordinates. Solve the corresponding ordinary differential equation to determine the exact shape of the traveling wave that satisfies $n(-\infty) = 1$ and $n(+\infty) = 0$.

EXAMPLE The Nagumo equation

The Nagumo or bistable equation,

$$\frac{\partial u}{\partial t} = u(u - a)(1 - u) + \frac{\partial^2 u}{\partial x^2}, \qquad (18.90)$$

is another reaction-diffusion equation that has been used to model the spread of invading organisms (Lewis and Kareiva, 1993). An organism governed by the Fisher equation does well at low densities. In contrast, an organism governed by the Nagumo equation does poorly at low densities because of a strong Allee effect. This population may still spread, however, if surplus reproduction and diffusion from regions of high density pushes the population past its threshold in regions of low density. This may or may not happen, depending on the size of the threshold value for growth. If the threshold is too high, waves of advance turn into waves of retreat. If the traveling wave solutions for the Fisher equation are 'pulled' waves, those for the Nagumo equation may be thought of as 'pushed' waves.

Despite the differences between the Fisher equation and the Nagumo equation, we may still look for a rightward moving traveling wave of the form

$$u(x, t) = u(x - ct) = u(z). \qquad (18.91)$$

Following this assumption, the Nagumo equation reduces to the second-order differential equation

$$u'' + c u' + u(u - a)(1 - u) = 0, \qquad (18.92)$$

which we may rewrite as the system

$$u' = v, \qquad (18.93a)$$

$$v' = -c v - u(u - a)(1 - u). \qquad (18.93b)$$

We are, once again, searching for a nonnegative heteroclinic orbit that

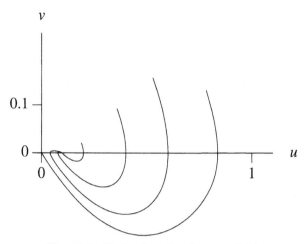

Fig. 18.8. Undershooting for $c = 0.56$.

satisfies

$$\lim_{z \to -\infty} [u(z),\ v(z)] = (1,\ 0), \tag{18.94a}$$

$$\lim_{z \to +\infty} [u(z),\ v(z)] = (0,\ 0). \tag{18.94b}$$

Is such a heteroclinic connection possible?

The phase portrait for the Nagumo equation differs from the phase portrait for the Fisher equation. There are now three equilibria in the phase plane: one at $(0, 0)$, another at $(a, 0)$, and a last equilibrium at $(1, 0)$. More importantly, $(0, 0)$ and $(1, 0)$ are both saddle points. Saddle–saddle connections are quite rare for nonconservative systems. For most values of c we may expect a separatrix emanating from one saddle point to either undershoot or overshoot the other equilibrium point. This is seen clearly in Figures 18.8 and 18.9 (for $a = 0.1$). In these figures I am shooting backwards along the separatrix from the origin. For $c = 0.56$, I undershoot the equilibrium at $(1, 0)$. For $c = 0.57$, I overshoot this equilibrium. One does not pick up an entire range of speeds c as for the Fisher equation.

To say more about the Nagumo equation, we may multiply equation (18.92) by $u'(z)$ and integrate in terms of z from minus infinity to positive infinity to get

$$c = \frac{\displaystyle\int_0^1 u\,(u - a)\,(1 - u)\ du}{\displaystyle\int_{-\infty}^{+\infty} [u'(z)]^2\ dz}. \tag{18.95}$$

If a traveling wave solution exists, the sign of the speed c is the same as the sign of the area that one gets by integrating the reaction term between

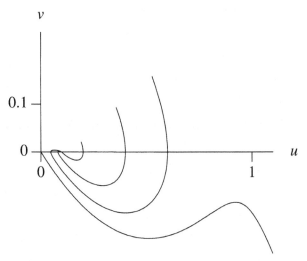

Fig. 18.9. Shooting too far for $c = 0.57$.

$u = 0$ and $u = 1$. If this area is positive, the wave advances; if it is negative, the wave retreats. For our special growth function, the wave will advance for $a < \frac{1}{2}$.

Remarkably, one can actually guess the shape of the heteroclinic connection in the traveling-coordinate phase plane for the Nagumo equation. The heteroclinic connection looks parabolic, and if one guesses that it is parabolic,

$$v = b\,u(u - 1), \tag{18.96}$$

one finds that

$$u'' = v' = b(2u - 1)\,u' = b^2 u(u - 1)(2u - 1) \tag{18.97}$$

so that equation (18.92) reduces to

$$u(u - 1)\,[b^2(2u - 1) + bc - (u - a)] = 0 \tag{18.98}$$

or

$$(bc + a - b^2) + (2b^2 - 1)\,u = 0. \tag{18.99}$$

It quickly follows that

$$b = \frac{1}{\sqrt{2}}, \quad c = \sqrt{2}\left(\frac{1}{2} - a\right). \tag{18.100}$$

Since

$$\frac{du}{dz} = \frac{1}{\sqrt{2}}u(u - 1), \tag{18.101}$$

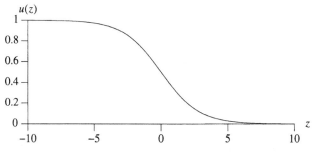

Fig. 18.10. Nagumo wave.

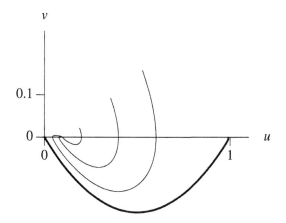

Fig. 18.11. Heteroclinic connection for $c = 0.5656854249$.

one can show that

$$u(z) = \frac{1}{2} \left[1 - \tanh \left(\frac{z}{2\sqrt{2}} \right) \right] \qquad (18.102)$$

(see Figure 18.10).

Figure 18.11 illustrates the heteroclinic connection for the Nagumo equation in the traveling-wave phase plane for $a = 0.1$. ◇

EXAMPLE The spread of rabies
Rabies is a horrible viral disease of the central nervous system. All warm-blooded animals (including birds) can be infected with rabies (Twisleton-Wykeham-Fiennes, 1978), but susceptibilities differ and we are most interested in the effects of rabies on humans and other mammals (especially carnivores). The disease itself is caused by a rod- or bullet-shaped virus of the rhabdovirus group (Shope, 1991; Patterson, 1993). The virus is present in high concentration in the saliva of rabid animals; it is transmitted to new

hosts in bite wounds. During an incubation period that may vary from days to many months, the virus multiplies in the muscle close to the site of entry. It then travels along nerves (the virus is neurotrophic) centripetally to the brain and to the rest of the central nervous system, where it causes extensive damage. There is no cure for the disease and, indeed, there are only three well-documented instances of humans recovering from rabies (Shope, 1991; Alvarez *et al.*, 1994). However, there do exist extremely effective vaccines. The earliest was introduced by Louis Pasteur in the 1880s; the most recent was released in the USA in 1988. These vaccines must be given if there is any reasonable chance of infection. The USA has only a few cases per year, but 20 000–30 000 people are treated annually with the vaccine at a cost of $400 per individual. In contrast, 20 000 cases occurred annually in India in the early 1980s and almost three million individuals were treated for animal bites each year during that time (Patterson, 1993).

Rabies occurs in many parts of the world. It is absent from Britain and Ireland, Sweden, Norway, Japan, Australia, New Zealand, and many islands of the Pacific and Caribbean (Patterson, 1993). Rabies tends to occur in irregular waves that may spread thousands of kilometres over a few decades. There was a major epizootic of rabies in Europe in the early part of the 18th century. Rabies declined in Europe in the early part of the 20th century and was eliminated from Britain in 1922. However, at the start of World War II, a fox epizootic developed along the Polish–Russian border. It has since spread westward at 30 to 60 km per year, reaching East Germany by 1948, West Germany by 1950, and France by 1968 (Murray, 1989; Shope, 1991; Patterson, 1993). Denmark has protected itself with a 'fox-free zone' north of Germany.

In the New World, the earliest reports of rabies are from Mexico in 1709, Cuba in 1719, Barbados in 1741, Virginia in 1753, North Carolina in 1762, New England in 1768, Jamaica in 1783, and Peru in 1803 (Patterson, 1993). There are more recent epizootics. An outbreak in racoons spread slowly from Florida into Georgia and South Carolina in the 1960s and 1970s. Sportsmen then transported infected individuals to Virginia and West Virginia for hunting. Between 1977 and 1982, racoon rabies spread rapidly through Washington, DC, Maryland, Virginia, and Pennsylvania (Patterson, 1993). The epizootic is now advancing towards Delaware and New Jersey at 25–50 miles (40–80 km) per year (Shope, 1991). Another epizootic, amongst skunks, has been spreading slowly from two foci in the Midwest for several decades and now extends across much of central North America (Patterson, 1993).

Monoclonal antibody studies suggest at least five rabies strains in the USA and southern Canada: the antigenic variety for racoons in the Southeast; one in skunk in the North Central States, and Manitoba; another in skunk in the south central USA; a focus in red fox in the Northeast; and a focus in gray fox in Arizona. There are several other antibody patterns in bats. In Alaska, the Arctic fox is the reservoir. In South and Central America rabies of dogs and cattle is widespread. Vampire bats are the important reservoir. It is estimated that up to a million cattle die each year of vampire-transmitted rabies in Latin America; losses are estimated at \$250 000 000 annually (Shope, 1991).

The spatial spread of rabies is a complex process. One approach to modeling this spread is to start with a simple epidemiology model and to add diffusion (Kallen *et al.*, 1985; Murray *et al.*, 1986). We imagine two groups of foxes, susceptible fox (S) and infective fox (I). Infective fox encounter susceptible fox and transmit rabies at a rate proportional to their densities. Infective fox die at a constant (proportional) rate from rabies. Susceptible fox are territorial and hence do not diffuse. If the rabies virus enters the spinal cord of an infective fox, it induces paralysis. If it enters the limbic system, it may induce aggression and a loss of sense of territory; the infective fox may wander about in a random manner. We thus include diffusion for the infective fox. The model, at this stage, does not allow for reproduction or the movement of young fox in search of territories. The model x we are considering takes the form

$$\frac{\partial S}{\partial t} = -r I S, \tag{18.103a}$$

$$\frac{\partial I}{\partial t} = r I S - a I + D \frac{\partial^2 I}{\partial x^2}. \tag{18.103b}$$

It is straightforward to nondimensionalize this system. If we let

$$\tilde{I} \equiv \frac{I}{S_0}, \quad \tilde{S} \equiv \frac{S}{S_0}, \tag{18.104a}$$

$$\tilde{x} \equiv \sqrt{\frac{r S_0}{D}}\, x, \quad \tilde{t} \equiv r S_0 t, \quad m \equiv \frac{a}{r S_0}, \tag{18.104b}$$

with S_0 the initial number of susceptibles, we obtain the dimensionless system

$$\frac{\partial S}{\partial t} = -S I, \tag{18.105a}$$

$$\frac{\partial I}{\partial t} = S I - m I + \frac{\partial^2 I}{\partial x^2}, \tag{18.105b}$$

where the tildes have been dropped for notational convenience.

We now seek a traveling wave solution to this system of the form

$$S(x, t) = S(x - ct) \equiv S(z), \tag{18.106a}$$
$$I(x, t) = I(x - ct) \equiv I(z). \tag{18.106b}$$

This leads us to the two ordinary differential equations

$$c\,S' = I\,S, \tag{18.107a}$$
$$I'' + c\,I' + (S - m)\,I = 0. \tag{18.107b}$$

We anticipate that ahead of the wave we should have the boundary conditions

$$S(+\infty) = 1, \quad I(+\infty) = 0. \tag{18.108}$$

This implies that there are no infectious fox before the epizootic. After the wave passes, we expect that there will again be no infectives (since rabies is almost always fatal) and that there will be, in addition, a constant but uncertain number of susceptibles,

$$S'(-\infty) = 0, \quad I(-\infty) = 0. \tag{18.109}$$

If we insert equation (18.107a) into (18.107b), we get

$$I'' + c\,I' + \frac{c\,S'\,(S - m)}{S} = 0. \tag{18.110}$$

Integration gives us

$$I' + c\,I + c\,S - c\,m\,\ln S = \text{constant} \tag{18.111}$$

and the boundary conditions (as z approaches infinity) imply that the constant is simply equal to c. If we now let z approach minus infinity and again apply our boundary conditions, we see that

$$S(-\infty) - m\,\ln\,S(-\infty) = 1 \tag{18.112}$$

or

$$\frac{S(-\infty) - 1}{\ln S(-\infty)} = m = \frac{a}{r\,S_0}. \tag{18.113}$$

This allows us to determine the fraction of the original susceptible fox population that survives the eipzootic wave (see Figure 18.12).

The parameter m is a measure of the severity of the epidemic, with lower values of m signifying fewer surviving susceptibles. We must have $m < 1$ for the onset of the epizootic. If $m > 1$, so that $a > r\,S_0$, the mortality rate amongst infected fox is greater than the recruitment rate of new infectives and rabies will not be able to persist. The value $m = 1$ also provides us with

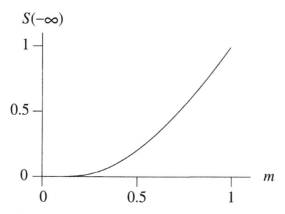

Fig. 18.12. Fraction surviving as a function of m.

a critical density of susceptible fox, $S_0 = a/r$, that must be present for an epidemic.

Equations (18.107a) and (18.111) may be combined as

$$S' = \frac{1}{c} S I, \tag{18.114a}$$

$$I' = -c(S + I) + mc \ln S + c. \tag{18.114b}$$

This system can be analyzed in the usual manner. The susceptible fox zero-growth isoclines are given by

$$S = 0 \tag{18.115}$$

and

$$I = 0. \tag{18.116}$$

The infective fox zero-growth isocline is given by

$$I = m \ln S - S + 1 \tag{18.117}$$

(see Figure 18.13).

Two equilibria are of interest. One is at $[S(-\infty), 0]$. The other is at $(1, 0)$. The Jacobian matrix is

$$J = \begin{pmatrix} \frac{I}{c} & \frac{S}{c} \\ -c + \frac{mc}{S} & -c \end{pmatrix}. \tag{18.118}$$

At $[S(-\infty), 0]$, the Jacobian matrix reduces to

$$J = \begin{pmatrix} 0 & \frac{S(-\infty)}{c} \\ -c + \frac{mc}{S(-\infty)} & -c \end{pmatrix}. \tag{18.119}$$

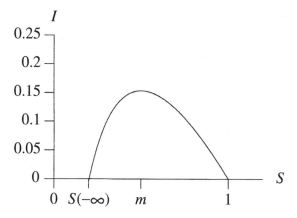

Fig. 18.13. (S, I) phase plane.

The corresponding characteristic equation,

$$\lambda^2 + c\lambda - [m - S(-\infty)] = 0, \tag{18.120}$$

has eigenvalues of opposite sign; $(S(-\infty), 0)$ is a saddle point.

At $(1, 0)$, the Jacobian matrix reduces to

$$J = \begin{pmatrix} 0 & \frac{1}{c} \\ c(m - 1) & -c \end{pmatrix}, \tag{18.121}$$

leaving us with the characteristic equation

$$\lambda^2 + c\lambda + (1 - m) = 0. \tag{18.122}$$

Thus,

$$\lambda = \frac{-c \pm \sqrt{c^2 - 4(1 - m)}}{2}. \tag{18.123}$$

For

$$c < 2\sqrt{1 - m}, \tag{18.124}$$

this equilibrium is a stable focus and there cannot be a heteroclinic connection with a nonnegative number of infectives (see Figure 18.14).

If we instead assume that

$$c \geq 2\sqrt{1 - m}, \tag{18.125}$$

the equilibrium at $(1, 0)$ is a stable node and we can have a positive heteroclinic connection (see Figure 18.15). This inequality gives us the minimum wave speed for the traveling wave.

The traveling wave that we have just discussed has a monotonically

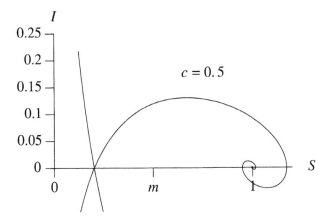

Fig. 18.14. Saddle-focus connection for $c < 2\sqrt{1-m}$.

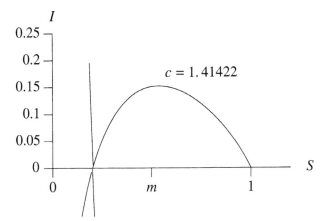

Fig. 18.15. Saddle-node connection for $c \geq 2\sqrt{1-m}$.

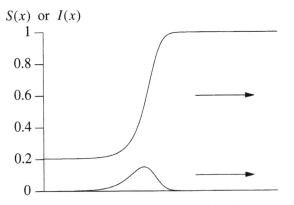

Fig. 18.16. Traveling wave of rabies.

decreasing susceptible population and a single peak in infectives (see Figure 18.16). In contrast, data show oscillations in the numbers of infectives and susceptibles behind the traveling wave. After the passage of the wavefront, the susceptible population begins to increase again. Unfortunately, equations (18.103a) and (18.103b) were intended only as a model for the initial traveling wave. Equation (18.103a) does not contain any growth terms for the susceptible fox population and cannot describe ripples. ◇

EXAMPLE   The spread of rabies with population growth

To amend this situation, we will consider a system in which the fox population grows logistically,

$$\frac{\partial S}{\partial t} = -r I S + B S \left(1 - \frac{S}{S_0} \right),$$ (18.126a)

$$\frac{\partial I}{\partial t} = r I S - a I + D \frac{\partial^2 I}{\partial x^2}.$$ (18.126b)

The change of variables

$$\tilde{I} \equiv \frac{r I}{B S_0}, \quad \tilde{S} \equiv \frac{S}{S_0},$$ (18.127a)

$$\tilde{x} \equiv \sqrt{\frac{B}{D}} \, x, \quad \tilde{t} \equiv B t,$$ (18.127b)

$$m \equiv \frac{a}{r S_0}, \quad b \equiv \frac{r S_0}{B}$$ (18.127c)

allows us to rewrite this system, omitting tildes, as

$$\frac{\partial S}{\partial t} = S (1 - S - I),$$ (18.128a)

$$\frac{\partial I}{\partial t} = b I (S - m) + \frac{\partial^2 I}{\partial x^2}.$$ (18.128b)

As usual, we will look for a traveling wave solution,

$$S(z) \equiv S(x - c t),$$ (18.129a)

$$I(z) \equiv I(x - c t).$$ (18.129b)

By assuming a solution of this form, we may reduce equations (18.128a) and (18.128b) to the ordinary differential equations

$$-c S' = S (1 - S - I),$$ (18.130a)

$$-c I' = b I (S - M) + I'',$$ (18.130b)

which we may further rewrite as a system of first-order differential equations,

$$S' = -\frac{1}{c} S (1 - S - I), \tag{18.131a}$$

$$I' = Q, \tag{18.131b}$$

$$Q' = -cQ - bI(S - m). \tag{18.131c}$$

Ahead of the wave, we expect that

$$\lim_{z \to +\infty} (S, I, Q) = (1, 0, 0). \tag{18.132}$$

Behind the wave we anticipate the coexistence of susceptibles and infectives,

$$\lim_{z \to -\infty} (S, I, Q) = (m, 1 - m, 0). \tag{18.133}$$

The Jacobian for our system of three ordinary differential equations is

$$J = \begin{pmatrix} \frac{-(1 - 2S - I)}{c} & \frac{S}{c} & 0 \\ 0 & 0 & 1 \\ bI & b(S - m) & -c \end{pmatrix}. \tag{18.134}$$

At $(1, 0, 0)$, the Jacobian reduces to

$$J = \begin{pmatrix} \frac{1}{c} & \frac{1}{c} & 0 \\ 0 & 0 & 1 \\ 0 & -b(1 - m) & -c \end{pmatrix}. \tag{18.135}$$

This has the characteristic equation

$$(\lambda - 1/c) [\lambda^2 + c\lambda + b(1 - m)] = 0 \tag{18.136}$$

and the eigenvalues

$$\lambda = \frac{1}{c}, \frac{-c \pm \sqrt{c^2 - 4b(1 - m)}}{2}. \tag{18.137}$$

We have a one-dimensional unstable manifold and two-dimensional stable manifold. Along the stable direction, $(1, 0, 0)$ may act like a node or a focus. The equilibrium $(1, 0, 0)$ needs to be saddle node for a positive heteroclinic connection. Thus,

$$c \geq 2\sqrt{b(1 - m)} \tag{18.138}$$

is a necessary condition for a traveling wave.

At $(m, 1 - m, 0)$, the Jacobian reduces to

$$J = \begin{pmatrix} \frac{m}{c} & \frac{m}{c} & 0 \\ 0 & 0 & 1 \\ -b(1 - m) & 0 & -c \end{pmatrix}. \tag{18.139}$$

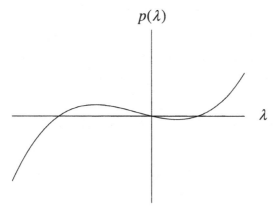

Fig. 18.17. Cubic characteristic polynomial.

The eigenvalues of this Jacobian are just the roots of the characteristic equation

$$p(\lambda) = \lambda^3 + \left(c - \frac{m}{c}\right)\lambda^2 - m\lambda + \frac{b}{c}m(1 - m) = 0. \qquad (18.140)$$

For $b = 0$, $p(\lambda)$ has three real roots,

$$\lambda = 0, \quad \frac{-\left(c - \frac{m}{c}\right) \pm \sqrt{\left(c - \frac{m}{c}\right)^2 + 4m}}{2}, \qquad (18.141)$$

and a graph that looks like Figure 18.17. The effect of increasing b is to raise the entire curve. If we do this, we go from having two positive real roots and a single negative real root (a saddle node) to having two complex roots (with positive real parts) and one negative real root (a saddle focus). Oscillations about $(m, 1 - m, 0)$ are not forbidden; as we increase b we pick up oscillations in susceptible and infectious fox behind the initial wave front. ◇

Mathematical meanderings

All of my examples have contained simple Fickian diffusion. I'd now like to look at an example with nonlinear density-dependent diffusion,

$$\frac{\partial n}{\partial t} = \frac{\partial}{\partial x}\left[D(n)\frac{\partial n}{\partial x}\right]. \qquad (18.142)$$

We considered an example with density-dependent diffusion (and advection) at the end of Chapter 17. We saw that this example gave rise to sharp boundaries in the distribution of a swarm of insects. What is the corresponding effect of density-dependent diffusion on the rate of spread of a diffusing population?

When we solved the heat equation on an infinite domain, we exploited the fact that the heat equation is a linear equation. We had separation of variables, Fourier transforms, and the superposition principle at our disposal. Now, however, we are dealing with a partial differential equation that is fundamentally nonlinear. Fortunately we will be able to convert our partial differential equation into an ordinary differential equation using a similarity method, the *method of stretchings* (see Logan, 1987, 1994), that was developed by Birkhoff (1950). This method is built on Lie's group theory of differential equations (Dresner, 1999).

Let us take our partial differential equation to be representative of a class of equations of the form

$$G(x, t, n, n_x, n_t, n_{xx}, n_{xt}, n_{tt}) = 0. \tag{18.143}$$

Now, consider a *one-parameter family of stretching transformations* of the form

$$\bar{x} = \epsilon^a x, \quad \bar{t} = \epsilon^b t, \quad \bar{n} = \epsilon^c n, \tag{18.144}$$

where a, b, and c are constants and ϵ is a real parameter that we restrict to some open interval containing $\epsilon = 1$. (The transformation that I have just described also automatically imposes a stretching transformation on the partial derivatives.) I will consider partial differential equation (18.143) to be *invariant* under this transformation if there exists a smooth function $f(\epsilon)$ such that

$$G(\bar{x}, \bar{t}, \bar{n}, \bar{n}_{\bar{x}}, \bar{n}_{\bar{t}}, \bar{n}_{\bar{x}\bar{x}}, \bar{n}_{\bar{x}\bar{t}}, \bar{n}_{\bar{t}\bar{t}}) = f(\epsilon) \, G(x, t, n, n_x, n_t, n_{xx}, n_{xt}, n_{tt}). \tag{18.145}$$

For an invariant second-order partial differential equation, the transformation[†]

$$n = t^{c/b} u(z), \quad z = \frac{x}{t^{a/b}} \tag{18.146}$$

reduces the partial differential equation to a second-order ordinary differential equation of the form

$$g(z, u, u', u'') = 0. \tag{18.147}$$

[†] Let us rewrite equation (18.143) as

$$G(x, t, n) = 0$$

for succinctness. Since this equation is invariant, we know that

$$G(\epsilon^a x, \epsilon^b t, \epsilon^c n) = 0.$$

If we differentiate this last equation with respect to ϵ and then set $\epsilon = 1$, we obtain the linear first-order partial differential equation

$$ax \frac{\partial G}{\partial x} + bt \frac{\partial G}{\partial t} + cn \frac{\partial G}{\partial n} = 0.$$

Our transformation comes from applying the method of characteristics to this first-order partial differential equation.

EXAMPLE   Nonlinear diffusion
Consider the nonlinear diffusion equation

$$\frac{\partial n}{\partial t} = \frac{\partial}{\partial x}\left(n\, \frac{\partial n}{\partial x} \right)$$

(18.148)

subject to the initial condition of a unit point release at the origin,

$$n(x,0) = \delta(x).$$

(18.149)

Since no organisms are being born or are dying, we will require that

$$\int_{-\infty}^{+\infty} n(x,\, t)\; dx = 1$$

(18.150)

for all $t > 0$. We will also assume that

$$\lim_{x \to \pm\infty} n(x,\, t) = 0$$

(18.151)

for all $t > 0$.
   Is equation (18.148) invariant? Well,

$$\frac{\partial \bar{n}}{\partial \bar{t}} - \frac{\partial}{\partial \bar{x}}\left(\bar{n}\, \frac{\partial \bar{n}}{\partial \bar{x}} \right) = \epsilon^{(c-b)}\frac{\partial n}{\partial t} - \epsilon^{(2c-2a)}\frac{\partial}{\partial x}\left(n\, \frac{\partial n}{\partial x} \right)$$

(18.152)

and we have invariance if

$$c = 2a - b.$$

(18.153)

This, in turn, dictates the similarity transformation

$$n = t^{(2a-b)/b}\, u(z), \quad z = \frac{x}{t^{a/b}}.$$

(18.154)

We can now get our ordinary differential equation. However, to simplify matters, let's avail ourselves of condition (18.150),

$$t^{(2a-b)/b}\int_{-\infty}^{+\infty} u(x/t^{a/b})\; dx = t^{(3a-b)/b}\int_{-\infty}^{+\infty} u(z)\; dz = 1.$$

(18.155)

For the left-hand side of equation (18.155) to be independent of time, we require that

$$b = 3a$$

(18.156)

so that our similarity transformation simplifies to

$$n = t^{-1/3}\, u(z), \quad z = x\, t^{-1/3}.$$

(18.157)

With this transformation, our partial differential equation reduces to

$$3\,(u u')' + u + z\, u' = 0,$$

(18.158)

which has the obvious first integral

$$3\, u u' + z\, u = \text{constant}.$$

(18.159)

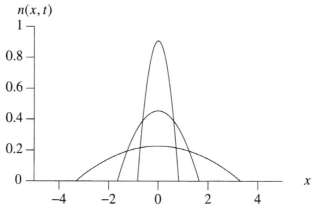

Fig. 18.18. Nonlinear diffusion.

We are interested in symmetric solutions with $u'(0) = 0$ and so we take the constant as zero,

$$3uu' + zu = 0. \tag{18.160}$$

After dividing by $3u$ and integrating, we obtain

$$u(z) = \frac{A^2 - z^2}{6}, \tag{18.161}$$

where A is a constant of integration. You should note, however, that in dividing by $3u$ we have thrown away the zero solution. Also, we expect our solution to be zero, not negative, at infinity. We will thus take our solution to be

$$u(z) = \begin{cases} \dfrac{A^2 - z^2}{6}, & |z| < A, \\ 0, & |z| > A. \end{cases} \tag{18.162}$$

The constant A may be determined from the fact that we had a unit point release and from the observation that there are no births or deaths,

$$1 = \int_{-\infty}^{+\infty} u(z)\,dz = \int_{-A}^{+A} u(z)\,dz = \frac{2}{9}A^3. \tag{18.163}$$

Thus

$$A = \left(\frac{9}{2}\right)^{\frac{1}{3}}. \tag{18.164}$$

In terms of our original coordinates, we conclude that

$$n(x,\,t) = \begin{cases} \dfrac{1}{6t}(A^2 t^{2/3} - x^2), & |x| < At^{1/3} \\ 0, & |x| > At^{1/3} \end{cases} \tag{18.165}$$

(see Figure 18.18).

This solution clearly has a sharp wave front at

$$x_f = A t^{1/3}. \tag{18.166}$$

The rate of spread is

$$\frac{dx_f}{dt} = \frac{1}{3} A t^{-2/3}; \tag{18.167}$$

density dependence causes the wave front to slow down as time increases. ◇

Problem 18.2 *The fundamental solution revisited*

Use the method of stretchings to derive the fundamental solution

$$u(x, t) = \frac{1}{2\sqrt{\pi D t}} e^{-x^2/4Dt} \tag{18.168}$$

for the heat equation.

Recommended readings

Logan (1994) provides a readable introduction to traveling-wave and similarity solutions. Consult Murray (1989) for other traveling-wave examples.

Section E
AGE-STRUCTURED MODELS

19 An overview of linear age-structured models

Age structure can have a profound effect on the growth of a population. It can, for example, lead to lags and oscillations.

The literature on age structure is often difficult to follow because of the existence of four different mathematical formulations for the equations of growth (Keyfitz, 1985). The equations (see Table 19.1) may be discrete or continuous in time and they may emphasize the birth rate or the age structure. Despite their profoundly dissimilar appearance, the four different models are all closely related. In this chapter, I will provide a quick overview of the formulation of these models and highlight some of their intrinsic similarities. I will follow this up (in subsequent chapters) with detailed analyses of each of the models.

For now, I will focus on the females in the population. We will turn to two-sex models later.

The Lotka integral equation

Lotka's model (Sharpe and Lotka, 1911; Lotka, 1939) is a continuous-time model that tracks a population's birth rate. Let

$$B(t)\, dt \equiv \text{number of female births during}$$
$$\text{the time interval } t \text{ to } t + dt. \tag{19.1}$$

Also, let

$$n(a, t)\, da \equiv \text{number of females of age } a \text{ to } a + da \text{ at time } t. \tag{19.2}$$

To determine these quantities, we must provide information about the age-specific survivorship and fertility. Let

$$l(a) \equiv \text{fraction of newborn females surviving to age } a. \tag{19.3}$$

Table 19.1. *Age-structured models*

	Births	Age structure
Continuous-time	Lotka integral equation	McKendrick–von Foerster PDE
Discrete-time	Difference equation	Leslie matrix

PDE, partial differential equation.

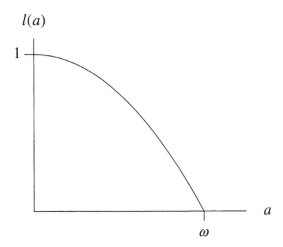

Fig. 19.1. Age-specific survivorship.

For now, we will assume that $l(a)$ is continuous and piecewise smooth. The function $l(a)$ should, of course, be nonincreasing. We will also assume that there is some maximum age of survivorship ω (see Figure 19.1).
In a similar manner, let

$$m(a)\, da \; \equiv \; \text{number of females born, on average, to a female}$$
$$\text{during the ages } a \text{ to } a + da. \qquad (19.4)$$

The age-specific rate $m(a)$ is sometimes referred to as a *maternity function*. We will assume that $m(a)$ is continuous and piecewise smooth and that there is a minimum age of reproduction α (menarche) and a maximum age of reproduction β (menopause). The maternity function $m(a)$ might look something like Figure 19.2.
Lotka's basic model for the rate of births can now be written as

$$B(t) \; = \; \int_0^t B(t - a)\, l(a)\, m(a)\, da \; + \; G(t), \qquad (19.5)$$

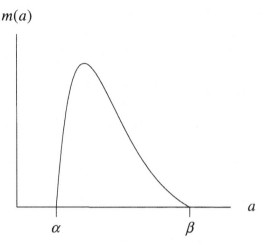

Fig. 19.2. Maternity function.

where

$$G(t) \;=\; \int_0^{\omega - t} n(a, 0)\, \frac{l(a + t)}{l(a)}\, m(a + t)\, da. \qquad (19.6)$$

The first integral in equation (19.5) describes that portion of the total birth rate attributable to females no older than t (i.e., born since $t = 0$). We take the production of females a years ago, weight those females by their odds of surviving a years and by the age-specific birth rate $m(a)$ for a-year-olds, and integrate over all relevant ages $0 \le a \le t$. $G(t)$, in turn, is the contribution from all females already present at $t = 0$. Here, we weight the density of females of age a at $t = 0$ by the probability that they will live an *additional* t years, multiply by the age-specific birth rate for $(a + t)$-year-olds, and integrate over all relevant ages. Equation (19.5) is a nonhomogeneous Volterra integral equation (of the second kind).

The difference equation

The use of difference equations and of discrete recurrence equations arguably goes back to Fibonacci in 1202. However, this approach was developed further in the 1930s, 1940s, and 1950s by Thompson (1931), Lotka (1948), and Cole (1954), among others.

By analogy with our previous discussion, we now let

$$B_t \equiv \text{number of female births at time } t, \qquad (19.7)$$

$$n_{a, t} \equiv \text{number of females of age } a \text{ at time } t, \qquad (19.8)$$

$$l_a \equiv \text{fraction of females surviving from birth to age } a, \qquad (19.9)$$

$$m_a \equiv \text{number of females born, on average,}$$

$$\text{to a female of age } a. \qquad (19.10)$$

Note that a and t are now integer-valued.

A discrete renewal equation may now be written

$$B_t = \sum_{a=1}^{t} B_{t-a} \, l_a \, m_a + G_t, \qquad (19.11)$$

where

$$G_t = \sum_{a=1}^{\omega - t} n_{a,0} \frac{l_{a+t}}{l_a} m_{a+t}. \qquad (19.12)$$

The first sum in equation (19.11) is the number of births due to the younger females (born at or since $t = 0$). G_t, in turn, is the contribution of the older females (nonneonates) from the initial population that are still reproductively active. Equation (19.11) is a nonhomogeneous difference equation.

EXAMPLE   Fibonacci's rabbits

A single newborn rabbit ($B_0 = 1$) initiates a population with the following parameters:

$$l_a = 1, \ a = 1, 2, \ldots, \qquad (19.13)$$

$$m_a = \begin{cases} 1, & a = 1, 2, \\ 0, & a > 2. \end{cases} \qquad (19.14)$$

The number of births is governed by the homogeneous second-order difference equation

$$B_t = B_{t-1} + B_{t-2}, \qquad (19.15)$$

which, in turn, generates the sequence of births (starting with $t = 0$)

$$B_t = 1, 1, 2, 3, 5, 8, \ldots. \qquad (19.16)$$

These are the well-known *Fibonacci numbers*.   ◇

The Leslie matrix

The Leslie matrix is a discrete-time model that emphasizes the age distribution (the number of females of each age) rather than the birth rate. It was introduced by Whelpton (1936) and formalized by Bernardelli (1941), Lewis (1942), and Leslie (1945, 1948).

If we start with the age distribution $n_{a,t}$ at time t, it is easy enough to project to time $t + 1$ using the notation that we have already introduced in our discussion of the discrete recurrence equation. Births occur at time $t + 1$ if females (of one or more ages) survive from time t and reproduce. These neonates also make up the youngest age class:

$$n_{0,t+1} = l_1 m_1 n_{0,t} + \frac{l_2}{l_1} m_2 n_{1,t} + \ldots + \frac{l_\omega}{l_{\omega-1}} m_\omega n_{\omega-1,t}. \tag{19.17}$$

In turn, females of older ages a (for $a > 0$) will be found at time $t + 1$ simply if females of age $a - 1$ present at time t manage to survive:

$$n_{a,t+1} = \frac{l_a}{l_{a-1}} n_{a-1,t}, \quad a = 1, 2, \ldots, \omega - 1. \tag{19.18}$$

Equations (19.17) and (19.18) are correct but somewhat unwieldy. There are several things that we can do to simplify the notation. First, we can follow Leslie (1945) and introduce year-to-year survival probabilities,

$$P_a \equiv \frac{l_{a+1}}{l_a}. \tag{19.19}$$

Secondly, we can define the *fertility* for each age class,

$$F_a \equiv P_a m_{a+1}, \tag{19.20}$$

as the number offspring produced in the next year, discounted by the survival probability for that age class. Finally, we can use the succinct vector and matrix notation offered by linear algebra. Employing all three of these remedies allows us to rewrite equations (19.17) and (19.18) as

$$
\begin{pmatrix} n_0 \\ n_1 \\ \cdot \\ \cdot \\ n_{\omega-1} \end{pmatrix}_{t+1}
=
\begin{pmatrix}
F_0 & F_1 & F_2 & \cdot & F_{\omega-1} \\
P_0 & 0 & 0 & \cdot & 0 \\
0 & P_1 & 0 & \cdot & 0 \\
\cdot & \cdot & \cdot & \cdot & \cdot \\
0 & 0 & \cdot & P_{\omega-2} & 0
\end{pmatrix}
\begin{pmatrix} n_0 \\ n_1 \\ \cdot \\ \cdot \\ n_{\omega-1} \end{pmatrix}_{t}
\tag{19.21}
$$

or, even better, as

$$\boldsymbol{n}_{t+1} = \boldsymbol{L}\boldsymbol{n}_t, \tag{19.22}$$

where $\boldsymbol{L}$ is the *Leslie matrix*.

The McKendrick–von Foerster partial differential equation

A fourth approach had its origins in the work of McKendrick (1926) and was popularized by von Foerster (1959). The McKendrick–von Foerster partial differential equation is a continuous-time model that emphasizes the age distribution.

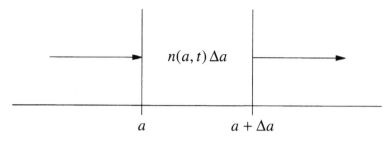

Fig. 19.3. A small cohort of females.

Once again, let

$$n(a, t)\, da \equiv \text{number of females of age } (a, a + da) \text{ at time } t. \qquad (19.23)$$

If we consider the rate of change of the number of females in a given age interval (see Figure 19.3), we may write

$$\frac{\partial}{\partial t} [n(a, t)\, \Delta a] = \left[\begin{array}{l} + \text{ rate of entry at } a \\ - \text{ rate of departure at } (a + \Delta a) \\ - \text{ deaths} \end{array} \right] \qquad (19.24)$$

or

$$\frac{\partial n}{\partial t} \Delta a = J(a, t) - J(a + \Delta a, t) - \mu(a, t)\, n(a, t)\, \Delta a, \qquad (19.25)$$

where $\mu(a, t)$ is the per capita mortality rate for individuals of age a at time t and $J(a, t)$ is the 'flux' of individuals of age a at time t. Dividing by Δa gives us

$$\frac{\partial n}{\partial t} = - \left[\frac{J(a + \Delta a, t) - J(a, t)}{\Delta a} \right] - \mu(a, t)\, n. \qquad (19.26)$$

Taking the limit, in turn, as Δa approaches zero gives us a 'conservation law' for the density of individuals:

$$\frac{\partial n}{\partial t} = -\frac{\partial J}{\partial a} - \mu(a, t)\, n. \qquad (19.27)$$

To progress any more, we must say something about the flux of individuals $J(a, t)$. This is not a flux in space but rather the movement of individuals in age. All individuals age, so we should expect this flux to be proportional to the density of individuals, with some characteristic velocity $v(a, t)$ of aging:

$$J(a, t) \equiv n(a, t)\, v(a, t). \qquad (19.28)$$

In the simplest instances, aging is just the passage of time:

$$v = \frac{da}{dt} = 1. \qquad (19.29)$$

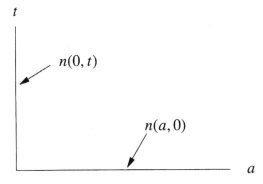

Fig. 19.4. Age and time domains.

This all seems eminently sensible. However, it should be noted that in size-structured (rather than age-structured) populations the rate at which organisms increase in size may vary with time and that equation (19.29) may no longer be correct. In any case, equation (19.27) now reads

$$\frac{\partial n}{\partial t} + \frac{\partial n}{\partial a} = -\mu(a, t)\, n. \tag{19.30}$$

Where do births enter into all this? In the McKendrick–von Foerster model, the birth rate ($B(t)$ in our discussion of the Lotka integral equation) occurs as a *boundary condition* at age zero:

$$n(0, t) = \int_0^\omega n(a, t)\, m(a, t)\, da. \tag{19.31}$$

The maternity function $m(a, t)$ is, in general, allowed to vary with time as well as with age, and ω is, as before, the maximum attainable age. To complete the model, we must also specify an initial age distribution,

$$n(a, 0) = n_0(a) \tag{19.32}$$

(see Figure 19.4).

Equation (19.30) is a first-order partial differential equation. The differential equation itself is not so bad. However, boundary condition (19.31) is rather nasty.

Similarities and differences

The four models that I have outlined may appear rather disparate. After all, we have gone from an integral equation to a difference equation, to a system of difference equation and, finally, to a partial differential equation.

However, first impressions can be deceiving. All four models have several features in common. Some of the obvious similarities are the inclusion of:

(1)   an initial age distribution,
(2)   a detailed description of age-specific mortality, and
(3)   a detailed description of age-specific natality.

There are also subtle similarities that appear only after detailed analyses. In particular, we shall see that each model (typically) possesses:

(1)   a characteristic equation,
(2)   a 'dominant' eigenvalue that provides us with the intrinsic rate of growth of the population, and
(3)   a positive right eigenvector that provides us with a stable age distribution.

Similarly, there are subtle differences between these models that make them particularly well suited to one or another application and that have led (along with historical happenstance) to the dominance of each model in one or more areas of mathematical ecology.

Recommended readings

Pollard (1973), Keyfitz (1985), and Caswell (2001) provide readable introductions to age-structured models. Smith and Keyfitz (1977) contains many classic papers.

20 The Lotka integral equation

In the previous chapter, I introduced Lotka's model for population growth. I wrote this as

$$B(t) = \int_0^t B(t - a)\, l(a)\, m(a)\, da + G(t), \tag{20.1}$$

where

$$G(t) = \int_0^{\omega - t} n(a, 0)\, \frac{l(a + t)}{l(a)}\, m(a + t)\, da, \tag{20.2}$$

$B(t)\, dt$ is the number of female births in the interval t to $t + dt$, $n(a, t)\, da$ is the number of females of age a to $a + da$ at time t, $l(a)$ is the probability that a newborn female will survive to age a, and $m(a)\, da$ is the expected number of female offspring to a female during the ages a to $a + da$. The product $l(a)\, m(a)$ is frequently referred to as the *net maternity function*.

There are two well-known approaches to solving this equation. The first, favored by Lotka, uses elementary methods, starts out quickly, but ends awkwardly. In contrast, Feller's (1941) analysis requires greater mathematical sophistication, but is ultimately more satisfying for many applied mathematicians. Let us start with Lotka's approach.

Elementary approach

We have already assumed that all reproduction takes place between menarche ($a = \alpha$) and menopause ($a = \beta$). Once the youngest of the initial females reaches menopause, nonhomogeneity $G(t)$ vanishes,

$$G(t) = 0, \ t \geq \beta, \tag{20.3}$$

and equation (20.1) simplifies to a *homogeneous* integral equation,

$$B(t) = \int_\alpha^\beta B(t - a)\, l(a)\, m(a)\, da. \tag{20.4}$$

What can we do with a linear, homogeneous, integral equation? Well, by analogy with linear, homogeneous, differential equations, I will try an exponential solution of the form

$$B(t) = Q e^{rt}. \tag{20.5}$$

Direct substitution gives us

$$\int_\alpha^\beta e^{-ra} l(a) m(a) \, da = 1. \tag{20.6}$$

This equation is called the *Euler equation*, the *Lotka equation*, or the *Euler–Lotka equation* and is the characteristic equation that I promised in Chapter 19. For convenience, I will define the function

$$\psi(r) \equiv \int_\alpha^\beta e^{-ra} l(a) m(a) \, da \tag{20.7}$$

so that the characteristic equation may be written succinctly as

$$\psi(r) = 1. \tag{20.8}$$

What can we can say about the roots of this (typically) transcendental equation? Quite a bit:

- Equation (20.6) has exactly one *real* root $r = r^*$.

To see this, note that

$$\lim_{r \to -\infty} \psi(r) = \infty, \tag{20.9}$$

and that

$$\lim_{r \to +\infty} \psi(r) = 0. \tag{20.10}$$

Moreover,

$$\frac{d\psi}{dr} = -\int_\alpha^\beta a e^{-ra} l(a) m(a) \, da < 0 \tag{20.11}$$

and (what the heck)

$$\frac{d^2\psi}{dr^2} = \int_\alpha^\beta a^2 e^{-ra} l(a) m(a) \, da > 0 \tag{20.12}$$

so that $\psi(r)$ is a continuous, strictly decreasing, concave-up function of r that takes on all positive values (see Figure 20.1). Since $\psi(r)$ is strictly decreasing, it can cross any given horizontal line in the upper half-plane, including the line $\psi(r) = 1$, just once. It follows that there is only one real solution $r = r^*$ of $\psi(r) = 1$.

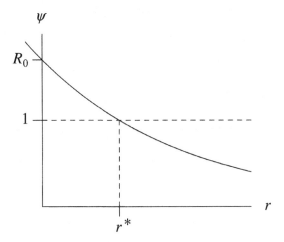

Fig. 20.1. The function $\psi(r)$.

The function $\psi(r)$ crosses the ordinate at

$$R_0 \equiv \psi(0) = \int_\alpha^\beta l(a)\,m(a)\,da. \tag{20.13}$$

R_0 is the *net reproductive rate*†. It is the lifetime reproductive potential of a female, corrected for mortality. It thus gives the average number of daughters born to a female in her lifetime. R_0 must exceed unity for r^* to be positive. If $R_0 = 1$, $r^* = 0$. Finally, if females cannot replace themselves ($R < 1$), $r^* < 0$.

- All other roots $\{r_j\}$ of equation (20.6) occur as complex conjugate pairs.

Suppose that $r_j = u + iv$ is a complex root. Since this root satisfies equation (20.6), we may write

$$\int_\alpha^\beta e^{-ua}\,[\cos(-va) + i\sin(-va)]\,l(a)\,m(a)\,da = 1. \tag{20.14}$$

Equating real and imaginary parts gives us

$$\int_\alpha^\beta e^{-ua}\cos(va)\,l(a)\,m(a)\,da = 1, \tag{20.15a}$$

$$\int_\alpha^\beta e^{-ua}\sin(va)\,l(a)\,m(a)\,da = 0. \tag{20.15b}$$

However, since these two equations are unchanged if v is replaced by $-v$,

---

† R_0 is called the net reproductive rate in ecology. Demographers, however, refer to it as the net reproduction rate.

$\bar{r}_j = u - iv$, the complex conjugate of our original root, is also a complex root of the Euler–Lotka equation.

- The real root r^* dominates all the complex roots: $r^* > \mathrm{Re}\, r_j$.

Because $\cos(va) < 1$ for some range of ages in the range of integration of equation (20.15a), we have that

$$\int_\alpha^\beta e^{-ua}\, l(a)\, m(a) > 1. \tag{20.16}$$

However,

$$\int_\alpha^\beta e^{-r^*a}\, l(a)\, m(a) = 1. \tag{20.17}$$

A direct comparison of these two integrals implies that $u < r^*$ or that $r^* > \mathrm{Re}\, r_j$.

The real root r^* and the complex roots r_j each give rise to solutions of the form (20.5). Since integral equation (20.4) is linear, you might expect that its general solution is merely the linear combination of these simpler solutions,

$$B(t) = Q^* e^{r^*t} + \sum_j Q_j e^{r_j t}, \quad t \geq \beta. \tag{20.18}$$

This is certainly so when all of the roots are simple. However, nothing prohibits the complex roots from having multiplicity greater than 1. A complex root r_j of multiplicity 2, say, can arise when

$$\psi(r_j) = 1, \quad \psi'(r_j) = 0. \tag{20.19}$$

If some of the roots do have multiplicity greater than one, the general solution will take the form

$$B(t) = Q^* e^{r^*t} + \sum_j \sum_{k=1}^{\gamma(j)} Q_{jk}\, t^{k-1} e^{r_j t}, \quad t \geq \beta, \tag{20.20}$$

where $\gamma(j)$ is the multiplicity of the j-th complex root.†

The complex roots at first lead to oscillations in the birth rate. Asymptotically, however, the birth rate is governed by the dominant real eigenvalue. Indeed, equation (20.18) may be rewritten

$$B(t) = Q^* e^{r^*t} \left[1 + \sum_j \frac{Q_j}{Q^*} e^{(r_j - r^*)t} \right], \tag{20.21}$$

---

† We are writing $B(t)$ as an *infinite* series. Hadwiger (1940) has shown that the continuous Euler–Lotka equation has infinitely many roots when the fertile age interval is finite. See Lopez (1961) for details.

whence

$$\lim_{t \to \infty} B(t) \sim Q^* e^{r^* t}. \tag{20.22}$$

After the death of the founder females, the total population is given by

$$N(t) = \int_0^\omega B(t - a) l(a) \, da \tag{20.23}$$

so that,

$$\lim_{t \to \infty} N(t) \sim Q^* e^{r^* t} \int_0^\omega e^{-r^* a} l(a) \, da. \tag{20.24}$$

Since the last integral is a constant, the population will eventually grow (or decay) exponentially with rate r^*.

Similarly, the number of females in some small age range is given by

$$n(a, t) \, da = B(t - a) l(a) \, da \tag{20.25}$$

so that,

$$\lim_{t \to \infty} n(a, t) \, da \sim Q e^{r^* t} [e^{-r^* a} l(a)] \, da. \tag{20.26}$$

Eventually, the number of females in some small age range will also grow (or decay) exponentially with rate r^*. The practical import of this can be seen by noting that the fraction of the population in some small age range can be written

$$c(a, t) \equiv \frac{n(a, t) \, da}{N(t)} \tag{20.27}$$

and that

$$c^*(a) \equiv \lim_{t \to \infty} c(a, t) = \frac{e^{-r^* a} l(a) \, da}{\int_0^\omega e^{-r^* a} l(a) \, da} \tag{20.28}$$

is asymptotically independent of t for large t. In other words, the population tends, asymptotically, towards a *stable age distribution*.

Equation (20.28) can be put another way. Asymptotically, the relative proportion of females of age a to newborns is

$$\frac{c^*(a)}{c^*(0)} = e^{-r^* a} l(a). \tag{20.29}$$

For a stationary population ($r^* = 0$), this ratio is simply the age-specific survivorship. For $r^* > 0$ (< 0), the age structure has an excess of young (old) females.

Problem 20.1 *The growth of Mexico* (Keyfitz and Beekman, 1984)
In the 1960 census of Mexico, the count of females of age 25–29 years was
1 314 000 while the count of females of age 50–54 was 538 000. A life table
that tracked the survivorship of a large cohort of females showed 408 000
and 351 000 for those two ages. Assume that the Mexican population was
growing stably. Use equation (20.29) to estimate the intrinsic rate of growth
r^* of the population. What is the doubling time of the population?

Problem 20.2 *The burden of bigamy*
In a hypothetical community, men marry at age 40 years, women marry at
age 20, and each man aspires to having two wives. Is such a society possible?
Assume that the population is growing stably, that the sex ratio at birth is
unity, that male and female survivorships are equal, and that all individuals
experience a uniform death rate of 1.5% at all ages, $l(a) = e^{-0.015a}$. How
fast does the community have to increase for each man to have two wives?

Until now, our elementary approach has brought us everything we could
hope for, rather painlessly. We have found the Euler–Lotka equation, the
asymptotic rate of growth, and the stable age distribution. What more could
we want? Well we do not have the solution for $t < \beta$. Moreover, we do not
really have the solution for $t \geq \beta$, since we have yet to evaluate Q^* and
the Q_i. These constants can be evaluated by elementary methods (Lotka,
1939; Keyfitz, 1968; Frauenthal, 1986), but the procedure is both tedious
and unintuitive. Rather, it makes sense to follow the lead of Feller (1941)
and to avail ourselves of the Laplace transform.

Solution *à la* Laplace

Given a function $f(t)$, we will write its Laplace transform as

$$\hat{f}(r) = \mathscr{L}[f(t)] = \int_0^\infty e^{-rt} f(t)\, dt. \tag{20.30}$$

Here r can be either a real variable or a complex quantity. Taking a Laplace
transform may be thought of as moving from a t space, where a problem
is difficult to solve, to an r space, where it is easy (see Figure 20.2). Of
course, not all functions have Laplace transforms. However, if a function is

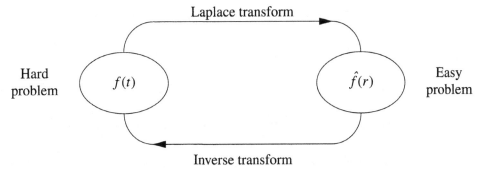

Fig. 20.2. Laplace transform pair.

piecewise continuous and grows no faster than a simple exponential it will
be Laplace transformable.

For example, let

$$f(t) = e^{at}. \tag{20.31}$$

Then

$$\hat{f}(r) = \int_0^\infty e^{-rt} e^{at} \, dt = \int_0^\infty e^{-(r-a)t} \, dt,$$

$$= \frac{-1}{(r-a)} e^{-(r-a)t} \Big|_0^\infty,$$

$$= \frac{1}{(r-a)}, \quad \text{Re}(r) > a. \tag{20.32}$$

The integral transform diverges for $\text{Re}(r) \le a$ and converges for $\text{Re}(r) > a$.
Such restrictions are important in developing the theory of Laplace transforms, but are of little consequence in most applications. We will be quite
happy in knowing that the transform exists for *some* values of r. Note that
equation (20.32) also implies that the inverse transform of $1/(r-a)$ is e^{at}.

Before moving on, let us look at one more example. Renewal equation
(20.1) contains a convolution of the birth rate and of the net maternity
function. Let us look at the Laplace transform of a convolution integral.

$$\mathscr{L}[f * g(t)] \equiv \mathscr{L}\left[\int_0^t f(t-a) g(a) \, da\right]$$

$$= \int_0^\infty e^{-rt} \int_0^t f(t-a) g(a) \, da \, dt$$

$$= \int_0^\infty \int_0^t e^{-rt} f(t-a) g(a) \, da \, dt. \tag{20.33}$$

We can interpret this as an iterated integral over the region $0 \le a \le t$,

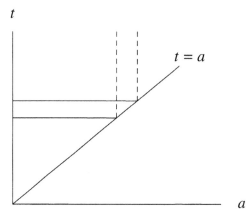

Fig. 20.3. Domain of integration.

$0 \le t < \infty$. If we change the order of integration (see Figure 20.3), we find that the region of integration is $a \le t < \infty, 0 \le a < \infty$,

$$\mathscr{L}\,[f * g(t)] \;=\; \int_0^\infty \int_a^\infty e^{-rt} f(t - a)\, g(a)\, dt\, da. \qquad (20.34)$$

The change of variables $x \equiv t - a$ (or $t = x + a$) then leads to

$$
\begin{aligned}
\mathscr{L}\,[f * g(t)] \;&=\; \int_0^\infty \int_0^\infty e^{-r(x+a)} f(x)\, g(a)\, dx\, da \\
&=\; \left(\int_0^\infty e^{-rx} f(x)\, dx \right) \left(\int_0^\infty e^{-ra} g(a)\, da \right) \\
&=\; \hat{f}(r)\, \hat{g}(r). \qquad (20.35)
\end{aligned}
$$

In other words, the Laplace transform of a convolution is simply the product of two Laplace transforms. We are now ready to solve Lotka's model using the Laplace transform.

Let $\phi(a)$ represent the net maternity function. Upon taking the Laplace transform of each side of equation (20.1), we obtain

$$\hat{B}(r) \;=\; \hat{B}(r)\, \hat{\phi}(r) + \hat{G}(r). \qquad (20.36)$$

It is now easy to solve for the transformed birth rate. A little bit of algebra yields

$$\hat{B}(r) \;=\; \frac{\hat{G}(r)}{1 - \hat{\phi}(r)}. \qquad (20.37)$$

I have transformed our equation and solved for the transformed birth rate. This was the easy part. Now comes the tough part. I must determine the inverse transform of the transformed birth rate. Note, however, that

$$1 - \hat{\phi}(r) = 1 - \int_0^\infty e^{-ra} l(a) m(a) \, da, \tag{20.38}$$

and that the roots of equation (20.38) are the same as those of Euler–Lotka characteristic equation. Let us assume, for the moment, that all of these roots are simple. It may well be that the transformed birth rate has a partial fractions expansion in terms of the roots of the Euler–Lotka equation:

$$\hat{B}(r) = \frac{\hat{G}(r)}{1 - \hat{\phi}(r)} = \frac{Q^*}{r - r^*} + \sum_j \frac{Q_j}{r - r_j}. \tag{20.39}$$

In this case,

$$B(t) = Q^* e^{r^* t} + \sum_j Q_j e^{r_j t}. \tag{20.40}$$

(Remember that the inverse transform of $1/(r - r_j)$ is simply $e^{r_j t}$.) Moreover, by the usual method for finding the coefficients of a partial fraction expansion,

$$Q^* = \lim_{r \to r^*} \left[\frac{(r - r^*) \hat{G}(r)}{1 - \hat{\phi}(r)} \right] = \left[\frac{\hat{G}(r)}{-d\hat{\phi}/dr} \right]_{r = r^*}$$

$$= \frac{\int_0^\beta e^{-r^* t} G(t) \, dt}{\int_\alpha^\beta a e^{-r^* a} l(a) m(a) \, da}. \tag{20.41}$$

Similarly,

$$Q_j = \frac{\int_0^\beta e^{-r_j t} G(t) \, dt}{\int_\alpha^\beta a e^{-r_j a} l(a) m(a) \, da}. \tag{20.42}$$

If you have taken complex variables, you probably know that the inverse Laplace transform can also be expressed directly as the complex integral

$$B(t) = \mathcal{L}^{-1}[\hat{B}(r)] = \frac{1}{2\pi i} \int_{c - i\infty}^{c + i\infty} e^{rt} \hat{B}(r) \, dr \tag{20.43}$$

with, in our case, $c > r^*$. The usual method of evaluating such a complex integral is to close the contour off in the left half-plane and to use the residue theorem. In this case, equations (20.41) and (20.42) are simply the residues from the simple poles of $\hat{B}(r)$. How would you handle a second-order pole (corresponding to a double root)?

EXAMPLE The momentum of population growth
Consider an idealized population that possesses a net reproductive rate in excess of unity,

$$R_0 = \int_0^\infty l(a) m(a) \, da > 1, \tag{20.44}$$

and that grows in a stable manner,

$$B(t) = Q^* e^{r^* t}, \tag{20.45}$$

$$N(t) = Q^* e^{r^* t} \int_0^\omega e^{-r^* a} l(a) \, da. \tag{20.46}$$

Let us imagine that one night (at exactly $t = 0$) all of the females abruptly reduce their reproduction, in a manner ultimately consistent with zero population growth,

$$\bar{m}(a) = \frac{1}{R_0} m(a). \tag{20.47}$$

The new net reproductive rate is simply

$$\bar{R}_0 = \int_0^\infty l(a) \frac{m(a)}{R_0} \, da = 1. \tag{20.48}$$

Eventually,

$$\bar{r}^* = 0, \tag{20.49}$$

$$\bar{B}(t) = \bar{Q}^* e^{\bar{r}^* t} = \bar{Q}^*, \tag{20.50}$$

$$\bar{N}(t) = \bar{Q}^* \int_0^\omega l(a) \, da = N_\infty. \tag{20.51}$$

However, there is a transient. It takes a while for the other components of the solution to die out and for the shift to the new stable age distribution. During this time, the population may continue to grow. What then is the relationship between Q^* and $\bar{Q}^*$?

We spent some time deriving equation (20.41), so let us use it.

$$\bar{Q}^* = \frac{\int_0^\beta e^{-\bar{r}^* t} \bar{G}(t) \, dt}{\int_\alpha^\beta a e^{-\bar{r}^* a} l(a) \bar{m}(a) \, da} = \frac{1}{\mu} \int_0^\beta \bar{G}(t) \, dt, \tag{20.52}$$

where

$$\mu \equiv \int_\alpha^\beta a l(a) \bar{m}(a) \, da = \frac{\int_\alpha^\beta a l(a) m(a) \, da}{\int_\alpha^\beta l(a) m(a) \, da}. \tag{20.53}$$

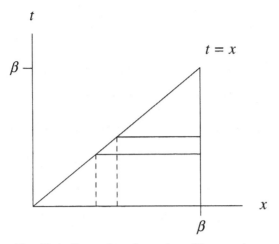

Fig. 20.4. Reversing the order of integration.

The constant μ has a simple interpretation. It is simply the average age of childbearing in the population (both before and after the change in the maternity function). Okay, how about $\bar{G}(t)$? That is simply

$$\bar{G}(t) = \int_0^{\omega-t} n(a, 0) \frac{l(a + t)}{l(a)} \bar{m}(a + t) \, da. \tag{20.54}$$

And $n(a, 0)$? Well, initially the females are arrayed along the stable age distribution for the old growth rate,

$$n(a, 0) = Q^* e^{-r^*a} l(a). \tag{20.55}$$

We can combine equations (20.52), (20.54), and (20.55) to get

$$\bar{Q}^* = \frac{Q^*}{\mu} \int_0^{\beta} \int_0^{\omega-t} e^{-r^*a} l(a + t) \bar{m}(a + t) \, da \, dt, \tag{20.56}$$

or

$$\frac{\bar{Q}^*}{Q^*} = \frac{1}{\mu R_0} \int_0^{\beta} \int_0^{\omega-t} e^{-r^*a} l(a + t) m(a + t) \, da \, dt. \tag{20.57}$$

Letting $x \equiv a + t$ in the inner integral gives us

$$\frac{\bar{Q}^*}{Q^*} = \frac{1}{\mu R_0} \int_0^{\beta} \int_t^{\omega} e^{-r^*(x-t)} l(x) m(x) \, dx \, dt. \tag{20.58}$$

To get any further, we must reverse the order of integration (see Figure 20.4; I have assumed that $\beta < \omega$):

$$\frac{\bar{Q}^*}{Q^*} = \frac{1}{\mu R_0} \int_0^{\beta} l(x) m(x) e^{-r^*x} \left(\int_0^x e^{r^*t} dt \right) dx, \tag{20.59}$$

$$\frac{\bar{Q}^*}{Q^*} = \frac{1}{\mu R_0 r^*} \int_0^\beta l(x) m(x) (1 - e^{-r^*x}) dx, \tag{20.60}$$

$$\frac{\bar{Q}^*}{Q^*} = \frac{R_0 - 1}{\mu R_0 r^*}. \tag{20.61}$$

If we look at Ecuador in 1965 as an example (Keyfitz and Beekman, 1984), $R_0 = 2.59$, $r^* = 0.0331$, and $\mu = 29.41$ so that expression (19.65) is 0.63, even though $1/R_0 = 0.386$.

The ratio of the ultimate to the initial population size is given by

$$\frac{N_\infty}{N_0} = \frac{\bar{Q}^* \int_0^\omega l(a) da}{Q^* \int_0^\omega e^{-r^*a} l(a) da} = \frac{R_0 - 1}{\mu R_0 r^*} b \mathring{e}, \tag{20.62}$$

where the integral in the numerator is the 'expectation of life at birth', $\mathring{e}$, and the integral in the denominator is the reciprocal of the intrinsic birth rate b. All of the quantities in equation (20.62) refer to the population prior to the shift in maternity. Returning again to Ecuador in 1965, $\mathring{e} = 60.16$ and $b = 0.0482$ so that expression (20.62) reduces to 1.69, corresponding to a 69% increase in the total population after a sudden shift in maternity. Comparable data for the USA from 1966 suggest an increase of about a third (Keyfitz, 1971; Pollard, 1973). ◇

The Lotka integral equation is a rigorous framework for studying age-structured population growth. The real root of the Euler–Lotka equation can be found with great precision, thereby allowing accurate estimates of the asymptotic rate of growth. However, finding the complex roots, the roots that describe oscillations, is anything but pleasant. Fortunately, this is less of a problem with the difference equation formulation that we are about to consider.

Recommended readings

I first learned this material from Frauenthal (1986). This chapter is modeled after and borrows heavily from Frauenthal (1986) and from the earlier book by Keyfitz (1968).

21 The difference equation

In Chapter 19, I introduced the model

$$B_t = \sum_{a=1}^{t} B_{t-a} l_a m_a + G_t, \qquad (21.1)$$

where

$$G_t = \sum_{a=1}^{\omega-t} n_{a,0} \frac{l_{a+t}}{l_a} m_{a+t}, \qquad (21.2)$$

B_t is the number of female births at time t, $n_{a,t}$ is the number of females of age a at time t, l_a is the fraction of females surviving from birth to age a, and m_a is the number of females born, on average, to a female of age a.

Equation (21.1) is a difference equation rather than an integral equation, but the parallels between this difference equation and the Lotka integral equation are striking. I will begin, therefore, by trying to reproduce the main results of Chapter 20. This raises some interesting mathematical issues. What, for example, is the discrete-time analog of a Laplace transform? While retracing our steps, we will discover subtle differences between this difference equation and the Lotka integral equation.

Elementary approach

For large times, nonhomogeneity G_t vanishes,

$$G_t = 0, \quad t \geq \beta, \qquad (21.3)$$

and we are left with the homogeneous difference equation

$$B_t = \sum_{a=\alpha}^{\beta} B_{t-a} l_a m_a. \qquad (21.4)$$

Since this is a linear difference equation, we will try a solution of the form

$$B_t = c \lambda^t. \tag{21.5}$$

Direct substitution into equation (21.4) gives us

$$\sum_{a=\alpha}^{\beta} \lambda^{-a} l_a m_a = 1. \tag{21.6}$$

It should come as no surprise that this equation is named the Euler equation, the Lotka equation, or the Euler–Lotka equation. This equation allows us to determine the rate of growth for a population.

EXAMPLE The Northern Spotted Owl (Lande, 1988; Bulmer, 1994)
The Spotted Owl (*Strix occidentalis*) has the following population parameters:

$$m_a = \begin{cases} 0, & a < 3, \\ m, & a \geq 3, \end{cases} \tag{21.7}$$

$$m = 0.24, \tag{21.8}$$

$$l_3 = 0.0722, \tag{21.9}$$

$$P = \frac{l_{a+1}}{l_a} = 0.942, \quad a \geq 3. \tag{21.10}$$

This looks like a forbidding problem. There is an infinite number of reproductive age classes! However, it is not really so bad. All we need to do is avail ourselves of the binomial expansion:

$$\sum_{a=3}^{\infty} \lambda^{-a} l_a m_a = \frac{l_3 m}{\lambda^3} + \frac{l_3 P m}{\lambda^4} + \frac{l_3 P^2 m}{\lambda^5} + \cdots = 1, \tag{21.11}$$

$$\frac{l_3 m}{\lambda^3} \sum_{a=0}^{\infty} \left(\frac{P}{\lambda}\right)^a = 1, \tag{21.12}$$

$$\frac{l_3 m}{\lambda^3} \frac{1}{1 - (P/\lambda)} = 1, \tag{21.13}$$

$$\lambda^3 - P \lambda^2 - l_3 m = 0. \tag{21.14}$$

Thus,

$$\lambda_0 = 0.96, \quad \lambda_{1,2} = -0.009 \pm 0.134 \, i. \tag{21.15}$$

We are after the first root. It is the only one that is bigger than P in magnitude. It is also close enough to unity to make one believe that there is no great decline within the owl's habitat. Loss of habitat, however, is another matter.

The sensitivity of λ with respect to any of the demographic parameters may be found by implicit differentiation of equation (21.14). For example, if I designate the left-hand side of equation (21.14) as $f(\lambda, P, l_3, m)$ and think of λ as the dependent variable and the three demographic parameters as three independent variables, then

$$\frac{\partial f}{\partial P} + \frac{\partial f}{\partial \lambda} \frac{\partial \lambda}{\partial P} = 0, \tag{21.16}$$

so that

$$\frac{\partial \lambda}{\partial P} = -\frac{\partial f / \partial P}{\partial f / \partial \lambda} = \frac{\lambda^2}{3\lambda^2 - 2P\lambda} = 0.962 \tag{21.17}$$

(for $\lambda = \lambda_0$). Similarly,

$$\frac{\partial \lambda}{\partial l_3} = 0.251, \tag{21.18}$$

and

$$\frac{\partial \lambda}{\partial m} = 0.075. \tag{21.19}$$

The geometric growth rate of the population is most sensitive to changes in adult annual survival, less so to survival to breeding age, and only lastly to average reproductive rate. ◇

What can we say *in general* about the roots of equation (21.6)? If you read the last section, you can undoubtedly guess some of the answers.

- Equation (21.6) has exactly one positive real root, $\lambda = \lambda_0$, of algebraic multiplicity one.

For convenience, let us rewrite equation (21.6) as

$$\psi(\lambda) = \frac{l_\alpha m_\alpha}{\lambda^\alpha} + \frac{l_{\alpha+1} m_{\alpha+1}}{\lambda^{\alpha+1}} + \cdots + \frac{l_\beta m_\beta}{\lambda^\beta} = 1. \tag{21.20}$$

Since all of the coefficients are nonnegative, and all of the terms are connected by plus signs,

$$\lim_{\lambda \to 0^+} \psi(\lambda) = \infty, \tag{21.21}$$

and

$$\lim_{\lambda \to \infty} \psi(\lambda) = 0. \tag{21.22}$$

It is also clear that $\psi(\lambda)$ is a strictly decreasing function since

$$\frac{d\psi}{d\lambda} = -\sum_{a=\alpha}^{\beta} a\lambda^{-a-1} l_a m_a < 0. \tag{21.23}$$

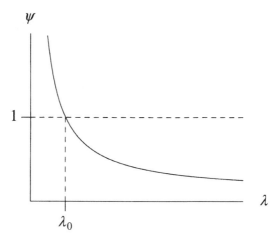

Fig. 21.1. Plot of $\psi\,(r)$.

Since $\psi(\lambda)$ is continuous and strictly decreasing (see Figure 21.1), it can cross any horizontal line in the upper half-plane, including the line $\psi(r) = 1$, just once. It follows that there is only one positive real solution $\lambda = \lambda_0$ of $\psi(r) = 1$.

All other roots λ_j are either negative or complex. If these roots are all simple, we may write the general solution of equation (21.4) as

$$B_t = c_0 \lambda_0^t + \sum_{j=1} c_j \lambda_j^t, \ t \geq \beta. \tag{21.24}$$

Now, I know what you are thinking. You are thinking that the positive real root must dominate all the negative and complex roots, by analogy with what happened in Chapter 20. However, life is not so generous. We must settle for a weaker result.

For a given maternity schedule m_a, let d be the greatest common divisor of the ages of positive reproduction. If $d > 1$, we say that the maternity function is *periodic* with period d. Otherwise, m_a is *aperiodic*. It then follows that:

- No other root λ_j can be greater than λ_0 in modulus. However, if the maternity function is periodic with period d, there will be $d - 1$ other eigenvalues with the same modulus as λ_0.

To see this, let us suppose that

$$\lambda_j = |\lambda_j| e^{i\theta}, \ \theta \neq 2n\pi, \tag{21.25}$$

is a negative or complex root of the discrete Euler–Lotka equation. Since this root satisfies equation (21.6), we may write

$$\sum_{a=\alpha}^{\beta} |\lambda_j|^{-a} e^{-ia\theta} l_a m_a = 1. \tag{21.26}$$

If we equate the real and the imaginary parts of both sides of equation (21.26), we get

$$\sum_{a=\alpha}^{\beta} |\lambda_j|^{-a} \cos(a\theta) l_a m_a = 1, \tag{21.27a}$$

$$\sum_{a=\alpha}^{\beta} |\lambda_j|^{-a} \sin(a\theta) l_a m_a = 0. \tag{21.27b}$$

If the maternity function is periodic with period d, we can satisfy these equations with $|\lambda_j| = \lambda_0$ and $\theta = 2\pi n/d$, $n = 1, 2, \ldots, d$ (i.e., with the d complex roots of λ_0) since we will then have $\cos(a\theta) = 1$ and $\sin(a\theta) = 0$ whenever $m_a \neq 0$. The last root is just λ_0. The remaining $d - 1$ roots are complex or negative. All d roots have the same modulus.

If the maternity schedule is aperiodic, life is simpler. There is then a range of ages for which $\cos(a\theta) < 1$. Hence,

$$\sum_{a=\alpha}^{\beta} |\lambda_j|^{-a} l_a m_a > 1. \tag{21.28}$$

However,

$$\sum_{a=\alpha}^{\beta} \lambda_0^{-a} l_a m_a = 1. \tag{21.29}$$

A direct comparison of these two summations implies that $|\lambda_j| < \lambda_0$.

EXAMPLE   Semelparous reproduction (Bernardelli, 1941)

$$m_a = \begin{cases} 0, & a = 1, \\ 0, & a = 2, \\ 6, & a = 3, \end{cases} \quad l_a = \begin{cases} 1, & a = 1, \\ 1/2, & a = 2, \\ 1/6, & a = 3. \end{cases} \tag{21.30}$$

All reproduction occurs in year 3. The greatest common divisor of 3 is, of course, 3. So, the maternity function is periodic with period 3. The Euler–Lotka equation now reduces to

$$\frac{1}{\lambda^3} = 1, \tag{21.31}$$

which has the three roots

$$\lambda_0 = 1, \quad \lambda_{1,2} = -\frac{1}{2} \pm \frac{\sqrt{3}}{2} i. \tag{21.32}$$

All three roots are of modulus 1. ◇

EXAMPLE The effect of immature reproduction

$$m_a = \begin{cases} 1/4, & a = 1, \\ 0, & a = 2, \\ 6, & a = 3, \end{cases} \quad l_a = \begin{cases} 1, & a = 1, \\ 1/2, & a = 2, \\ 1/6, & a = 3. \end{cases} \tag{21.33}$$

Reproduction occurs at ages 1 and 3. The greatest common divisor of 1 and 3 is 1; this is an aperiodic maternity schedule. The Euler–Lotka equation for this problem is

$$\frac{1}{4}\frac{1}{\lambda} + \frac{1}{\lambda^3} = 1 \tag{21.34}$$

or

$$\lambda^3 - \frac{1}{4}\lambda^2 - 1 = 0. \tag{21.35}$$

I computed the roots and their moduli numerically. They are

$$\lambda_0 = 1.09, \quad |\lambda_0| = 1.09, \tag{21.36a}$$
$$\lambda_{1,2} = -0.42 \pm 0.86\,i, \quad |\lambda_{1,2}| = 0.957. \tag{21.36b}$$

The single positive root is now dominant. ◇

For an aperiodic maternity schedule, we can proceed much as we did in the case of the Lotka integral equation. Equation (21.24) may be rewritten

$$B_t = c_0 \lambda_0^t \left[1 + \sum_{j=1}^{} \frac{c_j}{c_0} \left(\frac{\lambda_j}{\lambda_0} \right)^t \right], \tag{21.37}$$

and, since λ_0 is the dominant eigenvalue,

$$\lim_{t \to \infty} B_t \sim c_0 \lambda_0^t. \tag{21.38}$$

After the death of the founder females, the number of females in each age class is given by

$$n_{a,t} = B_{t-a} l_a \tag{21.39}$$

so that,

$$\lim_{t \to \infty} n_{a,t} \sim c_0 \lambda_0^t (\lambda_0^{-a} l_a). \tag{21.40}$$

The term in parentheses gives the relative proportion of a-year-olds to newborns for the stable age distribution.

However, for a periodic maternity schedule, there is no stable age distribution (except perhaps in the average). The birth rate and the age distribution will both continue to oscillate.

Problem 21.1 *The Eastern Screech-Owl* (Gill, 1995)

The Eastern Screech-Owl (*Otus asio*) is a small, cavity-nesting owl found throughout eastern North America. Frank Gehlbach has prepared life tables for two study populations in Texas, one in the suburbs, the other in rural woodlands.

Age	Suburban		Rural	
a	l_a	m_a	l_a	m_a
0	1.00	0.0	1.00	0.0
1	0.49	0.8	0.30	0.8
2	0.18	1.3	0.11	1.1
3	0.10	1.5	0.06	1.6
4	0.06	1.6	0.04	1.0
5	0.04	1.3	0.02	1.0
6	0.03	1.3		
7	0.02	1.3		
8	0.02	1.3		
9	0.01	1.3		
10	0.01	1.3		

Compare and contrast these two populations, discussing such quantities as the net reproductive rate, the geometric growth rate, the average age of reproduction, and the stable age distribution.

Solution by *Z*-transform

We are left with the unpleasant task of finding the solution for $t < \beta$ and of finding c_0 and the c_j in equation (21.24). We solved a similar problem in the last chapter by applying the Laplace transform. Is there a comparable transform for discrete-time problems?

One of the oldest methods for solving difference equations, dating back to de Moivre (1730a,b, 1756), uses generating functions. A closely related approach uses the Z-transform. For a sequence of real numbers f_t, the (one-sided) Z-transform is

$$\hat{f}(\lambda) = Z(f_t) = \sum_{t=0}^{\infty} f_t \lambda^{-t}. \qquad (21.41)$$

(Note our use of λ instead of the traditional z.) The Z-transform looks and acts much like a Laplace transform.

Consider, for example,

$$f_t = a^t. \qquad (21.42)$$

Then

$$\hat{f}(\lambda) = \sum_{t=0}^{\infty} \left(\frac{a}{\lambda}\right)^t = \frac{1}{1 - a/\lambda}. \qquad (21.43)$$

Since equation (21.1) contains a convolution of the birth rate and the net maternity function, it is also worth our while to consider the Z-transform of a convolution:

$$\begin{aligned} Z(f * g) &= \sum_{t=0}^{\infty} \left(\sum_{a=0}^{t} f_{t-a}\, g_a\right) \lambda^{-t} \\ &= \sum_{t=0}^{\infty} \sum_{a=0}^{t} f_{t-a}\, \lambda^{-(t-a)}\, g_a\, \lambda^{-a} \\ &= \sum_{a=0}^{\infty} g_a\, \lambda^{-a} \sum_{t=a}^{\infty} f_{t-a}\, \lambda^{-(t-a)} \\ &= \sum_{a=0}^{\infty} g_a\, \lambda^{-a} \sum_{b=0}^{\infty} f_b\, \lambda^{-b} \\ &= Z(f_t)\, Z(g_t). \qquad (21.44) \end{aligned}$$

This extremely powerful result says that Z-transform of a convolution sum can be replaced by a simple product of Z-transforms.

Events now follow the course of Chapter 20. Let ϕ_a be the net maternity function. Taking the Z-transform of each side of equation (21.1), we obtain

$$\hat{B}(\lambda) = \hat{B}(\lambda)\,\hat{\phi}(\lambda) + \hat{G}(\lambda). \qquad (21.45)$$

It follows that

$$\hat{B}(\lambda) = \frac{\hat{G}(\lambda)}{1 - \hat{\phi}(\lambda)}. \qquad (21.46)$$

But the denominator is simply

$$1 - \hat{\phi}(\lambda) = 1 - \sum_{a=\alpha}^{\beta} \lambda^{-a} l_a m_a. \tag{21.47}$$

Thus the roots of equation (21.47) are identical to those of the (discrete) Euler–Lotka equation. To progress further (in the case of simple roots) we now assume that the Z-transformed birth rate has a partial fractions expansion, in terms of the roots of Euler–Lotka equation, of the form:

$$\hat{B}(\lambda) = \frac{c_0}{\left(1 - \frac{\lambda_0}{\lambda}\right)} + \sum_{j=1}^{\infty} \frac{c_j}{\left(1 - \frac{\lambda_j}{\lambda}\right)}. \tag{21.48}$$

Since the inverse transform of $1/(1 - \lambda_j/\lambda)$ is simply λ_j^t, this implies that

$$B_t = c_0 \lambda_0^t + \sum_{j=1}^{\infty} c_j \lambda_j^t. \tag{21.49}$$

Moreover, a direct comparison of equations (21.46) and (21.49) implies that

$$c_j = \lim_{\lambda \to \lambda_j} \left[\left(1 - \frac{\lambda_j}{\lambda}\right) \frac{\hat{G}(\lambda)}{1 - \hat{\phi}(\lambda)} \right] = \left[\frac{\hat{G}(\lambda)}{-\lambda \frac{d\hat{\phi}(\lambda)}{d\lambda}} \right]_{\lambda=\lambda_j} \tag{21.50}$$

and hence that

$$c_j = \frac{\displaystyle\sum_{t=0}^{\beta-1} \lambda_j^{-t} G_t}{\displaystyle\sum_{a=\alpha}^{\beta} a \lambda_j^{-a} l_a m_a}. \tag{21.51}$$

To use equation (21.51), we must know G_0. So that equation (21.1) may be consistent, we shall simply define $G_0 = B_0$.

As an alternative to equation (21.49), if $\hat{B}(\lambda)$ is given in closed form as an algebraic expression and if its domain of analyticity is known, then its inverse is given by the contour integral

$$B_t = \frac{1}{2\pi i} \oint_C \hat{B}(\lambda) \lambda^{t-1} d\lambda, \tag{21.52}$$

where C is a closed contour, surrounding the origin, in the domain of analyticity of $\hat{B}(\lambda)$.

EXAMPLE Fibonacci's rabbits

You will remember that Fibonacci's rabbits had the following parameters:

$$l_a = 1, \quad a = 1, 2, \ldots, \tag{21.53}$$

$$m_a = \begin{cases} 1, & a = 1, 2, \\ 0, & a > 2. \end{cases} \tag{21.54}$$

The Euler–Lotka equation reduces, in this instance, to

$$\sum_{a=1}^{2} \lambda^{-a} l_a m_a = \frac{1}{\lambda} + \frac{1}{\lambda^2} = 1 \tag{21.55}$$

or

$$\lambda^2 - \lambda - 1 = 0. \tag{21.56}$$

This has the two roots

$$\lambda_0 = \frac{1 + \sqrt{5}}{2} \approx +1.618, \tag{21.57a}$$

$$\lambda_1 = \frac{1 - \sqrt{5}}{2} \approx -0.618. \tag{21.57b}$$

If we start the population off with a single newborn rabbit ($N_0 = G_0 = 1$),

$$c_0 = \frac{1}{\left(\frac{1}{\lambda_0} + \frac{2}{\lambda_0^2} \right)} = \frac{5 + \sqrt{5}}{10}, \tag{21.58a}$$

$$c_1 = \frac{1}{\left(\frac{1}{\lambda_1} + \frac{2}{\lambda_1^2} \right)} = \frac{5 - \sqrt{5}}{10}, \tag{21.58b}$$

and

$$B_t = \left(\frac{5 + \sqrt{5}}{10} \right) \left(\frac{1 + \sqrt{5}}{2} \right)^t + \left(\frac{5 - \sqrt{5}}{10} \right) \left(\frac{1 - \sqrt{5}}{2} \right)^t. \tag{21.59}$$

This generates the sequence

$$B_t = 1, 1, 2, 3, \ldots. \tag{21.60}$$

However, if we start the population off with one newborn rabbit ($B_0 = G_0 = 1$) and one 1-year-old rabbit,

$$G_1 = n_{1,0} \frac{l_2}{l_1} m_2 = 1. \tag{21.61}$$

Then,

$$c_0 = \frac{1 + \frac{1}{\lambda_0}}{\left(\frac{1}{\lambda_0} + \frac{2}{\lambda_0^2}\right)} = \frac{5 + 3\sqrt{5}}{10}, \tag{21.62a}$$

$$c_1 = \frac{1 + \frac{1}{\lambda_1}}{\left(\frac{1}{\lambda_1} + \frac{2}{\lambda_1^2}\right)} = \frac{5 - 3\sqrt{5}}{10}, \tag{21.62b}$$

and

$$B_t = \left(\frac{5 + 3\sqrt{5}}{10}\right)\left(\frac{1 + \sqrt{5}}{2}\right)^t + \left(\frac{5 - 3\sqrt{5}}{10}\right)\left(\frac{1 - \sqrt{5}}{2}\right)^t. \tag{21.63}$$

This generates the sequence

$$B_t = 1, 2, 3, 5, \ldots, \tag{21.64}$$

an answer that, with hindsight, is obvious. ◇

Problem 21.2 Double-trouble

a	l_a	m_a
1	0.50	2.0
2	0.25	8.0

Fibonacci suddenly finds himself confronted with a new set of rabbits with the population parameters shown above. Assume an initial female population of one newborn rabbit and one 1-year-old rabbit. Solve for the number of births in each year. What is the number of births, population size, and age structure for years 3, 4, and 5? How close is the age structure in years 3, 4, and 5 to the predicted stable age distribution?

For the discrete formulation, finding the roots of the Euler–Lotka equation is straightforward. However, the constants c_j are most unpleasant; projecting is still awkward.

Recommended readings

Charlesworth (1994) covers material in this chapter from a population genetic perspective. Caswell (2001) provides a thorough introduction to sensitivity analysis.

22 The Leslie matrix

Let me remind you of the ingredients of a Leslie matrix. We start with a vector of age-class abundances,

$$\boldsymbol{n}_t = \begin{pmatrix} n_0 \\ n_1 \\ \cdot \\ \cdot \\ n_{\omega-1} \end{pmatrix}_t , \tag{22.1}$$

a set of year-to-year survival probabilities,

$$P_a = \frac{l_{a+1}}{l_a}, \tag{22.2}$$

and a corresponding set of age-class fertilities,

$$F_a = P_a m_{a+1}, \tag{22.3}$$

that we combine into a simple projection scheme,

$$n_{0,t+1} = F_0\, n_{0,t} + F_1\, n_{1,t} + \ldots + F_{\omega-1}\, n_{\omega-1,t}, \tag{22.4a}$$

$$n_{1,t+1} = P_0\, n_{0,t}, \tag{22.4b}$$

$$n_{2,t+1} = P_1\, n_{1,t}, \tag{22.4c}$$

$$\vdots$$

$$n_{\omega-1,t+1} = P_{\omega-2}\, n_{\omega-2,t}. \tag{22.4d}$$

This may be written succinctly as

$$\begin{pmatrix} n_0 \\ n_1 \\ \cdot \\ \cdot \\ n_{\omega-1} \end{pmatrix}_{t+1} = \begin{pmatrix} F_0 & F_1 & F_2 & \cdot & F_{\omega-1} \\ P_0 & 0 & 0 & \cdot & 0 \\ 0 & P_1 & 0 & \cdot & 0 \\ \cdot & \cdot & \cdot & & \cdot \\ 0 & 0 & \cdot & P_{\omega-2} & 0 \end{pmatrix} \begin{pmatrix} n_0 \\ n_1 \\ \cdot \\ \cdot \\ n_{\omega-1} \end{pmatrix}_{t} \tag{22.5}$$

or, even better, as

$$n_{t+1} = L n_t, \tag{22.6}$$

where L is a *Leslie matrix* or a *projection matrix*.

In one sense, we are all done. If you want to know the vector of age-class abundances in any given year, you simply multiply by the Leslie matrix the appropriate number of times,

$$n_t = L^t n_0. \tag{22.7}$$

However, there are some interesting subtleties. Say we want to find the stable age distribution. We want a vector in which each age class grows by the same factor λ each year,

$$n_{t+1} = \lambda n_t. \tag{22.8}$$

Equations (22.6) and (22.8) together imply that the stable age distribution must satisfy

$$L n_t = \lambda n_t. \tag{22.9}$$

We must, therefore, solve for the eigenvalues and eigenvectors of the Leslie matrix.

There are two rather different approaches to finding the eigenvalues. The direct approach is to rewrite equation (22.9) as

$$(L - \lambda I) n_t = 0. \tag{22.10}$$

Since we are looking for a nontrivial solution to this homogeneous set of equations, we require that the matrix on the left-hand side be singular, and that the determinant of this matrix satisfy

$$|L - \lambda I| = 0. \tag{22.11}$$

Solving the resulting characteristic equation provides us with the desired eigenvalues.

A subtler approach is to begin by noting that stable-growth equation (22.8) implies that

$$n_{1,t+1} = \lambda n_{1,t}, \tag{22.12}$$

and that this has to be consistent with equation (22.4b),

$$n_{1,t+1} = P_0 n_{0,t}. \tag{22.13}$$

Thus,

$$n_{1,t} = \frac{P_0}{\lambda} n_{0,t}. \tag{22.14}$$

Similarly

$$n_{2,t} = \frac{P_1}{\lambda} n_{1,t} = \frac{P_1}{\lambda} \frac{P_0}{\lambda} n_{0,t}. \tag{22.15}$$

We can continue like this all the way through

$$n_{\omega-1,t} = \frac{P_0 P_1 \dots P_{\omega-2}}{\lambda^{\omega-1}} n_{0,t}. \tag{22.16}$$

Now consider the neonates. According to our Leslie matrix,

$$n_{0,t+1} = F_0 n_{0,t} + F_1 n_{1,t} + \cdots + F_{\omega-1} n_{\omega-1,t}, \tag{22.17}$$

so that

$$\lambda n_{0,t} = \left(F_0 + F_1 \frac{P_0}{\lambda} + \cdots + F_{\omega-1} \frac{P_0 P_1 \dots P_{\omega-2}}{\lambda^{\omega-1}} \right) n_{0,t}. \tag{22.18}$$

If we divide this equation by its left-hand side and substitute equation (22.3) for the fertilities, we obtain

$$\left(\frac{P_0 m_1}{\lambda} + \frac{P_0 P_1 m_2}{\lambda^2} + \cdots + \frac{P_0 P_1 \dots P_{\omega-1} m_\omega}{\lambda^\omega} \right) = 1. \tag{22.19}$$

Since

$$l_a = P_0 P_1 \dots P_{a-1}, \tag{22.20}$$

equation (22.19) can be rewritten

$$\sum_{a=1}^{\omega} \lambda^{-a} l_a m_a = 1. \tag{22.21}$$

Finally, since reproduction is typically over the ages α to β, this equation (22.21) reduces to the familiar form

$$\sum_{a=\alpha}^{\beta} \lambda^{-a} l_a m_a = 1. \tag{22.22}$$

This is just the discrete Euler–Lotka equation. For a Leslie matrix, we can immediately jump to the Euler–Lotka equation for our characteristic equation.

EXAMPLE   Semelparous reproduction (Bernardelli, 1941)
The Euler–Lotka equation for the example shown in the table on p. 380 reduces to

$$\sum_{a=\alpha}^{\beta} \lambda^{-a} l_a m_a = \frac{1}{\lambda^3} = 1, \tag{22.23}$$

Table 22.1. *Semelparous maternity schedule*

a	l_a	m_a	P_a	F_a
0	1	—	1	0
1	1	0	1/2	0
2	1/2	0	1/3	2
3	1/6	6	—	—

and the roots of this characteristic equation are simply

$$\lambda_0 = 1, \quad \lambda_1 = -\frac{1}{2} + \frac{\sqrt{3}}{2} i, \quad \lambda_2 = -\frac{1}{2} - \frac{\sqrt{3}}{2} i. \tag{22.24}$$

The Leslie matrix is

$$L = \begin{pmatrix} 0 & 0 & 2 \\ 1 & 0 & 0 \\ 0 & \frac{1}{2} & 0 \end{pmatrix}. \tag{22.25}$$

If we look for the eigenvalues directly, we have

$$\begin{vmatrix} -\lambda & 0 & 2 \\ 1 & -\lambda & 0 \\ 0 & \frac{1}{2} & -\lambda \end{vmatrix} = -\lambda^3 + 1 = 0, \tag{22.26}$$

with the same eigenvalues as before. ◇

With the eigenvalues in hand, we must still solve for the eigenvectors. We know from, Chapter 21, that the discrete Euler–Lotka equation possesses a single positive real root λ_0. The corresponding eigenvector has positive components. To see this, note that the matrix on the left hand of equation (22.10) is not of full rank for $\lambda = \lambda_0$. (This is just another way of saying that the matrix is singular.) This means that we are free to choose one component of the eigenvector. Okay, let us choose

$$n_{0,t} = 1. \tag{22.27}$$

Then, by equations (22.14), (22.15), etc.,

$$n_{1,t} = \frac{P_0}{\lambda} n_{0,t} = l_1 \lambda^{-1}, \tag{22.28a}$$

$$n_{2,t} = \frac{P_1}{\lambda} n_{1,t} = l_2 \lambda^{-2}, \tag{22.28b}$$

$$\vdots$$

$$n_{a,t} = l_a \lambda^{-a}. \tag{22.28c}$$

Fig. 22.1. Simple graph.

This sequence should look familiar. It is simply the stable age distribution that we derived in Chapter 21. The stable age distribution is an eigenvector of the Leslie matrix.

We would like

$$|\lambda_j| < \lambda_0, \quad j \neq 0. \tag{22.29}$$

Why ? In the case of ω distinct eigenvalues, we could then write the general solution as

$$\boldsymbol{n}_t = c_0 \lambda_0^t \boldsymbol{v}_0 + \sum_{j=1}^{\omega-1} c_j \lambda_j^t \boldsymbol{v}_j \tag{22.30}$$

or

$$\boldsymbol{n}_t = \lambda_0^t \left[c_0 \boldsymbol{v}_0 + \sum_{j=1}^{\omega-1} c_j \left(\frac{\lambda_j}{\lambda_0} \right)^t \boldsymbol{v}_j \right]. \tag{22.31}$$

If inequality (22.29) is true, the positive real eigenvalue will dominate the solution in the sense that

$$\lim_{t \to \infty} \boldsymbol{n}_t \sim c_0 \lambda_0^t \boldsymbol{v}_0. \tag{22.32}$$

However, we know, from Chapter 21, that if the maternity function is periodic, with period d, inequality (22.29) is not true and there will be $d - 1$ more eigenvalues with the same modulus as λ_0.

This characterization is perfectly adequate for Leslie matrices. However, we are really dealing with a special case of a more general theorem. The general theorem is useful for analyzing matrices based on stage as well as on age. Indeed, it can be applied to all nonnegative matrices. Before introducing this theorem, I must present two new properties: irreducibility and primitivity.

Irreducibility and primitivity are properties of matrices. Even so, one can recognize these properties by looking at *graphs* and *digraphs*. I mean graph in the graph-theoretic sense. A graph is a finite set of *vertices*, a finite set of *edges*, and a rule that specifies which edges join which pairs of vertices (see Figure 22.1). A digraph is a graph with *directed* edges (or arrows) (see Figure 22.2).

I can draw a digraph for every nonnegative matrix. For a Leslie matrix, a positive entry a_{ij} ($0 \le i, j < \omega$) implies a transition from age class j to age class i. The corresponding digraph has a directed edge from vertex j to vertex i.

EXAMPLE Annuals

The matrix

$$\begin{pmatrix} 0 & 2 \\ 0.6 & 0 \end{pmatrix} \tag{22.33}$$

has the digraph shown in Figure 22.2. Newborns mature to become 1-year-olds. One-year-olds produce more neonates.

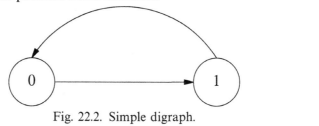

Fig. 22.2. Simple digraph.   ◇

Definition A nonnegative matrix is *irreducible* if and only if its digraph is *strongly connected*. A digraph is strongly connected if there is a path from every vertex to every other vertex.

EXAMPLE Irreducible matrix

The matrix

$$\begin{pmatrix} 1 & 0 & 1 \\ 1 & 0 & 0 \\ 0 & 1 & 0 \end{pmatrix} \tag{22.34}$$

is irreducible because its digraph, Figure 22.3, is strongly connected.

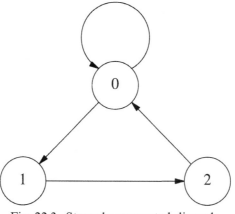

Fig. 22.3. Strongly connected digraph.   ◇

EXAMPLE   Reducible matrix
The matrix

$$\begin{pmatrix} 0 & 0 & 1 & 0 & 0 \\ 1 & 0 & 0 & 0 & 0 \\ 0 & 1 & 0 & 0 & 0 \\ 0 & 0 & 0 & 0 & 0 \\ 0 & 0 & 0 & 1 & 0 \end{pmatrix}$$ (22.35)

is reducible because its digraph, Figure 22.4, is not strongly connected.

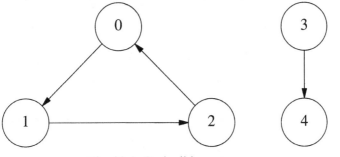

Fig. 22.4. Reducible system.   ◇

EXAMPLE   Postreproductive age classes
Any Leslie matrix that contains postreproductive age classes is reducible (see Figure 22.5).

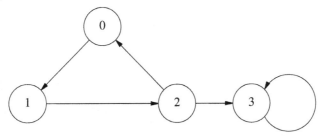

Fig. 22.5. Reducible system with postreproductive individuals.   ◇

Definition   A nonnegative matrix A is *primitive* if $A^m > 0$ for some m.

EXAMPLE   Imprimitive matrix
The matrix

$$A = \begin{pmatrix} 0 & 0 & 1 \\ 1 & 0 & 0 \\ 0 & 1 & 0 \end{pmatrix}$$ (22.36)

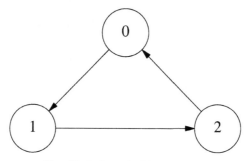

Fig. 22.6. Imprimitive system.

with digraph (see Figure 22.6) is imprimitive. To see this, consider the second, third, and fourth powers of this matrix:

$$A^2 = \begin{pmatrix} 0 & 1 & 0 \\ 0 & 0 & 1 \\ 1 & 0 & 0 \end{pmatrix},$$

(22.37)

$$A^3 = \begin{pmatrix} 1 & 0 & 0 \\ 0 & 1 & 0 \\ 0 & 0 & 1 \end{pmatrix} = I,$$

(22.38)

$$A^4 = A.$$

(22.39)

The system cycles and there is no integer power that will make all elements of A^n positive. ◇

EXAMPLE   Primitive matrix

The matrix

$$A = \begin{pmatrix} 1 & 0 & 1 \\ 1 & 0 & 0 \\ 0 & 1 & 0 \end{pmatrix},$$

(22.40)

with digraph (see Figure 22.7) is primitive. To see this, note that

$$A^4 = \begin{pmatrix} 3 & 1 & 2 \\ 2 & 1 & 1 \\ 1 & 1 & 1 \end{pmatrix}$$

(22.41)

has all positive elements.

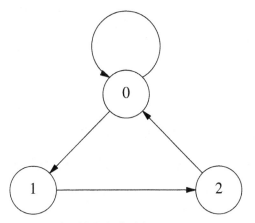

Fig. 22.7. Primitive system.   ◇

Primitivity implies that a number *m* can be found that allows one to go from any vertex of the digraph to any other vertex in exactly *m* moves.

Our definition of primitivity is couched in terms of A^m rather than A. However, the primitivity of a matrix can be easily recognized from its digraph (Demetrius, 1971):

Consider a nonnegative, irreducible matrix. This matrix is primitive if and only if the greatest common divisor of the lengths of all circuits (leading from a node back to itself) is one.

An irreducible imprimitive matrix is *cyclic*; its *index of imprimitivity* (*d*) is the greatest common divisor of the circuit lengths of its digraph.

EXAMPLE

(a)

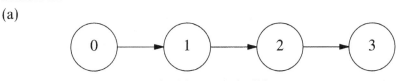

Fig. 22.8. A reducible system.

(b)

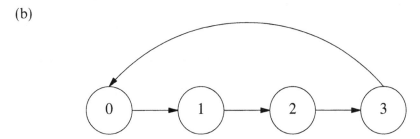

Fig. 22.9. Irreducible and imprimitive ($d = 4$).

(c)

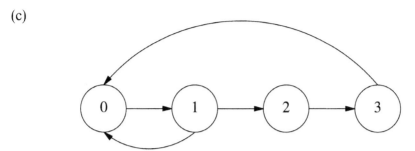

Fig. 22.10. Irreducible and imprimitive ($d = 2$).

(d)

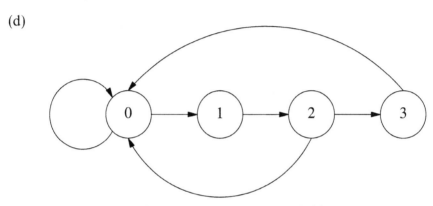

Fig. 22.11. Irreducible and primitive.　　　　　◇

I am finally ready to state the long-promised Perron–Frobenius theorem (see also Gantmacher, 1959; Karlin, 1966; and Caswell, 2001).

Theorem (Perron, 1907; Frobenius, 1912)

(I)　For a nonnegative, irreducible, *primitive* matrix:

    (A)　There exists a simple, positive, real λ_0 such that $\lambda_0 > |\lambda_j|$ for $j \neq 0$.

    (B)　The right eigenvector corresponding to λ_0 is real and strictly positive.

(II)　For a nonnegative, irreducible, but *imprimitive* (cyclic) matrix:

    (A)　There exists a simple, positive, real λ_0 such that $\lambda_0 \geq |\lambda_j|$ for $j \neq 0$.

    (B)　There are $d - 1$ (d being the index of imprimitivity) eigenvalues, $\lambda_j = \lambda_0 \exp(2n\pi i/d)$, $n = 1, 2, \ldots, d - 1$, equal in modulus to λ_0.

    (C)　The right eigenvector corresponding to λ_0 is real and strictly positive.

(III)　For a nonnegative but *reducible* matrix:

    (A)　There exists a nonnegative and real eigenvalue λ_0 such that $\lambda_0 \geq |\lambda_j|$ for $j \neq 0$.

    (B)　The right eigenvector corresponding to λ_0 is real and nonnegative.

EXAMPLE   Fibonacci's rabbits

a	l_a	m_a	P_a	F_a
0	1	—	1	1
1	1	1	1	1
2	1	1	1	0
3	1	0	—	—

$$L = \begin{pmatrix} 1 & 1 & 0 \\ 1 & 0 & 0 \\ 0 & 1 & 0 \end{pmatrix}. \qquad (22.42)$$

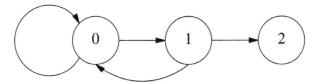

Fig. 22.12. Digraph for Fibonacci's rabbits.

As written, this Leslie matrix is reducible. So, we will reduce it by dropping the postreproductive age class:

$$L = \begin{pmatrix} 1 & 1 \\ 1 & 0 \end{pmatrix}. \qquad (22.43)$$

The simple cycles in Figure 22.13 have lengths 1 and 2. Since these two lengths are relatively prime, the Leslie matrix is both irreducible and primitive.

Direct approach

$$|L - \lambda I| = \begin{vmatrix} 1 - \lambda & 1 \\ 1 & -\lambda \end{vmatrix} = 0. \qquad (22.44)$$

$$\lambda^2 - \lambda - 1 = 0. \qquad (22.45)$$

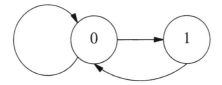

Fig. 22.13. Reduced digraph for Fibonacci's rabbits.

The two eigenvalues are

$$\lambda_0 = \frac{1 + \sqrt{5}}{2} \approx 1.6, \tag{22.46a}$$

$$\lambda_1 = \frac{1 - \sqrt{5}}{2} \approx -0.6. \tag{22.46b}$$

From the second row of the determinant, it is clear that, for each eigenvalue, the corresponding eigenvector $v = (v_0, v_1)$ satisfies

$$v_1 = \frac{1}{\lambda} v_0. \tag{22.47}$$

We may arbitrarily choose $v_0 = 1$. This gives us

$$\begin{pmatrix} n_0 \\ n_1 \end{pmatrix}_t = c_0 \begin{pmatrix} 1 \\ 1/\lambda_0 \end{pmatrix} \lambda_0^t + c_1 \begin{pmatrix} 1 \\ 1/\lambda_1 \end{pmatrix} \lambda_1^t. \tag{22.48}$$

Subtle approach

The alternative, of course, is to write the characteristic equation using the Euler–Lotka equation:

$$\sum_{a=1}^{2} \lambda_a l_a m_a = \frac{1}{\lambda} + \frac{1}{\lambda^2} = 1, \tag{22.49}$$

or

$$\lambda^2 - \lambda - 1 = 0. \tag{22.50}$$

The two eigenvalues are, once again, given by equations (22.46a) and (22.46b). Rather than solve for the eigenvectors directly, we can use the fact that the components of the eigenvectors are specified by the stable age distribution formula $l_a \lambda^{-a}$. Thus the general solution can, once again, be written as

$$\begin{pmatrix} n_0 \\ n_1 \end{pmatrix}_t = c_0 \begin{pmatrix} 1 \\ 1/\lambda_0 \end{pmatrix} \lambda_0^t + c_1 \begin{pmatrix} 1 \\ 1/\lambda_1 \end{pmatrix} \lambda_1^t. \tag{22.51}$$

$\diamond$

Lefkovitch (1965) was the first to argue that the Leslie matrix is a special case of a more general class of stage-structured matrix models. The exact age of an organism may be difficult to ascertain, whereas the size class or stage (e.g., egg, larva, pupa, and adult) may be easier to determine. Moreover, fecundity and survival are often directly related to stage and only indirectly to age. In these instances, it makes sense to subdivide the population into stages and to model the transition between the various stages.

The stages in a stage-structured model need not be of equal duration. Also, organisms may remain in a stage from one time to the next. An especially common form of the Lefkovitch matrix,

$$
L = \begin{pmatrix}
F_0 + P_0 & F_1 & F_2 & \cdot & \cdot & F_{\omega-1} \\
G_0 & P_1 & 0 & \cdot & \cdot & 0 \\
0 & G_1 & P_2 & \cdot & \cdot & 0 \\
\cdot & & 0 & G_2 & \cdot & \cdot & \cdot \\
\cdot & & & & \cdot & P_{\omega-2} & 0 \\
0 & 0 & 0 & \cdot & G_{\omega-2} & P_{\omega-1}
\end{pmatrix}, \tag{22.52}
$$

contains positive elements on the diagonal individuals that stay in the same stage, and positive elements on the subdiagonal for individuals that grow to the next stage. Lefkovitch matrices may, however, have positive elements anywhere; a positive ij-th element represents the contribution of the j-th stage to the i-th stage over some predefined interval.

For Lefkovitch matrices, the maternity schedule may no longer characterize the dominance of the Perron root. Fortunately, the Perron–Frobenius theorem is still quite applicable. Note, however, that the Lefkovitch matrix, unlike the Leslie matrix, may have more than one real, positive root.

EXAMPLE   Loggerhead sea turtles (Crouse *et al.*, 1987)

Let

$$eh \equiv \text{eggs, hatchlings,} \tag{22.53a}$$
$$sj \equiv \text{small juveniles,} \tag{22.53b}$$
$$lj \equiv \text{large juveniles,} \tag{22.53c}$$
$$sa \equiv \text{subadults,} \tag{22.53d}$$
$$nb \equiv \text{novice breeders,} \tag{22.53e}$$
$$re \equiv \text{1st year remigrants,} \tag{22.53f}$$
$$mb \equiv \text{mature breeders.} \tag{22.53g}$$

Crouse *et al.* (1987) showed that the stage-structured model for loggerhead

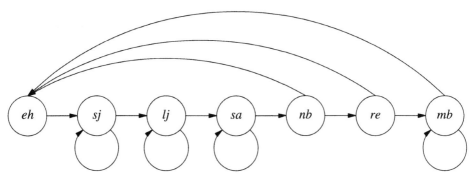

Fig. 22.14. Sea turtle digraph.

sea turtles can be written

$$
\begin{pmatrix} eh \\ sj \\ lj \\ sa \\ nb \\ re \\ mb \end{pmatrix}_{t+1} = \begin{pmatrix} 0 & 0 & 0 & 0 & 127 & 4 & 80 \\ 0.6747 & 0.737 & 0 & 0 & 0 & 0 & 0 \\ 0 & 0.0486 & 0.6610 & 0 & 0 & 0 & 0 \\ 0 & 0 & 0.0147 & 0.6907 & 0 & 0 & 0 \\ 0 & 0 & 0 & 0.0518 & 0 & 0 & 0 \\ 0 & 0 & 0 & 0 & 0.8091 & 0 & 0 \\ 0 & 0 & 0 & 0 & 0 & 0.8091 & 0.8089 \end{pmatrix} \begin{pmatrix} eh \\ sj \\ lj \\ sa \\ nb \\ re \\ mb \end{pmatrix}_t .
$$

(22.54)

The spectrum of eigenvalues is:

$$\lambda_0 = 0.945,$$ (22.55a)

$$\lambda_1 = 0.372,$$ (22.55b)

$$\lambda_2 = 0.265,$$ (22.55c)

$$\lambda_{3,4} = 0.746 \pm 0.213\, i,$$ (22.55d)

$$\lambda_{5,6} = -0.088 \pm 0.12\, i.$$ (22.55e)

A sensitivity analysis of this model highlights the importance of enhancing juvenile survivorship (using turtle excluder devices or TEDs), rather than simply protecting eggs on nesting-beaches. ◇

Recommended readings

Caswell (2001) is the definitive reference on matrix population models.

23 The McKendrick–von Foerster PDE

It has been a while since I introduced the McKendrick–von Foerster partial differential equation (PDE). So, let me remind you of the important components of this model.

We began by considering the rate of change of the number of females in some age interval (see Figure 23.1). We then derived a simple 'conservation law' that accounted for changes in the density of females with aging and mortality,

$$\frac{\partial n}{\partial t} = -\frac{\partial J}{\partial a} - \mu(a, t)\, n. \qquad (23.1)$$

Since all females age, the flux J was assumed to be proportional to the density of females,

$$J = n(a, t)\, v, \qquad (23.2)$$

with a velocity v that describes the rate of change of age with time,

$$v = \frac{da}{dt} = 1. \qquad (23.3)$$

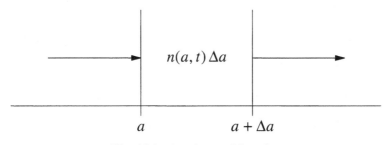

$$n(a, t)\, \Delta a$$

$$a \qquad\qquad a + \Delta a$$

Fig. 23.1. A cohort of females.

This led us to the McKendrick–von Foerster PDE,

$$\frac{\partial n}{\partial t} + \frac{\partial n}{\partial a} = -\mu(a, t)\, n. \tag{23.4}$$

Finally, we supplemented this partial differential equation with a boundary condition,

$$n(0, t) = \int_0^\omega n(a, t)\, m(a, t)\, da, \tag{23.5}$$

that accounted for births, and an initial condition,

$$n(a, 0) = n_0(a), \tag{23.6}$$

that captured the initial age distribution.

For now, we will restrict our attention to mortality functions that depend upon age but not upon time,

$$\mu(a, t) = \mu(a). \tag{23.7}$$

A special solution

The ingredients of this problem are similar to those for Lotka's integral equation. For Lotka's integral equation, the population asymptotically approached exponential growth and a stable age distribution. Taking that as our cue, we may try a solution of the form

$$n(a, t) = n^*(a)\, e^{rt}. \tag{23.8}$$

Substituting (23.8) into the McKendrick–von Foerster PDE, equation (23.4), gives us

$$r n^* e^{rt} + \frac{dn^*}{da} e^{rt} = -\mu(a)\, n^* e^{rt} \tag{23.9}$$

or

$$\frac{dn^*}{da} + [r + \mu(a)]\, n^* = 0. \tag{23.10}$$

This is a first-order, linear differential equation that you should be able to solve in your sleep. In particular,

$$n^*(a) = n^*(0)\, e^{-ra}\, e^{-\int_0^a \mu(\xi)\, d\xi}. \tag{23.11}$$

The last exponential may be interpreted as the survivorship, $l(a)$, so that

$$n^*(a) = n^*(0)\, e^{-ra}\, l(a) \tag{23.12}$$

or

$$\frac{n^*(a)}{n^*(0)} = e^{-ra} l(a). \tag{23.13}$$

This is the stable age distribution that we first met while analyzing Lotka's integral equation.

For this stable age distribution, there is a solution of the form

$$n(a, t) = n^*(0) e^{rt} [e^{-ra} l(a)] \tag{23.14}$$

with the corresponding birth rate

$$n(0, t) = n^*(0) e^{rt}. \tag{23.15}$$

However, this birth rate must satisfy boundary condition (23.5), so that

$$n^*(0) e^{rt} = \int_\alpha^\beta n^*(0) e^{rt} [e^{-ra} l(a) m(a)] da. \tag{23.16}$$

This simplifies to our old friend

$$\int_\alpha^\beta e^{-ra} l(a) m(a) da = 1, \tag{23.17}$$

the Euler–Lotka equation. We have thus recaptured the correct stable age distribution and characteristic equation. Of course, we have not really shown that these are asymptotic solutions to the McKendrick–von Foerster equation. We are also still left with the tougher problem of computing the age distribution for all times.

A general solution

To solve

$$\frac{\partial n}{\partial t} + \frac{\partial n}{\partial a} = -\mu(a) n, \tag{23.18a}$$

$$n(0, t) = \int_0^\omega n(a, t) m(a, t) da \equiv B(t), \tag{23.18b}$$

$$n(a, 0) = n_0(a) \tag{23.18c}$$

in full generality we must use the *method of characteristics*. As with all good methods for solving PDEs, the goal is to reduce our PDE into one or more ordinary differential equations (ODEs). We do this by imposing a change of variables,

$$n(a, t) = n(\xi, \eta). \tag{23.19}$$

The appropriate change will allow us to use the chain rule to simplify our PDE. To wit, we replace our PDE with the three ODEs

$$\frac{\partial a}{\partial \xi} = 1, \tag{23.20a}$$

$$\frac{\partial t}{\partial \xi} = 1, \tag{23.20b}$$

$$\frac{\partial n}{\partial \xi} = -\mu(a)\, n, \tag{23.20c}$$

since

$$\frac{\partial n}{\partial \xi} = \frac{\partial n}{\partial a}\frac{\partial a}{\partial \xi} + \frac{\partial n}{\partial t}\frac{\partial t}{\partial \xi} = -\mu(a)\, n. \tag{23.21}$$

Equation (23.20c) is still not as simple as we would like: there is a ξ on the left-hand side and an a on the right-hand side. We would like to make ξ and a as similar as possible. Since we are already constrained by equations (23.20a) and (23.20b), we can only tilt matters in our favor using the initial conditions for these two differential equations. (This is where η comes in.) We will require that a be ξ-like, at first, in that $\xi = 0$ implies that $a = 0$ and that t be η-like, at first, in that $\xi = 0$ implies that $t = \eta$.

Solving equations (23.20a) and (23.20b) gives us

$$a = \xi + c_1(\eta), \tag{23.22a}$$

$$t = \xi + c_2(\eta). \tag{23.22b}$$

Imposing the initial conditions

$$a(0, \eta) = 0, \tag{23.23a}$$

$$t(0, \eta) = \eta, \tag{23.23b}$$

now implies that

$$a = \xi, \tag{23.24a}$$

$$t = \xi + \eta, \tag{23.24b}$$

or that

$$\xi = a, \tag{23.25a}$$

$$\eta = t - a. \tag{23.25b}$$

Equation (23.20c) thus simplifies to

$$\frac{dn}{d\xi} = -\mu(\xi)\, n. \tag{23.26}$$

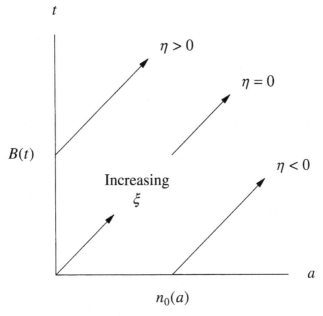

Fig. 23.2. Characteristics.

It is useful, at this point, to get a geometric feel for the effect of increasing ξ or η. If we increase ξ, for any fixed value of η, we generate a line or *characteristic* of slope 1 (see Figure 23.2). Equation (23.26) implies that we will be transmitting information along each of these characteristics. Biologically, we will be following cohorts as they age. The number in each cohort at some ξ (age) will also depend on the boundary condition (for $\eta \geq 0$ or $a \leq t$) or on the initial condition (for $\eta \leq 0$ or $a \geq t$). So, we will have to solve equation (23.26) separately for each of these two regions.

In each case, separating equation (23.26) gives us

$$\frac{1}{n} \, dn = -\mu(\xi) \, d\xi. \tag{23.27}$$

This is easy to integrate. The subtlety, however, is in how we handle the initial or boundary data. These initial and boundary data should be written solely in terms of η.

For $a \leq t$, $n(0, t) = n(0, \eta) = B(\eta)$, so that

$$\int_{B(\eta)}^{n(\xi, \eta)} \frac{1}{n} \, dn = -\int_0^\xi \mu(\xi) \, d\xi. \tag{23.28}$$

It follows that

$$n(\xi, \eta) = B(\eta) e^{-\int_0^\xi \mu(\xi)\,d\xi}. \tag{23.29}$$

Shifting back to a and t now yields

$$n(a, t) = B(t - a) e^{-\int_0^a \mu(\xi)\,d\xi}, \tag{23.30}$$

or

$$n(a, t) = n(0, t - a)\, l(a). \tag{23.31}$$

For $t \geq a$, we thus look at the number of births a years ago and multiply by the probability of surviving to age a.

For $a \geq t$, $n(a, 0) = n_0(a) = n_0(-\eta)$, so that

$$\int_{n_0(-\eta)}^{n(\xi, \eta)} \frac{1}{n}\, dn = -\int_{-\eta}^\xi \mu(\xi)\,d\xi. \tag{23.32}$$

It follows that

$$n(\xi, \eta) = n_0(-\eta) e^{-\int_{-\eta}^\xi \mu(\xi)\,d\xi}. \tag{23.33}$$

Reintroducing the original independent variables now gives us

$$n(a, t) = n_0(a - t) e^{-\int_{a-t}^a \mu(\xi)\,d\xi}, \tag{23.34}$$

or

$$n(a, t) = n(a - t, 0)\, \frac{l(a)}{l(a - t)}. \tag{23.35}$$

We therefore look at the females of age $a - t$ present at $t = 0$ and discount this number by the probability of surviving from age $a - t$ to age a.

We can now combine our two results into a single expression that describes the size of cohorts:

$$n(a, t) = \begin{cases} n(0, t - a)\, l(a), & a \leq t, \\ n(a - t, 0)\, \dfrac{l(a)}{l(a - t)}, & a \geq t. \end{cases} \tag{23.36}$$

Having solved for the density of females, we can guarantee consistency by substituting our expression for $n(a, t)$ from equation (23.26) into our boundary condition,

$$n(0, t) = \int_0^\omega n(a, t)\, m(a)\, da. \tag{23.37}$$

Thus

$$
n(0, t) = \int_0^t n(0, t-a)\, l(a)\, m(a)\, da
$$
$$
+ \int_t^\omega n(a-t, 0)\, \frac{l(a)}{l(a-t)}\, m(a)\, da. \tag{23.38}
$$

A minor change of variables in the second integral will bring this into a more familiar form. If we replace $a - t$ everywhere with a new a, equation (23.5) may be rewritten

$$
n(0, t) = \int_0^t n(0, t-a)\, l(a)\, m(a)\, da
$$
$$
+ \int_0^{\omega-t} n(a, 0)\, \frac{l(a+t)}{l(a)}\, m(a+t)\, da. \tag{23.39}
$$

The above is none other than Lotka's integral equation. To solve our problem, we must solve Lotka's integral equation for the birth rate. With this birth rate in hand, we can then use solution (23.36), derived using the method of characteristics, for the density of females of any particular age at any particular time.

I know what you are thinking. If we have to solve Lotka's integral equation to solve the our PDE, why do we not just start and stick with Lotka's integral equation. There are several reasons for preferring the McKendrick–von Foerster PDE as a starting point. In many practical situations, it is easier to construct models using the instantaneous mortality $\mu(a)$ rather than the cumulative survivorship $l(a)$. It is also easier to modify the PDE to accommodate stage structure and density dependence.

EXAMPLE   Age-dependent harvesting (Sanchez, 1978)
Consider an age-structured population in which a fraction δ, $0 < \delta < 1$, of the population of age $a \geq c$ is harvested. This harvesting occurs in addition to natural mortality. The McKendrick–von Foerster PDE can now be written

$$
\frac{\partial n}{\partial t} + \frac{\partial n}{\partial a} = -\mu(a)n - \delta H(a-c)n, \tag{23.40}
$$

where $H(a)$ is the Heaviside step function,

$$
H(a) = \begin{cases} 1, & a \geq 0, \\ 0, & a < 0. \end{cases} \tag{23.41}
$$

The right-hand side of equation (23.40) is still a form of age-specific mortality. Density (23.36), our result from using the method of characteristics, is still correct, if we make the appropriate substitutions into equations (23.30) and (23.34). The resulting density may be written

$$
n(a, t) = \begin{cases}
n(0, t - a) \, l(a), & a \le c, t, \\
n(0, t - a) \, l(a) \, e^{-\delta(a-c)}, & c < a < t, \\
n(a - t, 0) \dfrac{l(a)}{l(a-t)}, & t \le a < c, \\
n(a - t, 0) \dfrac{l(a)}{l(a-t)} \, e^{-\delta(a-c)}, & a \ge t, \, a \ge c, \, a \le t+c, \\
n(a - t, 0) \dfrac{l(a)}{l(a-t)} \, e^{-\delta t}, & a > c+t,
\end{cases}
$$

(23.42)

where $l(a)$ is the survivorship without harvesting.

Before, there were two distinct groups, those born since the start of the problem and those present at the start of the problem. Now, there are up to five distinct groups of individuals: females born since the start of the problem and too young to have been harvested, females born since the start of the problem and old enough to have been harvested for $a - c$ years, females present at the start of the problem but too young to have ever been harvested, females present at the start of the problem who have undergone $a - c$ years of harvesting, and females present at the start of the problem who have been harvested every year.

Two of the groups are incompatible. It is impossible for females born since the start of the problem to be old enough to have been harvested while females who were around at the start of the problem are still too young to have been harvested. Either $t \le c$ or $t > c$. We must drop either the second or the third row of equation (23.42) depending on which of these two inequalities holds.

Having solved for the density of the females, we must now guarantee consistency by substituting density (23.42) into boundary condition (23.37). I will also take this opportunity to change variables in the integrals involving the founder females. Thus, for $t > c$,

$$
n(0, t) = \int_0^c n(0, t - a) \, l(a) \, m(a) \, da + \int_c^t n(0, t - a) \, l(a) \, m(a) \, e^{-\delta(a-c)} \, da
$$
$$
+ \int_0^c n(a, 0) \frac{l(a+t)}{l(a)} \, m(a+t) \, e^{-\delta(a+t-c)} \, da
$$
$$
+ \int_c^{\omega - t} n(a, 0) \frac{l(a+t)}{l(a)} \, m(a+t) e^{-\delta t} \, da,
$$

(23.43)

while, for $t \leq c$,

$$
\begin{aligned}
n(0, t) = & \int_0^c n(0, t - a) \, l(a) \, m(a) \, da + \int_0^{c-t} n(a, 0) \frac{l(a + t)}{l(a)} m(a + t) \, da \\
& + \int_{c-t}^c n(a, 0) \frac{l(a + t)}{l(a)} m(a + t) \, e^{-\delta(a + t - c)} \, da \\
& + \int_c^{\omega - t} n(a, 0) \frac{l(a + t)}{l(a)} m(a + t) e^{-\delta t} \, da.
\end{aligned}
\tag{23.44}
$$

Clearly, these two equations are more complicated than our harvest-free solution.

Can age-dependent harvesting cause an growing population to collapse? How will this harvesting shift the stable age distribution? To answer these questions, we must analyze the asymptotic properties of equations (23.43) and (23.44).

We are interested in the behavior for large time – equation (23.43) rather than equation (23.44). If we wait long enough for the founder females to perish, equation (23.43) simplifies to

$$
n(0, t) = \int_\alpha^\beta n(0, t - a) \, E(a) \, l(a) \, m(a) \, da
\tag{23.45}
$$

with

$$
E(a) \equiv
\begin{cases}
1, & 0 \leq a \leq c, \\
e^{-\delta(a - c)}, & a \geq c.
\end{cases}
\tag{23.46}
$$

The associated characteristic equation is just

$$
\int_\alpha^\beta e^{-ra} E(a) \, l(a) \, m(a) \, da = 1.
\tag{23.47}
$$

You should compare this to Euler–Lotka equation (23.17).

Equation (23.47) has exactly one real root $r = r_h^*$. Moreover, since $0 < E(a) \leq 1$, it follows that $r_h^* \leq r^*$, where r^* is the solution to the Euler–Lotka equation without harvesting. We need more information about the net maternity if we are to say anything else about this root. $\diamond$

Problem 23.1 *Constant birth and death rates*

For the special case $m(a) = b$, $\mu(a) = d$, $\alpha = 0$, and $\beta = \infty$, show that

$$
r^* = b - d
\tag{23.48}
$$

while equation (23.47) gives us

$$\frac{b}{r+d+\delta}\, e^{-(r+d)c} \;-\; \frac{b}{r+d}\, [e^{-(r+d)c} - 1] \;=\; 1. \qquad (23.49)$$

Use a perturbation scheme to solve this equation for small δ. In particular, expand r in powers of delta,

$$r \;=\; r_0 + r_1 \delta + O(\delta^2), \qquad (23.50)$$

substitute this into equation (23.49), expand other terms that contain δ as appropriate, and equate different powers of δ to show that

$$r_h^* \;=\; b - d - e^{-bc}\, \delta + O(\delta^2). \qquad (23.51)$$

Recommended readings

Read Logan (1994) and Williams (1980) for introductions to the method of characteristics. Castillo-Chavez (1989) looks at applications of the McKendrick–von Foerster PDE.

24 Some simple nonlinear models

Throughout this book, I have emphasized the importance of nonlinearity. It is relatively easy to incorporate density dependence into age-structured models. Let us start with the McKendrick–von Foerster partial differential equation. Let

$$N(t) \equiv \int_0^\infty n(a, t) \, da \qquad (24.1)$$

be the population size at time t. If the birth and death rates depend, in the simplest instance, on age and population size, we may write

$$\frac{\partial n}{\partial t} + \frac{\partial n}{\partial a} = -\mu(a, N) \, n, \qquad (24.2a)$$

$$n(0, t) = \int_0^\infty n(a, t) \, m(a, N) \, da, \qquad (24.2b)$$

$$n(a, 0) = n_0(a). \qquad (24.2c)$$

EXAMPLE   Age-independent birth and death rates
As a simple example of this class of models, consider mortality and natality that depend on population size but not on age:

$$\frac{\partial n}{\partial t} + \frac{\partial n}{\partial a} = -\mu(N) \, n, \qquad (24.3a)$$

$$n(0, t) = \int_0^\infty n(a, t) \, m(N) \, da, \qquad (24.3b)$$

$$n(a, 0) = n_0(a). \qquad (24.3c)$$

This example hearkens back to Chapter 1. Since age is no longer important, we should somehow be able to recapture a simple differential equation for the population size. And, we can!

Integrating equation (24.3a) over all ages produces

$$\int_0^\infty \frac{\partial n}{\partial t}\, da \; + \; [n(\infty, t) - n(0, t)] \; = \; -\mu(N) \int_0^\infty n(a, t)\, da \qquad (24.4)$$

or

$$\frac{dN}{dt} \; = \; n(0, t) - \mu(N)\, N. \qquad (24.5)$$

Because of boundary condition (24.3b), this simplifies to

$$\frac{dN}{dt} \; = \; [m(N) - \mu(N)]\, N. \qquad (24.6)$$

For simple density dependence,

$$\frac{dm}{dN} \leq 0, \quad \frac{d\mu}{dN} \geq 0. \qquad (24.7)$$

Since equation (24.6) is a simple, autonomous, first-order differential equation, we may expect that the population size $N(t)$ will approach an equilibrium N^*.

How about the age structure? As the population size tends towards a constant, the population density must, by equation (24.1), lose its dependence on time,

$$\int_0^\infty n(a, t)\, da \;\rightarrow\; \int_0^\infty n^*(a)\, da. \qquad (24.8)$$

However, any equilibrial density $n^*(a)$ must still satisfy partial differential equation (24.3a), so that

$$\frac{dn^*}{da} \; = \; -\mu(N^*)\, n. \qquad (24.9)$$

Hence,

$$n^*(a) \; = \; n^*(0)\, e^{-\mu(N^*)a}. \qquad (24.10)$$

Boundary condition (24.3b) now implies that

$$n(0, t) \; = \; m(N)\, N(t), \qquad (24.11)$$

so that the equilibrial number of births is just

$$n^*(0) \; = \; m(N^*)\, N^* \qquad (24.12)$$

and

$$n^*(a) \; = \; m(N^*)\, N^*\, e^{-\mu(N^*)a}. \qquad (24.13)$$

The stable age distribution is a decaying exponential. ◇

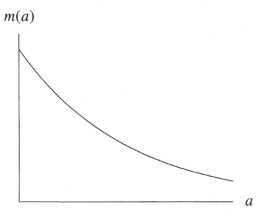

Fig. 24.1. Age-specific birth rate.

If the natality and mortality depend only on population size, we do not really need an age-structured model. Fortunately, there are models that contain both age and density dependence that can also be analyzed.

One interesting group of models is due to Gurtin and MacCamy (1974, 1979). Suppose that the mortality is age independent and that the natality is separable. Let

$$\mu(a, N) = \mu(N), \tag{24.14a}$$
$$m(a, N) = \beta(N)\, a^n\, e^{-\gamma a}. \tag{24.14b}$$

We can attack this class of problems using linear chain trickery.

EXAMPLE   The trouble with tribbles†

As a simple special case, consider a group of organisms, of potentially infinite longevity, that are born pregnant and whose age-specific birth rate decreases exponentially with age (see Figure 24.1). Thus

$$\frac{\partial n}{\partial a} + \frac{\partial n}{\partial t} = -\mu(N)\, n, \tag{24.15a}$$
$$n(0, t) = \int_0^\infty m_0\, e^{-\gamma a}\, n(a, t)\, da, \tag{24.15b}$$
$$n(a, 0) = n_0(a). \tag{24.15c}$$

For this model, we can follow a simple two-step procedure.

---

† For more information on tribbles, consult *Star Trek* episode 'The Trouble with Tribbles.'

First, integrate partial differential equation (24.15a) over all possible ages,

$$\int_0^\infty \frac{\partial n}{\partial t}\, da + [n(\infty, t) - n(0, t)] = -\mu(N) \int_0^\infty n(a, t)\, da, \qquad (24.16)$$

to produce

$$\frac{dN}{dt} = B(t) - \mu(N)\, N, \qquad (24.17)$$

where $B(t) = n(0, t)$.

Secondly, multiply partial differential equation (24.15a) by the maternity schedule and integrate over all possible ages:

$$m_0\, e^{-\gamma a} \frac{\partial n}{\partial a} + \frac{\partial}{\partial t} \left(m_0\, e^{-\gamma a}\, n \right) = -\mu(N)\, m_0\, e^{-\gamma a}\, n, \qquad (24.18)$$

$$m_0 \int_0^\infty e^{-\gamma a} \frac{\partial n}{\partial a}\, da + \frac{dB}{dt} = -\mu(N)\, B. \qquad (24.19)$$

The first integral may be handled by integration by parts,

$$\int_0^\infty e^{-\gamma a} \frac{\partial n}{\partial a}\, da = -n(0, t) + \gamma \int_0^\infty e^{-\gamma a}\, n\, da, \qquad (24.20)$$

so that equation (24.19) reduces to

$$-m_0\, B + \gamma B + \frac{dB}{dt} = -\mu(N)\, B \qquad (24.21)$$

or

$$\frac{dB}{dt} = [m_0 - \gamma - \mu(N)]\, B. \qquad (24.22)$$

This two-step procedure produces a coupled system of two first-order differential equations,

$$\frac{dN}{dt} = B - \mu(N)\, N, \qquad (24.23a)$$

$$\frac{dB}{dt} = [m_0 - \gamma - \mu(N)]\, B, \qquad (24.23b)$$

for the population size and the birth rate. These two variables are exactly what we need to integrate along characteristics (since the mortality μ depends on population size). If we could integrate equations (24.23a) and (24.23b) exactly, we could determine $n(a, t)$. However, our equations are nonlinear. We will have to settle for determining the asymptotic behavior of $N(t)$ and $B(t)$.

The nontrivial zero-growth isoclines for our $N - B$ system are

$$B = \mu(N)\, N, \qquad (24.24a)$$

$$\mu(N) = m_0 - \gamma. \qquad (24.24b)$$

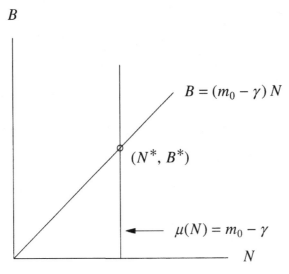

Fig. 24.2. Zero-growth isoclines.

A nontrivial equilibrium occurs (see Figure 24.2) at the intersection of the two lines

$$B = (m_0 - \gamma) N, \tag{24.25}$$

and

$$\mu(N) = m_0 - \gamma. \tag{24.26}$$

The first line is also an invariant manifold. To see this, let

$$x \equiv B - (m_0 - \gamma) N. \tag{24.27}$$

Now

$$\frac{dx}{dt} = \frac{dB}{dt} - (m_0 - \gamma) \frac{dN}{dt}, \tag{24.28}$$

so that

$$\frac{dx}{dt} = -\mu(N) [B - (m_0 - \gamma) N] \tag{24.29}$$

or

$$\frac{dx}{dt} = -\mu x. \tag{24.30}$$

If $x = 0$, $dx/dt = 0$. Furthermore, the distance from the $x = 0$ manifold decreases monotonically.

B

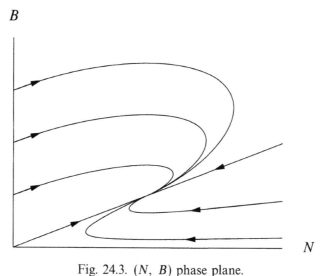

Fig. 24.3. (N, B) phase plane.

On the aforementioned invariant manifold, equation (24.23a) simplifies to

$$\frac{dN}{dt} = [(m_0 - \gamma) - \mu(N)]\, N. \tag{24.31}$$

Under straightforward biological assumptions, this will produce logistic-like growth. It is also clear that there cannot be closed orbits or damped oscillations around our nontrivial equilibrium, since these orbits would then cut the invariant manifold. We expect the nontrivial equilibrium to act like a node (see Figure 24.3). At this node, the stable age distribution reduces to

$$n^*(a) = n^*(0)\, e^{-\mu(N^*)\, a} \tag{24.32}$$

(compare this with equation (24.10)), where $n^*(0)$ is the equilibrial number of births. In light of equations (24.25) and (24.26),

$$n^*(a) = (m_0 - \gamma)\, N^*\, e^{-(m_0 - \gamma)\, a}. \tag{24.33}$$

◇

The fact that equations (24.23a) and (24.23b) lack oscillations is surprising. If we learned anything from our analysis of the linear Lotka integral equation and the linear McKendrick–von Foerster model, it was that the introduction of age structure can lead to oscillations. The early occurrence of reproduction in equations (24.23a) and (24.23b) appears to quench any propensity for oscillations. Does a delay in reproduction set off oscillations?

Problem 24.1 *More trickery*

Consider

$$\frac{\partial n}{\partial a} + \frac{\partial n}{\partial t} = -\mu(N)\, n, \tag{24.34a}$$

$$n(0, t) = \int_0^\infty m_0\, a\, e^{-\gamma a}\, n(a, t)\, da, \tag{24.34b}$$

$$n(a, 0) = n_0(a). \tag{24.34c}$$

Let

$$N(t) \equiv \int_0^\infty n(a, t)\, da, \tag{24.35a}$$

$$G(t) \equiv \int_0^\infty m_0\, e^{-\gamma a}\, n(a, t)\, da, \tag{24.35b}$$

$$B(t) \equiv \int_0^\infty m_0\, a\, e^{-\gamma a}\, n(a, t)\, da. \tag{24.35c}$$

Derive a system of three differential equations for $N(t)$, $G(t)$, and $B(t)$.

EXAMPLE   Juveniles and adults
Consider a population in which only adults (of age τ or older) breed. To keep things simple, let us assume that all individuals suffer constant mortality μ and that the maternity function depends on the number of adults. Thus

$$\frac{\partial n}{\partial a} + \frac{\partial n}{\partial t} = -\mu\, n, \tag{24.36a}$$

$$n(0, t) = \int_\tau^\infty m(N_A)\, n(a, t)\, da = m(N_A)\, N_A, \tag{24.36b}$$

$$n(a, 0) = n_0(a), \tag{24.36c}$$

where

$$N_A \equiv \int_\tau^\infty n(a, t)\, da \tag{24.37}$$

is the number of adults.

If we integrate partial differential equation (24.36a) over ages τ to infinity, we find that

$$\frac{dN_A}{dt} = n(\tau, t) - \mu\, N_A. \tag{24.38}$$

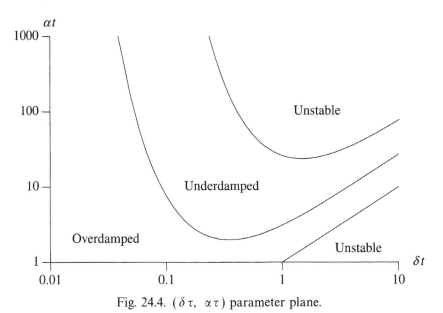

Fig. 24.4. ($\delta\tau$, $\alpha\tau$) parameter plane.

The first term on the right-hand side, $n(\tau, t)$, is the number of new adults (of age τ) at time t. For large times, the number of new adults is simply

$$n(\tau, t) = n(0, t - \tau)\, l(\tau) \tag{24.39}$$

or, in light of our assumptions,

$$n(\tau, t) = m[N_A(t - \tau)]\, N_A(t - \tau)\, e^{-\mu\tau}. \tag{24.40}$$

We thus have a delay-differential equation for the number of adults,

$$\frac{dN_A}{dt} = m[N_A(t - \tau)]\, N_A(t - \tau)\, e^{-\mu\tau} - \mu\, N_A(t), \tag{24.41}$$

with τ, the maturation time, acting as the delay.

Gurney *et al.* (1980), in their study of Nicholson's sheep blowflies, considered a maternity function that decayed exponentially with adult density,

$$m(N_A) = m_0\, e^{-\beta N_A}. \tag{24.42}$$

Equation (24.41) can now be written

$$\frac{dN_A}{dt} = \alpha N_A(t - \tau)\, e^{-\beta N_A(t - \tau)} - \delta\, N_A(t), \tag{24.43}$$

with

$$\alpha = m_0\, e^{-\mu\tau}, \quad \delta = \mu. \tag{24.44}$$

$m(a)$

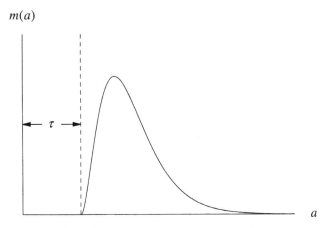

Fig. 24.5. Realistic maternity function.

Some readers may recognize this equation. It was a homework problem from when we first studied delay-differential equations (Chapter 5). This equation has a single nontrivial equilibrium. The equilibrium is not always stable. Small perturbations away from the equilibrium can grow monotonically, decay monotonically, decay in an oscillator manner, or grow in an oscillatory manner (see Figure 24.4) depending on the parameters. Delaying reproduction can set off oscillations. ◇

Greater reality can be achieved by combining the best features of the previous two examples. Nisbet and Gurney (1986) combined separate juvenile and adult stages with a gamma-distribution maternity function for the adults (see Figure 24.5) and obtained a set of delay-differential equations as the simplified age-structured equations.

Nonlinear Leslie matrices have also received attention. Caswell (2001) provides a useful review of this literature; the fishery models of Levin and Goodyear (1980) and DeAngelis *et al.* (1980) figure prominently in this review and in the following example. Density-dependent Leslie matrices are particularly useful when density effects are mediated by one or a few age classes.

EXAMPLE   A simple two-age-class fishery
A two-age-class Leslie matrix takes the form

$$\begin{pmatrix} n_0 \\ n_1 \end{pmatrix}_{t+1} = \begin{pmatrix} F_0 & F_1 \\ P_0 & 0 \end{pmatrix} \begin{pmatrix} n_0 \\ n_1 \end{pmatrix}_{t}. \tag{24.45}$$

We will assume density-dependent survivorship of the youngest individuals.

Let

$$\begin{pmatrix} n_0 \\ n_1 \end{pmatrix}_{t+1} = \begin{pmatrix} P_0 f(n_0) m_1 & P_1 m_2 \\ P_0 f(n_0) & 0 \end{pmatrix} \begin{pmatrix} n_0 \\ n_1 \end{pmatrix}_t. \tag{24.46}$$

P_0 is the density-independent component of the young age-class survivorship and

$$f(n_0) = e^{-\beta n_0} \tag{24.47}$$

is the density-dependent part.

If we abandon our matrix notation, equation (24.46) may be written, and thought of, as a coupled system of nonlinear difference equations,

$$n_{0,t+1} = P_0 f(n_{0,t}) m_1 n_{0,t} + P_1 m_2 n_{1,t}, \tag{24.48a}$$
$$n_{1,t+1} = P_0 f(n_{0,t}) n_{0,t}. \tag{24.48b}$$

This system may be analyzed using standard techniques (see Chapters 5 and 11).

Equations (24.48a) and (24.48b) possess a single nontrivial equilibrium at

$$n_0^* = \frac{1}{\beta} \ln R_0, \tag{24.49a}$$

$$n_1^* = \frac{P_0}{\beta} \frac{\ln R_0}{R_0}, \tag{24.49b}$$

where, by definition,

$$R_0 \equiv P_0 m_1 + P_0 P_1 m_2. \tag{24.50}$$

We must have $R_0 > 1$ for this equilibrium to be in the first quadrant.

We will ascertain the stability of this equilibrium in the usual way. Linearizing about the nontrivial equilibrium produces the Jacobian (community matrix)

$$J = \begin{pmatrix} P_0 m_1 (1 - \ln R_0)/R_0 & P_1 m_2 \\ P_0 (1 - \ln R_0)/R_0 & 0 \end{pmatrix} \tag{24.51}$$

and the characteristic equation

$$\lambda^2 - P_0 m_1 c \lambda - P_0 P_1 m_2 c = 0 \tag{24.52}$$

with

$$c = \frac{(1 - \ln R_0)}{R_0}. \tag{24.53}$$

For the nontrivial equilibrium to be asymptotically stable, both roots of this characteristic equation must be less than 1 in modulus. Equivalently, the characteristic equation must satisfy the Jury conditions.

The first Jury condition yields

$$1 - P_0 m_1 c - P_0 P_1 m_2 c > 0, \tag{24.54}$$

which reduces to

$$R_0 > 1. \tag{24.55}$$

The second Jury condition produces

$$1 + P_0 m_1 c - P_0 P_1 m_2 c > 0, \tag{24.56}$$

which reduces to

$$\ln R_0 < \frac{2 m_1}{m_1 - P_1 m_2} \tag{24.57}$$

if

$$P_1 < m_1/m_2. \tag{24.58}$$

The third Jury condition,

$$|P_0 P_1 m_2 c| < 1, \tag{24.59}$$

reduces to

$$\ln R_0 < 2 + \frac{m_1}{P_1 m_2} \tag{24.60}$$

in light of condition (24.55).

Figure 24.6 highlights the portions of parameter space that satisfy these conditions for the case $m_2 = 1.5 m_1$. For constant P_1, increasing R_0 is destabilizing. The presence of a cusp implies that increasing P_1, for constant R_0, may be stabilizing or destabilizing. Our results match those of Levin (1981), although our notation is like that of DeAngelis *et al.* (1980) in that we census the population after reproduction. Beyond the two boundaries in Figure 24.6, we may expect either a period-doubling bifurcation or a Hopf bifurcation. Various degenerate cases may also arise (Guckenheimer *et al.*, 1977). Figure 24.7 shows a typical bifurcation diagram, for $P_1 = 0.8$, $m_2 = 1.5$, $m_1 = 1.5$, and $\beta = 1$. The equilibrium loses stability via a Hopf bifurcation. Quasiperiodicity eventually yields to phase-locking and an assortment of other bifurcations.

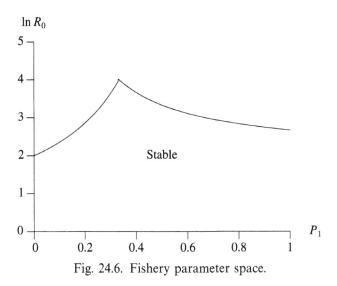

Fig. 24.6. Fishery parameter space.

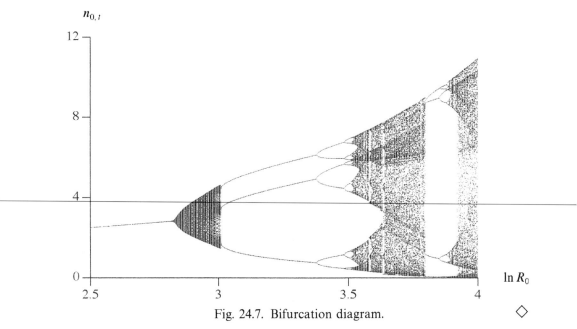

Fig. 24.7. Bifurcation diagram.

Recommended readings

The recent volume by Tuljapurkar and Caswell (1997) includes detailed case studies of a number of nonlinear, age-structured, partial differential equations, delay-differential equations, and projection matrices. A monograph by Cushing (1998) discusses many of the same topics.

Section F
SEX-STRUCTURED MODELS

25 Two-sex models

Populations are often sexually dimorphic for demographic traits. Caswell (2001) has done a good job of highlighting some of these differences. The life expectancy of women exceeds that of men by about 10%. The life expectancy of female Belding's ground squirrels is about 25% greater than that of males (Sherman and Morton, 1984). Female black widow spiders (Deevey and Deevey, 1945) typically live 170% longer than males. However, I do not want to give the impression that females always outlive males. Males of many species of monogamous birds live longer than females (Darwin, 1871; Breitwisch, 1989), for reasons that vary from differences in natal dispersal (Greenwood and Harvey, 1982) to differences in parental investment. Differences in survivorship frequently manifest themselves as a skewed sex ratio in the population.

Some species show sex-based differences in the age of maturity. Female sperm whales attain sexual maturity at 7 to 9 years. Males may become physiologically mature at 19 years, but do not reach social maturity for 25 to 27 years (Nowak, 1991). Most organisms have sex-based differences in fecundity.

As a result, life tables and demographic models for males and for females often give different predictions. Kuczynski (1931) calculated the male and female net reproductive rates for France for 1920–1923. The male rate was 1.194 and the female rate was 0.977. If one were to follow these calculations to their logical conclusion, one would now expect France to consist entirely of men ! Kuczynski attributed the difference in the rates to the skew in sex ratio caused by the First World War.

Male and female predictions can be reconciled by assuming that the limiting sex determines a population's dynamics. However, this approach has its own drawbacks and there are alternative approaches.

I will begin with two-sex models that assume no added structure, before turning to age-structured two-sex models.

Age-independent models

A time-honored approach, due to Kendall (1949) and Goodman (1953), is to write

$$\frac{dF}{dt} = -\mu_f F + b_f \Lambda(F, M), \tag{25.1a}$$

$$\frac{dM}{dt} = -\mu_m M + b_m \Lambda(F, M), \tag{25.1b}$$

where $F(t)$ is the number of females, $M(t)$ is the number of males, $b_f \Lambda(F, M)$ is the number of female births, $b_m \Lambda(F, M)$ is the number of male births. Males and females are assumed to have different birth and death rates. The primary sex ratio (sex ratio at birth), b_m/b_f, is close to 1.05 for most human populations.

The simplest version of this model is the symmetric (Kendall, 1949)

$$\frac{dF}{dt} = -\mu F + b \Lambda(F, M), \tag{25.2a}$$

$$\frac{dM}{dt} = -\mu M + b \Lambda(F, M). \tag{25.2b}$$

Subtraction gives

$$\frac{dF}{dt} - \frac{dM}{dt} = -\mu(F - M) \tag{25.3}$$

so that

$$F(t) - M(t) = [F(0) - M(0)] e^{-\mu t}. \tag{25.4}$$

Since male and female birth and death rates are equal, any disparity in the numbers of either sex quickly vanishes.

To say more, we must assume a form for Λ. It is tempting to assume the law of mass action,

$$\Lambda = FM, \tag{25.5}$$

so that equations (25.2a) and (25.2b) reduce to

$$\frac{dF}{dt} = -\mu F + bFM, \tag{25.6a}$$

$$\frac{dM}{dt} = -\mu M + bFM. \tag{25.6b}$$

However, the number of births per unit time then varies as the square of the population size and we may end up with both an Allee effect and blow-up in finite time. To see this, let $F(0) = M(0)$. Then

$$\frac{dF}{dt} = -\mu F \left(1 - \frac{b}{\mu} F\right). \tag{25.7}$$

For $F < \mu/b$, the population decays to extinction. However, for F large, equation (25.7) is approximated by

$$\frac{dF}{dt} \approx b F^2, \tag{25.8}$$

which has a solution,

$$F(t) = \frac{F_0}{1 - F_0 b t}, \tag{25.9}$$

that blows up in finite time.

These difficulties can be avoided if Λ is linear in the population size. We would then have a two-sex analog of exponential growth. There are many ways to achieve this goal:

Female dominance
Let

$$\Lambda(F, M) = F. \tag{25.10}$$

This is reasonable if females are limiting and there are always enough males to fertilize all females. This may be true for some species that lek.

Equations (25.1a) and (25.1b) now take the form

$$\frac{dF}{dt} = (b_f - \mu_f) F, \tag{25.11a}$$

$$\frac{dM}{dt} = b_m F - \mu_m M. \tag{25.11b}$$

The first of these two equations has the simple solution

$$F(t) = F(0) e^{(b_f - \mu_f) t} \tag{25.12}$$

so that equation (25.11b) reduces to

$$\frac{dM}{dt} + \mu_m M = b_m F(0) e^{(b_f - \mu_f) t}. \tag{25.13}$$

This last equation may be solved with an integrating factor. After we multiply through by $e^{\mu_m t}$, equation (25.13) may be rewritten

$$\frac{d}{dt} (M e^{\mu_m t}) = b_m F(0) e^{(b_f - \mu_f + \mu_m) t} \tag{25.14}$$

so that the male population size is

$$M(t) = \frac{b_m F_0}{b_f - \mu_f + \mu_m} e^{(b_f - \mu_f)t} + c e^{-\mu_m t}. \tag{25.15}$$

For large times, the sex ratio tends towards a constant,

$$\lim_{t \to \infty} \frac{M(t)}{F(t)} = \frac{b_m}{b_f - \mu_f + \mu_m}, \tag{25.16}$$

that depends on the various birth and death rates. Typically, we may expect a skew in the sex ratio.

In 1992, just over two million (2 175 613) Americans died. The crude death rates for males and females were 901.6 deaths per 100 000 males and 806.5 deaths per 100 000 females. In 1993, there were just over four million (4 000 240) births to some 131 983 000 females. The crude birth rate was thus 30.3 births per 1000 females. The primary sex ratio was 1.05. These facts imply that $b_f = 0.0148$, $b_m = 0.0155$, $\mu_f = 0.0081$, and $\mu_m = 0.0090$. Equation (25.16), predicts an asymptotic sex ratio of

$$\frac{M(\infty)}{F(\infty)} = \frac{0.0155}{0.0148 - 0.0081 + 0.0090} = 0.987. \tag{25.17}$$

The current sex ratio is 0.953. The asymptotic growth rate is

$$r = b_f - \mu_f = 0.0067 \text{ per year.} \tag{25.18}$$

This implies a doubling time of just over 103 years.

Male dominance
Let

$$\Lambda(F, M) = M. \tag{25.19}$$

This is reasonable if males are limiting. By symmetry, we now obtain

$$\lim_{t \to \infty} \frac{M(t)}{F(t)} = \frac{b_m - \mu_m + \mu_f}{b_f} \tag{25.20}$$

as the asymptotic sex ratio and

$$r = b_m - \mu_m \tag{25.21}$$

as the eventual asymptotic rate of growth.

There were roughly 125 800 000 males in America in 1993 when just over 4 million births occurred. The crude birth rate was 31.8 births per 1000 males. The primary sex ratio was 1.05. For $b_f = 0.0155$, $b_m = 0.0163$,

$\mu_f = 0.0081$, and $\mu_m = 0.0090$, equation (25.20) predicts an asymptotic sex ratio of

$$\frac{M(\infty)}{F(\infty)} = \frac{0.0163 - 0.0090 + 0.0081}{0.0155} = 0.994. \tag{25.22}$$

The asymptotic growth rate reduces to

$$r = 0.0163 - 0.0090 = 0.0073 \text{ per year}, \tag{25.23}$$

with a doubling time of just under 95 years.

Neither model is entirely satisfactory. The female-dominance model assumes that females can reproduce in the absence of males. The male-dominance model assumes that males can somehow reproduce without females.

Intermediate dominance

Various intermediate-dominance models require both sexes for reproduction. Here are three:

(1)   Geometric mean

$$\Lambda = \sqrt{F M}, \tag{25.24}$$

(2)   Minimum

$$\Lambda = \min(F, M), \tag{25.25}$$

(3)   Harmonic mean

$$\Lambda = \frac{2 F M}{F + M}. \tag{25.26}$$

Problem 25.1 *Geometric mean births*
Determine the asymptotic sex ratio for the model

$$\frac{dF}{dt} = -\mu_f F + b_f \sqrt{F M}, \tag{25.27a}$$

$$\frac{dM}{dt} = -\mu_m M + b_m \sqrt{F M}. \tag{25.27b}$$

Hint   Try the change of variables $R^2(t) = F(t)$, $S^2(t) = M(t)$.

Equation (25.26) may look strange. However, it is often considered the least objectionable of the intermediate-dominance models. It was introduced by Keyfitz (1972) in a paper entitled 'The mathematics of sex and marriage'. Keyfitz argued that one should take a weighted mean of males and females

$$\Lambda = DM + (1 - D)F, \tag{25.28}$$

and that the weighting or dominance should be determined by the relative abundance of females

$$D = \frac{F}{F + M}. \tag{25.29}$$

If females are rare, D is small and we tend towards female dominance. If males are rare, D is close to one and we have a male-dominance model. Equations (25.28) and (25.29) together give us the harmonic mean (25.26). At the same time, there has been very little testing of any intermediate-dominance model.

For the harmonic mean, equations (25.1a) and (25.1b) reduce to

$$\frac{dF}{dt} = -\mu_f F + b_f \frac{2FM}{F + M}, \tag{25.30a}$$

$$\frac{dM}{dt} = -\mu_m M + b_m \frac{2FM}{F + M}. \tag{25.30b}$$

Consider the ratio

$$x \equiv \frac{M}{F}. \tag{25.31}$$

As the sex ratio tends to a constant,

$$\frac{dx}{dt} = \frac{F\dot{M} - M\dot{F}}{F^2} \to 0 \tag{25.32}$$

so that

$$F\frac{dM}{dt} = M\frac{dF}{dt}. \tag{25.33}$$

Substituting equations (25.30a) and (25.30b) into equation (25.33) gives us (after some algebra)

$$\frac{M(\infty)}{F(\infty)} = \frac{2b_m - (\mu_m - \mu_f)}{2b_f - (\mu_f - \mu_m)}. \tag{25.34}$$

Let us call this ratio s. If the population has converged to this ratio,

$$M(t) = sF(t), \tag{25.35}$$

equation (25.30a) reduces to

$$\frac{dF}{dt} = -\mu_f F + b_f \frac{2s}{1 + s} F. \tag{25.36}$$

In other words, the asymptotic growth rate is

$$r = \frac{2s}{1 + s} b_f - \mu_f. \tag{25.37}$$

In 1993, there were close to 131 983 000 females and 125 800 000 males in the USA. The harmonic mean of these two counts is just over 128 817 000. The crude birth rate in that year was roughly 31 births per harmonic mean individual. With a primary sex ratio of 1.05, we thus have $b_f = 0.0151$, $b_m = 0.0159$, $\mu_f = 0.0081$, and $\mu_m = 0.0090$. The predicted asymptotic sex ratio is

$$s = \frac{M(\infty)}{F(\infty)} = \frac{2(0.0159) - 0.0090 + 0.0081}{2(0.0151) - 0.0081 + 0.0090} = 0.994 \tag{25.38}$$

and the resulting growth rate,

$$r = 0.007 \text{ per year}, \tag{25.39}$$

produces a doubling time of between 99 and 100 years.

We can extend these models in many ways:

(1) $\Lambda(F(t), M(t))$ should really be $\Lambda(F(t - \tau), M(t - \tau))$. We should include a gestation time delay in these models.

(2) We have ignored density dependence. We should perhaps consider a model of the form

$$\frac{dF}{dt} = -\mu_f F + b_f \Lambda(F, M) - \alpha_{ff} F^2 - \alpha_{fm} FM, \tag{25.40a}$$

$$\frac{dM}{dt} = -\mu_m M + b_m \Lambda(F, M) - \alpha_{mf} MF - \alpha_{mm} M^2. \tag{25.40b}$$

(3) We could follow the lead of Goodman (1953) and consider stochastic two-sex processes.

(4) We can also follow the lead of Kendall (1949) and consider more complicated social dynamics. For example, let F, M, and C be the number of unmarried females, unmarried males, and couples in a population. It then makes sense to consider

$$\frac{dF}{dt} = -\mu_f F + (b_f + \mu_m) C - W(F, M), \tag{25.41a}$$

$$\frac{dM}{dt} = -\mu_m M + (b_m + \mu_f) C - W(F, M), \tag{25.41b}$$

$$\frac{dC}{dt} = W(F, M) - (\mu_f + \mu_m) C. \tag{25.41c}$$

(5) Many populations are polygamous. Rosen (1983) has extended the harmonic-mean intermediate-dominance model to incorporate polygamous mating systems.

Lindström and Kokko (1998) have looked at the interactions between two of these extensions in the context of the discrete-time population model

$$F_{t+1} = \frac{2\,k\,F\,M}{h^{-1}\,F\,+\,M}\,e^{-\mu_f\,(F_t\,+\,M_t)}, \tag{25.42a}$$

$$M_{t+1} = \frac{2\,k\,F\,M}{h^{-1}\,F\,+\,M}\,e^{-\mu_m\,(F_t\,+\,M_t)}, \tag{25.42b}$$

where F_t and M_t are the number of females and males at time t. This model contains a harmonic birth function that has been modified to incorporate polygyny. The fecundity k is the number of female offspring per female; h is the average harem size. The primary sex ratio is 1:1. The model also contains strong, overcompensatory density dependence. One can study the interactions between sex, mating system, and density dependence using this simple model.

The authors first compare the dynamics of a monogamous ($h = 1$) version of their two-sex model (for $\mu_f = \mu_m = 1$) with a one-sex Ricker model. They conclude that the introduction of two sexes stabilizes dynamics in that bifurcations occur later in k for the monogamous two-sex model than for the one-sex Ricker model. However, the differences between the two models all but vanish if the harem size is increased to $h = 10$. If one introduces sex-specific differences in the response to population density by letting μ_f and μ_m differ, the dynamics can become quite complicated. Stabilization or destabilization can both occur for different choices of μ_f and μ_m.

Age-dependent models

Keyfitz (1972) introduced a continuous-time, age-structured, two-sex model that may be written

$$\frac{\partial f}{\partial t} + \frac{\partial f}{\partial a} = -\mu_f(a)\,f(a, t), \tag{25.43a}$$

$$\frac{\partial m}{\partial t} + \frac{\partial m}{\partial a} = -\mu_m(a)\,m(a, t), \tag{25.43b}$$

$$f(0, t) = \frac{1}{F(t) + M(t)}\int_0^\infty \int_0^\infty \beta(a, a')\,f(a, t)\,m(a', t)\,da\,da', \tag{25.43c}$$

$$m(0, t) = s\,f(0, t), \tag{25.43d}$$

$$f(a, 0) = f_0(a), \tag{25.43e}$$

$$m(a, 0) = m_0(a), \tag{25.43f}$$

where $f(a, t)$ is the density of females, $m(a, t)$ is the density of males, $F(t)$ is the total number of females, $M(t)$ is the total number of males, $\beta(a)$ is the maternity function, and s is the primary sex ratio at birth. Caswell and Weeks (1986) and Caswell (2001), in turn, introduced and studied two-sex Leslie matrices. These are good places to start for a general introduction to this literature. Since I am almost at the end of this book, I will instead focus on a recent applied problem.

EXAMPLE Population dynamics of alligators (Woodward and Murray, 1993)

Woodward and Murray developed a nonlinear age-structured population model for *Alligator mississippiensis*. Alligators are unusual in that they have a highly skewed sex ratio with as many as 10 females for every male. Ferguson and Joanen (1982), collected and incubated over 8000 eggs. They found a primary sex ratio of roughly five females for every male.

Alligators, like many lizards, turtles, and crocodilians, rely on environmental sex determination (ESD) rather than genetic or genotypic sex determination (GSD). In alligators, sex is determined by the incubation temperature. The artificial incubation of alligator eggs at low temperatures (29–31.5 °C) and at high temperatures (35 °C) yields female hatchlings. All young are male if the incubation temperature is 32.5–33 °C. Temperatures between (32 °C and from 33.5 °C to 34.5 °C) produce both sexes. One can often predict the sex of alligator offspring from the nest. Wet marsh nests are cool (30 °C) and hatch females while elevated and dry levee nests are quite warm (34 °C) and hatch males. Dry marsh nests have an intermediate temperature profile and produce an intermediate number of males and females. Marsh nests are far more common than levee nests (Ferguson and Joanen, 1982), consistent with the observed excess of females.

Woodward and Murray made several simplifying assumptions. They assumed that females prefer to nest in the same habitat in which they were raised and, failing that, that they prefer wet marsh to dry marsh to levee. (There is some evidence that alligators hatched from eggs incubated at low temperature have larger yolk reserves and grow faster than those hatched at high temperatures.) They also assumed that alligator populations are regulated by intraspecific competition for nesting sites. The partial differential equation part of their model is

$$\frac{\partial f_1}{\partial t} + \frac{\partial f_1}{\partial a} = -\mu(a) f_1(a, t), \tag{25.44a}$$

$$\frac{\partial f_2}{\partial t} + \frac{\partial f_2}{\partial a} = -\mu(a) f_2(a, t), \tag{25.44b}$$

$$\frac{\partial m_2}{\partial t} + \frac{\partial m_2}{\partial a} = -\mu(a)\, m_2(a, t), \tag{25.44c}$$

$$\frac{\partial m_3}{\partial t} + \frac{\partial m_3}{\partial a} = -\mu(a)\, m_3(a, t), \tag{25.44d}$$

where $f_1(a, t)$ is the density of females born in the wet marsh, $f_2(a, t)$ is the density of females born in the dry marsh, $m_2(a, t)$ is the density of males born in the dry marsh, and $m_3(a, t)$ is the density of males born on the levees. Each group is assumed to have the same age-specific mortality rate.

Things become more interesting when we look at the boundary conditions. Starting in the wet marsh, Woodward and Murray wrote

$$f_1(0, t) = \int_0^\omega f_1(a, t) \left[C\, S\, \beta(a)\, \frac{K_1}{K_1 + Q_1(t)} \right] da, \tag{25.45}$$

where C is the clutch size, S is the survival rate of eggs and hatchlings, $\beta(a)$ is the age-specific component of natality (sorry, $m(a)$ is already being used for male density), K_1 is the most nests that can be built in the wet marsh, and $Q_1(t)$ is the number of sexually mature females who were born in the wet marsh. The fraction $K_1/(K_1 + Q_1)$ is the relative proportion of sexually mature females, born in the wet marsh, that stay in the wet marsh. When Q_1 is small, this fraction is close to one. When Q_1 is large, this fraction is close to zero. The choice of this function was capricious. Indeed, when $Q_1 = K_1$, half the females leave the wet marsh.

We now move on to the production of females in the dry marsh. Woodward and Murray wrote

$$
\begin{aligned}
f_2(0, t) = {} & \frac{1}{2} \int_0^\omega f_1(a, t) \left[C\, S\, \beta(a)\, \frac{K_2}{K_2 + Q_1(t) + Q_2(t)}\, \frac{Q_1(t)}{K_1 + Q_1(t)} \right] da \\
& + \frac{1}{2} \int_0^\omega f_2(a, t) \left[C\, S\, \beta(a)\, \frac{K_2}{K_2 + Q_1(t) + Q_2(t)} \right] da.
\end{aligned} \tag{25.46}
$$

The second fraction in the first integral is the fraction of wet marsh females that have left the wet marsh. The first fraction in both integrals is supposed to be the fraction of females that are successful in nesting in the dry marsh. This is not quite right. There are not Q_1 wet marsh females trying to breed in the dry marsh. Therefore they should not all be counted against the carrying capacity. Finally, Woodward and Murray have assumed that half the alligators born in the dry marsh will be female. Hence the half before each integral.

It is worth noting that this is a female-dominance model in that births are proportional to females. This may seem peculiar, given the paucity of males. However, the mating system is polygamous; male alligators control harems of females. Male numbers are not limiting.

The male birth rate in the dry marsh is, by assumption, the same as the female birth rate:

$$m_2(0, t) = \frac{1}{2} \int_0^{\omega} f_1(a, t) \left[C\,S\,\beta(a) \frac{K_2}{K_2 + Q_1(t) + Q_2(t)} \frac{Q_1(t)}{K_1 + Q_1(t)} \right] da$$

$$+ \frac{1}{2} \int_0^{\omega} f_2(a, t) \left[C\,S\,\beta(a) \frac{K_2}{K_2 + Q_1(t) + Q_2(t)} \right] da. \qquad (25.47)$$

However, for the levee, we have

$$m_3(0, t) = \int_0^{\omega} f_1(a, t) \left[C\,S\,\beta(a) \frac{K_3}{K_3 + Q_1(t) + Q_2(t)} \right.$$

$$\times \left. \frac{Q_1(t) + Q_2(t)}{K_2 + Q_1(t) + Q_2(t)} \frac{Q_1(t)}{K_1 + Q_1(t)} \right] da$$

$$+ \int_0^{\omega} f_2(a, t) \left[CS\beta(a) \frac{K_3}{K_3 + Q_1(t) + Q_2(t)} \right.$$

$$\times \left. \frac{Q_1(t) + Q_2(t)}{K_2 + Q_1(t) + Q_2(t)} \right] da. \qquad (25.48)$$

Data for the age-specific natality and mortality were taken from Smith and Webb (1985). The carrying capacities were taken to be $K_1 : K_2 : K_3 = 79.7 : 13.6 : 6.7$.

This looks like a terribly complicated system. It should be. It includes spatial structure, age structure, and the existence of two sexes. Nevertheless, partial differential equation (25.44a) and its boundary condition, equation (25.45), decouple from the rest of the equations. It follows that the net reproductive rate in the wet marsh is simply

$$R_1(Q_1(t)) = \int_0^{\omega} l(a) \left[CS\beta(a) \frac{K_1}{K_1 + Q_1(t)} \right] da, \qquad (25.49)$$

where the survivorship,

$$l(a) = e^{-\int_0^a \mu(s)\,ds},$$

is obtained by integrating equation (25.44a). Woodward and Murray set this net reproductive rate equal to 1 to look for equilibria. They showed that 'in most problems of interest' there is at least one value of Q_1^* for which $R_1(Q_1^*) = 1$. They also observed that a stable solution of the female population in the wet marsh inevitably gives a stable solution for the males and females in the other two regions. They replaced the derivatives by finite differences and the integrals by quadrature formulae (Kostova, 1990) and numerically solve the full system to obtain an equilibrium sex ratio of just under $1:8$. Finally, they looked at the special case $K_2 = 0$, which implies that $Q_2 = 0$, and noted that the sex ratio can then be obtained as the ratio of the net reproductive rate in the levee to that in the wet marsh,

$$\frac{R_3(Q_1^*)}{R_1(Q_1^*)} = \int_0^\omega l(a) \left[CS\beta(a) \frac{K_3}{K_3 + Q_1^*} \frac{Q_1^*}{K_1 + Q_1^*} \right] da. \qquad (25.50)$$

Note that I have used the fact that $R_1(Q_1^*) = 1$. These are remarkably clean results for this complicated a model. ◇

The study of two-sex problems is a fascinating but oft-neglected part of mathematical ecology.

Recommended readings

Keyfitz (1968) and Pollard (1973) review much of the classical literature on two-sex models.

References

Abrams, P. 1983. The theory of limiting similarity. *Annual Review of Ecology and Systematics*, **14**, 359–376.

Adler, F. 1990. Coexistence of two types on a single resource in discrete time. *Journal of Mathematical Biology*, **28**, 695–713.

Ahmadjian, V. and Paracer, S. 1986. *Symbiosis: An Introduction to Biological Associations*. University Press of New England, Hanover, NH.

Allen, J. C. 1990. Chaos and phase-locking in predator–prey models in relation to functional response. *Florida Entomologist*, **73**, 100–110.

Alvarez, I., Fajardo, R., Lopez, E., Hemachudha, T., Kamolvarin, N., Cortes, G., and Baier, G. M. 1994. Partial recovery from rabies in a nine-year-old boy. *Pediatric Infectious Disease Journal*, **12**, 1154–1155.

Andrewartha, H. G. and Birch, L. C. 1954. *The Distribution and Abundance of Animals*. University of Chicago Press, Chicago, IL.

Andrews, J. F. 1968. A mathematical model for the continuous culture of microorganisms utilizing inhibitory substances. *Biotechnology and Bioengineering*, **10**, 707–723.

Andronov, A. and Leontovich, E. 1939. Some cases of the dependence of the limit cycles upon parameters. *Uchenye zapiski Gor'kovskogo Gosudarstvennogo Universiteta*, **6**, 3–24.

Aronson, D. G., Chory, M. A., Hall, G. R., and McGehee, R. P. 1980. A discrete dynamical system with subtly wild behavior. In *New Approaches to Nonlinear Problems in Dynamics*, P. J. Holmes, editor. SIAM, Philadelphia, PA, pp. 339–359.

Aronson, D. G., Hall, M. A., Chory, G. R., and McGehee, R. P. 1982. Bifurcations from an invariant circle for two-parameter families of maps of the plane: a computer assisted study. *Communications in Mathematical Physics*, **83**, 303–354.

Arrowsmith, D. K. and Place, C. M. 1992. *Dynamical Systems: Differential Equations, Maps, and Chaotic Behaviour*. Chapman & Hall, London, UK.

Bailey, N. T. 1964. *The Elements of Stochastic Processes*. John Wiley & Sons, New York.

Ball, P. 1999. *The Self-Made Tapestry: Pattern Formation in Nature*. Oxford University Press, Oxford.

Banks, R. B. 1994. *Growth and Diffusion Phenomena: Mathematical Frameworks and Applications*. Springer-Verlag, Berlin.

Bardi, M. 1981. Predator–prey models in periodically fluctuating environments. *Journal of Mathematical Biology*, **12**, 127–140.

Bazykin, A. D. 1998. *Nonlinear Dynamics of Interacting Populations*. World Scientific, Singapore.

Beattie, A. J. 1985. *The Evolutionary Ecology of Ant–Plant Mutualisms*. Cambridge University Press, Cambridge.

Beavis, B. and Dobbs, I. 1990. *Optimization and Stability Theory for Economic Analysis*. Cambridge University Press, Cambridge.

Beddington, J. R., Free, C. A., and Lawton, J. H. 1975. Dynamic complexity in predator–prey models framed in difference equations. *Nature*, **255**, 58–60.

Bell, G., Handford, P., and Dietz, C. 1977. Dynamics of an exploited population of Lake Whitefish (*Coregonus clupeaformis*). *Journal of the Fisheries Research Board of Canada*, **34**, 942–953.

Bellman, R. and Cooke, K. L. 1963. *Differential Difference Equations*. Academic Press, New York.

Bellman, R. and Harris, T. 1948. On the theory of age-dependent stochastic branching processes. *Proceedings of the National Academy of Sciences of the United States of America*, **34**, 601–604.

Bellman, R. and Harris, T. 1952. On age-dependent binary branching processes. *Annals of Mathematics*, **55**, 280–295.

Belt, T. 1874. *The Naturalist in Nicaragua*. J. Murray, London.

Bendixson, I. 1901. Sur les courbes définies par des équations différentielles. *Acta Mathematica*, **24**, 1–88.

Benoiston de Châteauneuf, L. F. 1847. Mémoire sur la durée des familes nobles de France. *Memoires de l'Académie royale des sciences morales et politique de l'Institut de France*, **5**, 753–794.

Berge, P., Pomeau, Y., and Vidal, C. 1984. *Order with Chaos: Towards a Deterministic Approach to Turbulence*. John Wiley & Sons, New York.

Bernardelli, H. 1941. Population waves. *Journal of the Burma Research Society*, **31**, 1–18.

Bernussou, J. 1977. *Point Mapping Stability*. Pergamon Press, New York.

~~Berryman, A. A. and Millstein, J. A. 1989. Are ecological systems chaotic—and if~~ not, why not? *Trends in Ecology and Evolution*, **4**, 26–28.

Beverton, R. J. H. and Holt, S. J. 1957. On the dynamics of exploited fish populations. *Fishery Investigations, Series II*, **19**, 1–533.

Bharucha-Reid, A. T. 1997. *Elements of the Theory of Markov Processes and Their Applications*. Dover Publications, Mineola, NY.

Bienaymé, I. J. 1845. De la loi de multiplication et de la durée des familles. *Société Philomatique de Paris Extraits*, **5**, 37–39.

Birkhoff, G. 1950. *Hydrodynamics*. Princeton University Press, Princeton, NJ.

Birkhoff, G. and Rota, G.-C. 1978. *Ordinary Differential Equations*. John Wiley & Sons, New York.

Bock, W. J., Balda, R. P., and Vander Wall, S. B. 1973. Morphology of the sublingual pouch and tongue musculature of the Clark's nutcracker. *Auk*, **90**, 491–519.

Boltyanskii, V. G. 1971. *Mathematics Methods of Optimal Control*. Holt, Rinehart and Winston, Inc., New York.

Boucher, D. H. 1985. *The Biology of Mutualism: Ecology and Evolution*. Oxford University Press, New York.

Braithwaite, R. W. and Lee, A. K. 1979. A mammalian example of semelparity. *American Naturalist*, **113**, 151–155.

Braun, M. 1978. *Differential Equations and Their Applications*. Springer-Verlag, New York.

Breitwisch, R. 1989. Mortality patterns, sex ratios, and parental investment in monogamous birds. *Current Ornithology*, **6**, 1–50.

Britton, N. F. 1986. *Reaction-Diffusion Equations and Their Applications to Biology*. Academic Press, London.

Bulmer, M. 1994. *Theoretical Evolutionary Ecology*. Sinauer Associates, Inc., Sunderland, MA.

Butler, G. J. and Waltman, P. 1981. Bifurcation from a limit cycle in a two prey one predator ecosystem modeled on a chemostat. *Journal of Mathematical Biology*, **12**, 295–310.

Campbell, B. and Lack, E. 1985. *Dictionary of Birds*. Buteo Books, Vermillion, SD.

Canale, R. P. 1970. An analysis of models describing predator–prey interaction. *Biotechnology and Bioengineering*, **12**, 353–378.

Candolle, A. de. 1873. *Histoire des Sciences et des Savants depuis Deux Siècles*. H. Georg, Geneva.

Canosa, J. 1973. On a nonlinear diffusion equation describing population growth. *IBM Journal of Research and Development*, **17**, 307–313.

Carleton, W. M. 1965. Food habits of two sympatric Colorado sciurids. *Journal of Mammalogy*, **47**, 91–103.

Castillo-Chavez, C. 1989. Some applications of structured models in population dynamics. In *Applied Mathematical Ecology*, S. A. Levin, T. G. Hallam, and L. J. Gross, editors. Springer-Verlag, Berlin, pp. 450–470.

Caswell, H. 1989. *Matrix Population Models*. Sinauer Associates, Inc., Sunderland, MA.

Caswell, H. 2001. *Matrix Population Models: Construction, Analysis, and Interpretation*. Sinauer Associates, Inc., Sunderland, MA.

Caswell, H., Fujiwara, M., and Brault, S. 1999. Declining survival probability threatens the North Atlantic right whale. *Proceedings of the National Academy of Sciences of the United States of America*, **96**, 3308–3313.

Caswell, H. and Weeks, D. E. 1986. Two-sex models: chaos, extinction, and other dynamic consequences of sex. *American Naturalist*, **128**, 707–735.

Causton, D. R. and Venus, J. C. 1981. *The Biometry of Plant Growth*. Edward Arnold, London.

Cesari, L. 1983. *Optimization—Theory and Applications*. Springer-Verlag, New York.

Charlesworth, B. 1994. *Evolution in Age-Structured Populations*. Cambridge University Press, Cambridge.

Chiang, A. C. 1992. *Elements of Dynamics Optimization*. McGraw-Hill, New York.

Chiang, C. L. 1980. *An Introduction to Stochastic Processes and Their Applications*. R. E. Krieger Publishing Company, Huntington, NY.

Clark, C. W. 1990. *Mathematical Bioeconomics: The Optimal Management of Renewable Resources*. John Wiley & Sons, New York.

Clark, C. W., Clarke, F. H., and Munro, G. R. 1979. The optimal exploitation of renewable resource stocks: problems of irreversible investment. *Econometrica*, **47**, 25–47.

Coddington, E. A. and Levinson, N. 1955. *Theory of Ordinary Differential Equations*. McGraw-Hill, New York.

Cole, L. C. 1954. The population consequences of life history phenomena. *Quarterly Review of Biology*, **19**, 103–137.

Cole, L. C. 1957. Sketches of general and comparative demography. *Cold Spring Harbor Symposium on Quantitative Biology*, **22**, 1–15.

Coleman, C. S. 1983. Biological cycles and the fivefold way. In *Differential Equation Models*, M. Braun, C. S. Coleman, and D. A. Drew, editors. Springer-Verlag, New York, pp. 251–278.

Collings, J. B. 1997. The effects of the functional response on the bifurcation behavior of a mite predator–prey interaction model. *Journal of Mathematical Biology*, **36**, 149–168.

Connell, J. H. 1978. Diversity of tropical rain forests and coral reefs. *Science*, **199**, 1302–1309.

Conrad, J. M. and Clark, C. W. 1987. *Natural Resource Economics: Notes and Problems*. Cambridge University Press, Cambridge, UK.

Conway, E. D. and Smoller, J. A. 1986. Global analysis of a systems of predator–prey equations. *SIAM Journal on Applied Mathematics*, **46**, 630–642.

Crawley, M. J. 1992. *Natural Enemies: The Population Biology of Predators, Parasites, and Disease*. Blackwell Scientific Publications, Oxford.

Crone, E. E. 1997. Delayed density dependence and the stability of interacting populations and subpopulations. *Theoretical Population Biology*, **51**, 67–76.

Crouse, D. T., Crowder, L. B., and Caswell, H. 1987. A stage-based population model for loggerhead sea turtles and implications for conservation. *Ecology*, **68**, 1412–1423.

Cunningham, A. and Nisbet, R. M. 1983. Transients and oscillations in continuous culture. In *Mathematics in Microbiology*, M. Bazin, editor. Academic Press, London, pp. 77–103.

Cushing, J. M. 1977a. *Integrodifferential Equations and Delay Models in Population Dynamics*. Springer-Verlag, New York.

Cushing, J. M. 1977b. Periodic time-dependent predator–prey systems. *SIAM Journal on Applied Mathematics*, **32**, 82–95.

Cushing, J. M. 1980. Two species competition in a periodic environment. *Journal of Mathematical Biology*, **10**, 384–400.

Cushing, J. M. 1982. Periodic Kolmogorov systems. *SIAM Journal on Mathematical Analysis*, **13**, 811–827.

Cushing, J. M. 1998. *An Introduction to Structured Population Dynamics*. SIAM, Philadelphia.

d'Alembert, J. le R. 1749. Recherches sur la courbe que forme une corde tendue mise en vibration. *Histoire de l'Académie Royale des Sciences et des Belles-Lettres de Berline*, **3** (1747/1749), 214–219.

d'Alembert, J. le R. 1752. "Addition" to 1749 paper. *Histoire de l'Académie Royale des Sciences et des Belles-Lettres de Berlin*, **6** (1750/1752), 355–360.

D'Ancona, U. 1926. *Dell'influenza della stasi peschereccia del periodo 1914–1918 sul patrimonio ittico dell'Alto Adriatico*. Regio Comitato Talassografico Italiano, Memoria CXXVI.

D'Ancona, U. 1954. *The Struggle of Existence*. E. J. Brill, Leiden.

Darwin, C. 1871. *The Descent of Man and Selection in Relation to Sex*. John Murray, London.

Dean, A. M. 1985. The dynamics of microbial commensalisms and mutualisms. In *The Biology of Mutualism: Ecology and Evolution*, D. H. Boucher, editor. Oxford University Press, New York, pp. 270–304.

DeAngelis, D. L., Svoboda, L. J., Cristensen, S. W., and Vaughan, D. S. 1980. Stability and return times of Leslie matrices with density-dependent survival: applications to fish populations. *Ecological Modeling*, **8**, 149–163.

DeBach, P. 1974. *Biological Control by Natural Enemies*. Cambridge University Press, Cambridge.

Deevey, G. B. and Deevey, Jr, E. S. 1945. A life table for the black widow. *Transactions of the Connecticut Academy of Arts and Sciences*, **36**, 115–134.

Demetrius, L. 1971. Primitivity conditions for growth matrices. *Mathematical Biosciences*, **12**, 53–58.

Dhondt, A. A. and Eyckerman, R. 1980. Competition between the great tit and the blue tit outside the breeding season in field experiments. *Ecology*, **61**, 1291–1296.

Doveri, F., Kuznetsov, Y., Muratori, S., Rinaldi, S., and Scheffer, M. 1992. Seasonality and chaos in a plankton–fish model. *Theoretical Population Biology*, **43**, 159–183.

Drazin, P. G. 1992. *Nonlinear Systems*. Cambridge University Press, Cambridge.

Dresner, L. 1999. *Applications of Lie's Theory of Ordinary and Partial Differential Equations*. Institute of Physics Publishing, Bristol.

Easton, D. M. 1995. Gompertz survival kinetics: fall in number alive or growth in number dead? *Theoretical Population Biology*, **48**, 1–6.

Ebenhoh, W. 1988. Coexistence of an unlimited number of algal species in a model system. *Theoretical Population Biology*, **34**, 130–144.

Edelstein-Keshet, L. 1988. *Mathematical Models in Biology*. Random House, New York.

Elaydi, S. N. 1996. *An Introduction to Difference Equations*. Springer-Verlag, New York.

El'sgol'ts, L. E. and Norkin, S. B. 1973. *Introduction to the Theory and Applications of Differential Equations with Deviating Arguments*. Academic Press, New York.

Elton, C. and Miller, R. S. 1954. The ecological survey of animal communities: with a practical system of classifying habitats by structural characters. *Journal of Ecology*, **42**, 460–496.

Emlen, J. M. 1984. *Population Biology: The Coevolution of Population Dynamics and Behavior*. MacMillan, New York.

Farkas, M. 1994. *Periodic Motions*. Springer-Verlag, New York.

Farlow, S. J. 1982. *Partial Differential Equations for Scientists and Engineers*. John Wiley & Sons, New York.

Feller, W. 1941. On the integral equation of renewal theory. *Annals of Mathematical Statistics*, **12**, 243–267.

Ferguson, M. W. J. and Joanen, T. 1982. Temperature of egg-incubation determines sex in *Alligator mississippiensis*. *Nature*, **296**, 850-853.

Fife, P. 1979. *Mathematical Aspects of Reacting and Diffusing Systems*. Springer-Verlag, New York.

Finch, C. E. 1990. *Longevity, Senescence, and the Genome*. University of Chicago Press, Chicago.

Fisher, R. A. 1922. On the dominance ratio. *Proceedings of the Royal Society of Edinburgh*, **42**, 321–341.

Fisher, R. A. 1930a. The distribution of gene ratios for rare mutations. *Proceedings of the Royal Society of Edinburgh*, **50**, 204–219.

Fisher, R. A. 1930b. *The Genetical Theory of Natural Selection*. Oxford University Press, Oxford.

Fisher, R. A. 1937. The wave of advance of advantageous genes. *Annals of Eugenics*, **7**, 355–369.

Foster, R. B. 1977. *Tachigalia versicolor* is a suicidal neotropical tree. *Nature*, **268**, 624–626.

Fourier, J. B. 1822. *Théorie Analytique de la Chaleur*. Didot, Paris. (English translation by A. Freeman: *The Analytical Theory of Heat*. Cambridge University Press, Cambridge, 1878. Reprinted by Dover Publications, Inc., New York, 1955.)

Fox, W. W. 1970. An exponential surplus yield model for optimizing in exploited fish populations. *Transactions of the American Fisheries Society*, **99**, 80–88.

Frauenthal, J. C. 1986. Analysis of age-structure models. In *Mathematical Ecology: An Introduction*, T. G. Hallam and S. A. Levin, editors. Springer-Verlag, Berlin, pp. 117–147.

Freedman, H. I. and Wolkowicz, G. S. K. 1986. Predator–prey systems with group defense: the paradox of enrichment revisited. *Bulletin of Mathematical Biology*, **48**, 493–508.

Frobenius, G. 1912. Über Matrizen aus nicht negativen Elementen. *Sitzungsberichte der Kgl. Preussicschen Akademie der Wissenschaften*, 456–477.

Fuller, E. 1987. *Extinct Birds*. Facts on File Publications, New York, New York, USA.

Funasaki, E. and Kot, M. 1993. Invasion and chaos in a periodically pulsed mass-action chemostat. *Theoretical Population Biology*, **44**, 203–224.

Furry, W. H. 1937. On fluctuation phenomena in the passage of high energy electrons through lead. *Physics Review*, **52**, 569–581.

Galton, F. 1873. *Educational Times*, **19**, 103–105.

Gantmacher, F. R. 1959. *Matrix Theory*. Chelsea Publishing Company, New York.

Gause, G. F. 1934. *The Struggle for Existence*. Hafner Press, New York.

Gause, G. F. 1935. Verifications experimentales de la théorie mathématique de la lutte pour la vie. *Actualités scientifiques et industrielles*, **277**, 1–63.

Gill, F. B. 1995. *Ornithology*. W. H. Freeman and Company, New York.

Gilpin, M. E. 1979. Spiral chaos in a predator–prey model. *American Naturalist*, **113**, 306–8.

Glendinning, P. 1994. *Stability, Instability, and Chaos: An Introduction to the Theory of Nonlinear Differential Equations*. Cambridge University Press, Cambridge.

Godfray, H. C. J. 1994. *Parasitoids: Behavioral and Evolutionary Ecology*. Princeton University Press, Princeton, NJ.

Gompertz, B. 1825. On the nature of the function expressive of the law of mortality, and on a new method of determining the value of life contingencies. *Philosophical Transactions of the Royal Society*, **27**, 513–585.

Goodman, L. A. 1953. Population growth of the sexes. *Biometrics*, **9**, 212–225.

Gopalsamy, K. 1992. *Stability and Oscillations in Delay Differential Equations of Population Dynamics*. Kluwer Academic Publishers, Dordrecht.

Graunt, J. 1662. *Nature and Political Observations Mentioned in a Following Index, and Made Upon the Bills of Mortality*. T. Roycroft, London.

Greenwood, P. J. and Harvey, P. H. 1982. The natal and breeding dispersal of birds. *Annual Review of Ecology and Systematics*, **13**, 1–21.

Grimmett, G. R. and Stirzaker, D. R. 1992. *Probability and Random Processes*. Oxford University Press, Oxford.

Grindrod, P. 1991. *Patterns and Waves: The Theory and Applications of Reaction-Diffusion Equations*. Oxford University Press, Oxford.

Grindrod, P. 1996. *The Theory and Applications of Reaction-Diffusion Equations: Patterns and Waves.* Oxford University Press, Oxford.

Grover, J. P. 1997. *Resource Competition.* Chapman & Hall, London, UK.

Guckenheimer, J. and Holmes, P. 1983. *Nonlinear Oscillations, Dynamical Systems, and Bifurcations of Vector Fields.* Springer-Verlag, New York.

Guckenheimer, J., Oster, G. F., and Ipaktchi, A. 1977. The dynamics of density dependent population models. *Journal of Mathematical Biology*, **4**, 101–147.

Gurney, W. S. C., Blythe, S. P., and Nisbet, R. M. 1980. Nicholson's blowflies revisited. *Nature*, **287**, 17–22.

Gurney, W. S. C. and Nisbet, R. M. 1975. The regulation of inhomogeneous populations. *Journal of Theoretical Biology*, **52**, 441–457.

Gurtin, M. E. and MacCamy, R. C. 1974. Non-linear age-dependent population dynamics. *Archive for Rational Mechanics and Analysis*, **54**, 281–300.

Gurtin, M. E. and MacCamy, R. C. 1979. Some simple models for nonlinear age-dependent population dynamics. *Mathematical Biosciences*, **43**, 199–211.

Haberman, R. 1983. *Elementary Applied Partial Differential Equations.* Prentice-Hall, Inc., Englewood Cliffs, NJ.

Hadwiger, H. 1940. Eine analytische Reproduktionsfunktion für biologische Gesamtheiten. *Skandinavisk Aktuarietidskrift*, **23**, 101–113.

Haeckel, E. 1866. *Generelle Morphologie der Organismen. Allgemeine Grundzüge der organischen Formen-Wissenschaft, mechanisch begründet durch die von Charles Darwin reformirte Descendenz-Theorie.* 2 vols. Georg Reimer, Berlin.

Haigh, J. and Maynard Smith, J. 1972. Can there be more predators than prey? *Theoretical Population Biology*, **3**, 290–299.

Haldane, J. B. S. 1927. A mathematical theory of natural and artificial selection, part V: Selection and mutation. *Proceedings of the Cambridge Philosophical Society*, **23**, 838–844.

Haldane, J. B. S. 1930. *Enzymes.* Longmans, London.

Haldane, J. B. S. 1939. The equilibrium between mutation and random extinction. *Annals of Eugenics*, **9**, 400–405.

Haldane, J. B. S. 1948. The theory of a cline. *Journal of Genetics*, **48**, 277–284.

Hale, J. K. 1977. *Theory of Functional Differential Equations.* Springer-Verlag, New York.

Hale, J. K. and Kocak, H. 1991. *Dynamics and Bifurcations.* Springer-Verlag, New York.

Hansen, S. R. and Hubbell, S. P. 1980. Single nutrient microbial competition: agreement between experimental and theoretical forecast outcomes. *Science*, **20**, 1491–1493.

Hanski, I. 1999. *Metapopulation Ecology.* Oxford University Press, New York.

Hardin, G. 1960. The competitive exclusion principle. *Science*, **131**, 1292–1297.

Harris, T. E. 1963. *The Theory of Branching Processes.* Springer-Verlag, New York.

Hassell, M. P. 1978. *The Dynamics of Arthropod Predator–Prey Systems.* Princeton University Press, Princeton, NJ.

Hassell, M. P., Lawton, J. H., and May, R. M. 1976. Patterns of dynamical behaviour in single-species populations. *Journal of Animal Ecology*, **45**, 471–486.

Hastings, A., Hom, C. L., Ellner, S., Turchin, P., and Godfray, H. C. J. 1993. Chaos in ecology: is mother nature a strange attractor? *Annual Review of Ecology and Systematics*, **24**, 1–33.

Hastings, A. and Powell, T. 1991. Chaos in a three-species food chain. *Ecology*, **72**, 896–903.

Heyde, C. C. and Seneta, E. 1972. The simple branching process, a turning point test and a fundamental inequality: a historical note on I. J. Bienaymé. *Biometrika*, **59**, 680–683.

Heyde, C. C. and Seneta, E. 1977. *I. J. Bienaymé: Statistical Theory Anticipated.* Springer-Verlag, New York.

Hirsch, M. W. 1982. Systems of differential equations which are competitive or cooperative: I. Limit sets. *SIAM Journal on Mathematical Analysis*, **13**, 167–179.

Hirsch, M. W. 1985. Systems of differential equations which are competitive or cooperative: II. Convergence almost everywhere. *SIAM Journal on Mathematical Analysis*, **16**, 423–439.

Hirsch, M. W. 1988. Systems of differential equations which are competitive or cooperative: III. Competing species. *Nonlinearity*, **1**, 51–71.

Hirsch, M. W. 1990. Systems of differential equations which are competitive or cooperative: IV. Structural stability in three dimensional systems. *SIAM Journal on Mathematical Analysis*, **21**, 1225–1234.

Holling, C. S. 1959a. The characteristics of simples types of predation and parasitism. *Canadian Entomologist*, **91**, 385–398.

Holling, C. S. 1959b. The components of predation as revealed by a study of small mammal predation of the European pine sawfly. *Canadian Entomologist*, **91**, 293–320.

Holling, C. S. 1965. The functional response of predators to prey density and its role in mimicry and population regulation. *Memoirs of the Entomological Society of Canada*, **45**, 1–60.

Holling, C. S. 1966. The functional response of invertebrate predators to prey density. *Memoirs of the Entomological Society of Canada*, **47**, 3–86.

Hoppensteadt, F. C. 1993. *Analysis and Simulation of Chaotic Systems.* Springer-Verlag, New York.

Hoppensteadt, F. C. and Izhikevich, E. M. 1997. *Weakly Connected Neural Networks.* Springer-Verlag, New York.

Hotelling, H. 1931. The economics of exhaustible resources. *Journal of Political Economy*, **39**, 137–175.

Howe, H. F. 1985. Gomphothere fruit: a critique. *American Naturalist*, **125**, 853–865.

Hsu, C. S. 1977. On nonlinear parametric excitation problems. *Advances in Applied Mechanics*, **17**, 245–301.

Hsu, C. S. 1987. *Cell-to-Cell Mapping.* Springer-Verlag, New York.

Hsu, S. B. 1978. Limiting behavior for competing species. *SIAM Journal on Applied Mathematics*, **34**, 760–763.

Hsu, S. B., Hubbell, S. P., and Waltman, P. 1977. A mathematical theory for single nutrient competition in continuous cultures of microorganisms. *SIAM Journal on Applied Mathematics*, **32**, 366–383.

Hsu, S. B., Hubbell, S. P., and Waltman, P. 1978. A contribution to the theory of competing predators. *Ecological Monographs*, **48**, 337–349.

Huffaker, C. B. 1958. Experimental studies on predation: dispersion factors and predator–prey oscillations. *Hilgardia*, **27**, 343–383.

Hutchinson, G. E. 1948. Circular causal systems in ecology. *Annals of the New York Academy of Sciences*, **50**, 221–246.

Hutchinson, G. E. 1980. *An Introduction to Population Ecology.* Yale University Press, New Haven, CT.

Inoue, M. and Kamifukumoto, H. 1984. Scenarios leading to chaos in a forced Lotka–Volterra model. *Progress in Theoretical Physics,* **71,** 930–937.

Intrilligator, M. D. 1971. *Mathematical Optimization and Economic Theory.* Prentice Hall, Englewood Cliffs, NJ.

Jackson, E. A. 1989. *Perspectives of Nonlinear Dynamics.* Cambridge University Press, Cambridge.

Jagers, P. 1975. *Branching Processes with Biological Applications.* John Wiley & Sons, New York.

Janzen, D. H. 1966. Coevolution of mutualism between ants and acacias in Central America. *Evolution,* **20,** 249–275.

Janzen, D. H. 1976. Why bamboos wait so long to flower. *Annual Review of Ecology and Systematics,* **7,** 347–391.

Janzen, D. H. 1979. How to be a fig. *Annual Review of Ecology and Systematics,* **10,** 13–51.

Janzen, D. H. 1985. The natural history of mutualisms. In *The Biology of Mutualism,* D. H. Boucher, editor. Oxford University Press, Oxford, pp. 40–99.

Janzen, D. H. 1986. Chihuahuan desert nopaleras: defaunated big mammal vegetation. *Annual Review of Ecology and Systematics,* **17,** 595–636.

Janzen, D. H. and Martin, P. 1982. Neotropical anachronisms: what the gomphotheres ate. *Science,* **215,** 19–27.

Kallen, A., Arcuri, P., and Murray, J. D. 1985. A simple model for the spatial spread and control of rabies. *Journal of Theoretical Biology,* **116,** 377–393.

Karlin, S. 1966. *A First Course in Stochastic Processes.* Academic Press, New York.

Keeler, K. H. 1989. Ant–plant interactions. In *Plant–Animal Interactions,* W. G. Abrahamson, editor. McGraw-Hill Book Company, New York, pp. 207–242.

Kendall, D. G. 1949. Stochastic processes and population growth. *Journal of the Royal Statistical Society B,* **11,** 230–264.

Kendall, D. G. 1975. The genealogy of genealogy: branching processes before (and after) 1873. *Bulletin of the London Mathematical Society,* **7,** 225–253.

Keyfitz, N. 1968. *Introduction to the Mathematics of Population.* Addison-Wesley, Reading, MA.

Keyfitz, N. 1971. On the momentum of population growth. *Demography,* **8,** 71–80.

Keyfitz, N. 1972. The mathematics of sex and marriage. *Proceedings of the 6th Berkeley Symposium of Mathematical Statistics and Probability,* **4,** 89–108.

Keyfitz, N. 1985. *Applied Mathematical Demography.* Springer-Verlag, New York.

Keyfitz, N. and Beekman, J. A. 1984. *Demography through Problems.* Springer-Verlag, New York.

Kierstead, H. and Slobodkin, L. B. 1953. The size of water masses containing plankton blooms. *Journal of Marine Research,* **12,** 141–147.

Kingsland, S. E. 1985. *Modeling Nature: Episodes in the History of Population Ecology.* University of Chicago Press, Chicago, IL.

Koch, A. L. 1974. Competitive coexistence of two predators utilizing the same prey under constant environmental conditions. *Journal of Theoretical Biology,* **44,** 387–95.

Kolmanovskii, V. B. and Nosov, V. R. 1986. *Stability of Functional Differential Equations.* Academic Press, New York.

Kolmogorov, A. 1938. Zur Losung einer biologischen Aufgabe. *Izvestiya nauchno-issledovatelskogo instituta matematiki i mechaniki pri Tomskom Gosudarstvennom Universitete*, **2**, 1–6.

Kostova, T. V. 1990. Numerical solutions to equations modelling nonlinearly interacting age-dependent populations. *Computers and Mathematics with Applications*, **19**, 95–103.

Kot, M., Sayler, G. S., and Schultz, T. W. 1992. Complex dynamics in a model microbial system. *Bulletin of Mathematical Biology*, **54**, 619–648.

Kuang, Y. 1993. *Delay Differential Equations with Applications to Population Dynamics*. Academic Press, San Diego, CA.

Kuczynski, R. R. 1931. *Fertility and Reproduction: Methods of Measuring the Balance of Births and Deaths*. MacMillan, New York.

Kuznetsov, Y. A. 1995. *Elements of Applied Bifurcation Theory*. Springer-Verlag, New York.

Kuznetsov, Y. A. and Rinaldi, S. 1995. Remarks on food chain dynamics. *Mathematical Biosciences*, **135**, 1–34.

Lack, D. L. 1954. *The Natural Regulation of Animal Numbers*. Oxford University Press, Oxford.

Lande, R. 1988. Demographic models of the northern spotted owl (*Strix occidentalis caurina*). *Oecologia*, **75**, 601–607.

Lanner, R. M. 1996. *Made for Each Other: A Symbiosis of Birds and Pines*. Oxford University Press, New York.

Lauwerier, H. A. 1986. Two-dimensional iterative maps. In *Chaos*, A. V. Holden, editor. Princeton University Press, Princeton, NJ, pp. 58–95.

Lauwerier, H. A. and Metz, J. A. 1986. Hopf bifurcation in host–parasitoid models. *IMA Journal of Mathematics Applied in Medicine and Biology*, **3**, 191–210.

Lee, I. H., Fredrickson, A. G., and Tsuchiya, H. M. 1974. Diauxic growth of *Propionibacterium shermanii*. *Applied Microbiology*, **28**, 831–835.

Lee, I. H., Fredrickson, A. G., and Tsuchiya, H. M. 1976. Dynamics of mixed cultures of *Lactobacillus plantarum* and *Propionibacterium shermanii*. *Biotechnology and Bioengineering*, **18**, 513–526.

Lefkovitch, L. P. 1965. The study of population growth in organisms grouped by stages. *Biometrics*, **21**, 1–18.

Lemieux, L. 1959. The breeding biology of the greater snow goose on Bylot Island, Northwest Territories. *Canadian Field Naturalist*, **73**, 117–128.

Lengeler, J. W., Drews, G., and Schlegel, H. G. 1999. *Biology of the Prokaryotes*. Blackwell Science, Stuttgart.

Leslie, P. H. 1945. On the use of matrices in certain population mathematics. *Biometrika*, **33**, 183–212.

Leslie, P. H. 1948. Some further notes on the use of matrices in population mathematics. *Biometrika*, **35**, 213–245.

Leven, R. W., Kock, B. P., and Markman, G. S. 1987. Periodic, quasiperiodic, and chaotic motion in a forced predator–prey ecosystem. In *Dynamical Systems and Environmental Models*, H. G. Bothe, W. Ebeling, A. B. Kurzhanski, and M. Peschel, editors. Akademie-Verlag, Berlin, pp. 95–104.

Levin, S. A. 1981. Age-structure and stability in multiple-age spawning populations. In *Renewable Resource Management*, T. L. Vincent and J. M. Skowronski, editors. Springer-Verlag, Heidelberg, pp. 21–45.

Levin, S. A. and Goodyear, C. P. 1980. Analysis of an age-structured fishery model. *Journal of Mathematical Biology*, **9**, 245–274.

Levin, S. A. and May, R. M. 1976. A note on difference-delay equations. *Theoretical Population Biology*, **9**, 178–187.

Lewis, E. G. 1942. On the generation and growth of a population. *Sankhya, Indian Journal of Statistics*, **6**, 93–96.

Lewis, M. A. and Kareiva, P. 1993. Allee dynamics and the spread of invading organisms. *Theoretical Population Dynamics*, **43**, 141–158.

Lindström, J. and Kokko, H. 1998. Sexual reproduction and population dynamics: the role of polygyny and demographic sex differences. *Proceedings of the Royal Society of London B*, **265**, 483–488.

Lloyd, M., Kritsky, C., and Simon, C. 1983. A simple Mendelian model for 13- and 17-year life cycles of periodical cicadas, with historical hybridization between them. *Evolution*, **37**, 1162–1180.

Lloyd, M. and White, J. 1987. Xylem feeding by periodical cicada nymphs on pine and grass roots, with new suggestions for pest control in conifer plantations and orchards. *Ohio Journal of Science*, **87**, 50–54.

Logan, J. D. 1987. *Applied Mathematics: A Contemporary Approach*. John Wiley & Sons, New York.

Logan, J. D. 1994. *An Introduction to Nonlinear Partial Differential Equations*. John Wiley & Sons, New York.

Lopez, A. 1961. *Problems in Stable Population Theory*. Princeton University Press, Princeton, NJ.

Lotka, A. J. 1931a. The extinction of families – I. *Journal of the Washington Academy of Sciences*, **21**, 377–380.

Lotka, A. J. 1931b. The extinction of families – II. *Journal of the Washington Academy of Sciences*, **21**, 453–459.

Lotka, A. J. 1932. The growth of mixed populations, two species competing for a common food supply. *Journal of the Washington Academy of Sciences*, **22**, 461–469.

Lotka, A. J. 1939. A contribution to the theory of self-renewing aggregates, with special reference to industrial replacement. *Annals of Mathematical Statistics*, **10**, 1–25.

Lotka, A. J. 1948. Application of recurrent series in renewal theory. *Annals of Mathematical Statistics*, **19**, 190–206.

Lotka, A. J. 1998. *Analytical Theory of Biological Populations*. Plenum Press, New York.

Ludwig, D., Aronson, D. G., and Weinberger, H. F. 1979. Spatial pattern of the spruce budworm. *Journal of Mathematical Biology*, **8**, 217–258.

Ludwig, D., Jones, D. D., and Holling, C. S. 1978. Qualitative analysis of insect outbreak systems: the spruce budworm and forest. *Journal of Animal Ecology*, **47**, 315–322.

MacArthur, R. H. 1958. Population ecology of some warblers of the northeastern coniferous forest. *Ecology*, **39**, 599–619.

MacDonald, N. 1978. *Time Lags in Biological Models*. Springer-Verlag, New York.

MacDonald, N. 1989. *Biological Delay Systems: Linear Stability Theory*. Cambridge University Press, Cambridge.

Malthus, T. R. 1798. *An Essay on the Principle of Population, as it Affects the Future Improvement of Society, with Remarks on the Speculations of Mr. Godwin, M. Condorcet, and Other Writers*. J. Johnson, London, England.

Margulis, L. 1970. *Origin of Eukaryotic Cells*. Yale University Press, New Haven.

Margulis, L. 1981. *Symbiosis in Cell Evolution*. W. H. Freeman, San Francisco, CA.

Marsden, J. E. and McCracken, M. 1976. *The Hopf Bifurcation and Its Applications.* Springer-Verlag, New York.

May, R. M. 1974. Biological populations with nonoverlapping generations: stable points, stable cycles, and chaos. *Science,* **186**, 645–647.

May, R. M. 1975. Biological populations obeying difference equations: stable points, stable cycles, and chaos. *Journal of Theoretical Biology,* **49**, 511–524.

May, R. M. 1976. Some mathematical models with very complicated dynamics. *Nature,* **261**, 459–467.

May, R. M. 1977. Threshold and breakpoints in ecosystems with a multiplicity of stable states. *Nature,* **269**, 471–477.

May, R. M. 1980. Mathematical models in whaling and fisheries management. In *Some Mathematical Questions in Biology,* R. C. DiPrima, editor. American Math Society, Providence, RI, pp. 1–64.

May, R. M. 1981. *Theoretical Ecology: Principles and Applications.* Sinauer Associates, Sunderland, MA.

May, R. M. and Leonard, W. 1975. Nonlinear aspects of competition between three species. *SIAM Journal on Applied Mathematics,* **29**, 243–252.

May, R. M. and Oster, G. F. 1976. Bifurcations and dynamic complexity in simple ecological models. *American Naturalist,* **110**, 573–599.

Maynard Smith, J. 1968. *Mathematical Ideas in Biology.* Cambridge University Press, New York.

Mazumdar, J. 1989. *An Introduction to Mathematical Physiology and Biology.* Cambridge University Press, Cambridge.

McClure, F. A. 1967. *The Bamboos—a Fresh Perspective.* Harvard University Press, Cambridge, MA.

McKendrick, A. G. 1926. Applications of mathematics to medical problems. *Proceedings of the Edinburgh Mathematical Society,* **40**, 98–130.

Meyer, J. S., Tsuchiya, H. M., and Fredrickson, A. G. 1975. Dynamics of mixed populations having complementary metabolism. *Biotechnology and Bioengineering,* **17**, 1065–1081.

Mills, N. J. and Getz, W. M. 1996. Modelling the biological control of insect pests: a review of host–parasitoid models. *Ecological Modelling,* **92**, 121–143.

Miura, Y., Tanaka, H., and Okazaki, M. 1980. Stability analysis of commensal and mutual relations with competitive assimilation in continuous mixed culture. *Biotechnology and Bioengineering,* **22**, 929–948.

Moivre, A. de 1730a. *Miscellanea Analytica de Seriebus et Quadraturis.* Tonson & Watts, London.

Moivre, A. de 1730b. *Miscellaneis Analyticis Supplementum.* Tonson & Watts, London.

Moivre, A. de 1756. *The Doctrine of Chances.* Millar, London.

Moran, P. A. P. 1962. *The Statistical Processes of Evolutionary Theory.* Clarendon Press, Oxford.

Murdoch, W. W. 1994. Population regulation in theory and practice. *Ecology,* **75**, 271–287.

Murray, J. D. 1989. *Mathematical Biology.* Springer-Verlag, Berlin.

Murray, J. D., Stanley, E. A., and Brown, D. L. 1986. On the spatial spread of rabies among foxes. *Proceedings of the Royal Society of London B,* **229**, 111–150.

Nagumo, J., Arimoto, S., and Yoshizawa, S. 1962. An active pulse transmission line simulating nerve axon. *Proceedings of the Institute of Radio Engineering*, **50**, 2061–2070.

Neave, F. 1953. Principles affecting the size of pink and chum salmon populations in British Columbia. *Journal of Fisheries Research Board of Canada*, **9**, 450–491.

Neubert, M. G. and Kot, M. 1992. The subcritical collapse of predator populations in discrete-time predator–prey models. *Mathematical Biosciences*, **110**, 45–66.

Nicholson, A. J. 1933. The balance of animal populations. *Journal of Animal Ecology*, **2**, 132–178.

Nicholson, A. J. 1954. An outline of the dynamics of animal populations. *Australian Journal of Zoology*, **2**, 9–65.

Nicholson, A. J. and Bailey, V. A. 1935. The balance of animal populations. Part I. *Proceedings of Zoological Society of London*, **3**, 551–598.

Nisbet, R. M. and Gurney, W. S. C. 1982. *Modelling Fluctuating Populations*. John Wiley & Sons, Chichester.

Nisbet, R. M. and Gurney, W. S. C. 1986. The formulation of age-structure models. In *Mathematical Ecology*, T. G. Hallam and S. A. Levin, editors. Springer-Verlag, Berlin, pp. 95–115.

Nowak, R. M. 1991. *Walker's Mammals of the World*. Johns Hopkins University Press, Baltimore, MD.

Okubo, A. 1978. Horizontal dispersion and critical scales of phytoplankton patches. In *Spatial Patterns in Plankton Communities*, J. H. Steele, editor. Plenum, New York, pp. 21–42.

Okubo, A. 1980. *Diffusion and Ecological Problems: Mathematical Models*. Springer-Verlag, Berlin.

Oliveira-Pinto, F. and Conolly, B. W. 1982. *Applicable Mathematics of Non-Physical Phenomena*. Ellis Horwood, Chichester.

Olsen, L. F. and Schaffer, W. M. 1990. Chaos versus noisy periodicity: alternative hypotheses for childhood epidemics. *Science*, **249**, 499–504.

Paine, R. T. 1966. Food web complexity and species diversity. *American Naturalist*, **100**, 65–75.

Paine, R. T. 1969. A note on trophic complexity and community stability. *American Naturalist*, **103**, 91–93.

Panikov, N. S. 1995. *Microbial Growth Kinetics*. Chapman Hall, London.

Park, T. 1954. Experimental studies of interspecific competition. II. Temperature, humidity, and competition in two species of Tribolium. *Physiological Zoology*, **27**, 177–238.

Parker, T. S. and Chua, L. O. 1989. *Practical Numerical Algorithms of Chaotic Systems*. Springer-Verlag, New York.

Patterson, K. D. 1993. Rabies. In *The Cambridge World History of Human Disease*, K. F. Kiple, editor. Cambridge University Press, Cambridge, pp. 962–967.

Pavlou, S. and Kevrekidis, I. G. 1992. Microbial predation in a periodically operated chemostat: a global study of the interaction between natural and externally imposed frequencies. *Mathematical Biosciences*, **108**, 1–55.

Pearl, R. and Reed, L. J. 1920. On the rate of growth of the population of the United States since 1790 and its mathematical representation. *Proceedings of the National Academy of Sciences of the United States of America*, **6**, 275–288.

Perko, L. 1991. *Differential Equations and Dynamical Systems*. Springer-Verlag, New York.

Perron, O. 1907. Zur Theorie der Über Matrizen. *Mathematische Annalen*, **64**, 248–263.

Peters, J. G., Peters, W. L., and Fink, T. J. 1987. Seasonal synchronization of emergence in *Dolania americana* (Ephemeroptera: Behningiidae). *Canadian Journal of Zoology*, **65**, 3177–3185.

Petty, W. 1683. *Another Essay in Political Arithmetick Concerning the Growth of the City of London: with the Measures, Periods, Causes, and Consequences Thereof.* Mark Pardoe, London, England.

Pielou, E. C. 1977. *Mathematical Ecology*. John Wiley & Sons, New York.

Pielou, E. C. 1981. The usefulness of ecological models: a stock-taking. *Quarterly Review of Biology*, **56**, 17–31.

Pollard, J. H. 1973. *Mathematical Models for the Growth of Human Populations.* Cambridge University Press, Cambridge.

Pontryagin, L. S., Boltyanskii, V. G., Gamkrelidze, R. V., and Mischenko, E. F. 1962. *The Mathematical Theory of Optimal Processes.* John Wiley & Sons, New York.

Powell, E. O. 1958. Criteria for the growth of contaminants and mutants in continuous culture. *Journal of General Microbiology*, **18**, 259–268.

Renshaw, E. 1991. *Modelling Biological Populations in Space and Time.* Cambridge University Press, Cambridge.

Ricciardi, L. M. 1986. Stochastic population theory: birth and death processes. In *Mathematical Ecology: An Introduction*, T. G. Hallam and S. A. Levin, editors. Springer-Verlag, Berlin, pp. 155–190.

Richards, A. J. 1997. *Plant Breeding Systems.* Chapman Hall, London.

Richards, J. A. 1983. *Analysis of Periodically Time-Varying Systems.* Springer-Verlag, Berlin.

Ricker, W. E. 1954. Stock and recruitment. *Journal of the Fisheries Research Board of Canada*, **11**, 559–623.

Rinaldi, S. and Muratori, S. 1993. Conditioned chaos in seasonally perturbed predator–prey models. *Ecological Modeling*, **69**, 79–97.

Rinaldi, S., Muratori, S., and Kuznetsov, Y. 1993. Multiple attractors, catastrophes and chaos in seasonally perturbed predator–prey communities. *Bulletin of Mathematical Biology*, **55**, 15–35.

Riordan, J. 1958. *An Introduction to Combinatorial Analysis.* John Wiley & Sons, New York.

Rosen, K. H. 1983. Mathematical models for polygamous mating systems. *Mathematical Modelling*, **4**, 27–39.

Rosenzweig, M. L. 1971. Paradox of enrichment: destabilization of exploitation ecosystems in ecological time. *Science*, **171**, 385–387.

Rosenzweig, M. L. and MacArthur, R. H. 1963. Graphical representation and stability conditions of predator–prey interactions. *American Naturalist*, **97**, 209–223.

Sabin, G. C. W. and Summers, D. 1992. Chaos in a periodically forced predator–prey ecosystem model. *Mathematical Biosciences*, **113**, 91–113.

Sagan, L. 1967. On the origin of mitosing cells. *Journal of Theoretical Biology*, **14**, 225–274.

Sanchez, D. A. 1978. Linear age-dependent population growth with harvesting. *Bulletin of Mathematical Biology*, **40**, 377–385.

Schaefer, M. B. 1954. Some aspects of the dynamics of populations important to the management of commercial marine fisheries. *Bulletin of the Inter-American Tropical Tuna Commission*, **1**, 25–56.

Schaffer, W. M. 1985. Can nonlinear dynamics elucidate mechanisms in ecology and epidemiology? *IMA Journal of Mathematics Applied in Medicine and Biology*, **2**, 221–252.

Schaffer, W. M. 1988. Perceiving order in the chaos of nature. In *Evolution of Life Histories of Mammals*, M. S. Boyce, editor. Yale University Press, New Haven, CT, pp. 313–350.

Schaffer, W. M., Ellner, S., and Kot, M. 1986. Effects of noise on some dynamical models in ecology. *Journal of Mathematical Biology*, **24**, 479–523.

Schaffer, W. M. and Kot, M. 1986a. Chaos in ecological systems: the coals that Newcastle forgot. *Trends in Ecology and Evolution*, **1**, 58–63.

Schaffer, W. M. and Kot, M. 1986b. Differential systems in ecology and epidemiology. In *Chaos*, A. V. Holden, editor. Princeton University Press, Princeton, New Jersey, USA, pp. 158–178.

Schoener, T. W. 1986. Resource partitioning. In *Community Ecology*, J. Kikkawa, editor. Blackwell Scientific Publications, Oxford, pp. 91–125.

Schuster, H. G. 1988. *Deterministic Chaos: An Introduction*. VCH, Weinheim.

Scuda, F. M. and Ziegler, J. R. 1978. *The Golden Age of Theoretical Ecology, 1923–1940: A Collection of Works by Volterra, Kostitzin, Lotka, and Kolmogorov*. Springer-Verlag, Berlin.

Sharpe, F. R. and Lotka, A. J. 1911. A problem in age-distribution. *Philosophical Magazine*, **21**, 435–438.

Sherman, P. W. and Morton, M. L. 1984. Demography of Belding's ground squirrels. *Ecology*, **65**, 1617–1628.

Shope, R. E. 1991. Rabies. In *Viral Infections of Humans: Epidemiology and Control*, A. S. Evans, editor. Plenum Medical Book Company, New York, pp. 509–523.

Silverman, E. and Kot, M. 2000. Rate estimation for a simple movement model. *Bulletin of Mathematical Biology*, **62**, 351–375.

Sinclair, A. R. E. 1989. Population regulation in animals. In *Ecological Concepts: The Contribution of Ecology to an Understanding of the Natural World*, J. M. Cherrett, editor. Blackwell Scientific Publications, Oxford, pp. 197–241.

Skellam, J. G. 1951. Random dispersal in theoretical populations. *Biometrika*, **38**, 196–218.

Smith, A. M. A. and Webb, G. J. W. 1985. *Crocodylus johnstoni* in the McKinlay river area, N. T. VII. A population simulation model. *Australian Wildlife Research*, **12**, 541-554.

Smith, C. C. and Follmer, D. 1972. Food preferences of squirrels. *Ecology*, **53**, 82–91.

Smith, D. and Keyfitz, N. 1977. *Mathematical Demography*. Springer-Verlag, Berlin.

Smith, H. L. and Waltman, P. 1995. *The Theory of the Chemostat: Dynamics of Microbial Competition*. Cambridge University Press, New York.

Smoller, J. 1983. *Shock Waves and Reaction-Diffusion Equations*. Springer-Verlag, New York.

Smoller, J. and Wasserman, A. 1981. Global bifurcation of steady-state solutions. *Journal of Differential Equations*, **39**, 269–290.

Solomon, M. E. 1949. The natural control of animal populations. *Journal of Animal Ecology*, **18**, 1–35.

Stakgold, I. 1979. *Green's Functions and Boundary Value Problems*. Wiley, New York.

Stapanian, M. A. and Smith, C. C. 1978. A model for seed scatterhoarding: coevolution of fox squirrels and black walnuts. *Ecology,* **59**, 884–896.

Stauffer, R. C. 1960. Ecology in the long manuscript version of Darwin's *Origin of Species* and Linnaeus's *Oeconomy of Nature. Proceedings of the American Philosophical Society,* **104**, 235–241.

Steffensen, J. F. 1930. Om sandsynligheden for at afkommet uddøor. *Matematisk Tidsskrift B,* **1**, 19–23.

Steffensen, J. F. 1932. Deux problèmes du calcul des probabilité. *Annales de l'Institut Henri Poincaré,* **3**, 319–344.

Stewart, F. M. and Levin, B. R. 1973. Partitioning of resources and the outcome of interspecific competition: a model and some considerations. *American Naturalist,* **107**, 171–198.

Strutt, R. J. 1905. On the radio-active minerals. *Proceedings of the Royal Society of London,* series A, **76**, 88–101.

Tamarin, R. H. 1978. *Population Regulation.* Dowden, Hutchinson, and Ross, Inc., Stroudsburg, PA.

Temple, S. A. 1977. Plant–animal mutualism: coevolution with dodo leads to near extinction of plant. *Science,* **197**, 885–886.

Thompson, J. M. T., Steward, H. B., and Ueda, Y. 1994. Safe, explosive, and dangerous bifurcations in dissipative dynamical systems. *Physical Review E,* **49**, 1019–1027.

Thompson, W. R. 1931. On the reproduction of organisms with overlapping generations. *Bulletin of Entomological Research,* **22**, 147–172.

Thomson, W. 1863. On the secular cooling of the earth. *Philosophical Magazine,* **25** (4th series), 1–14.

Thomson, W. 1864. On the secular cooling of the earth. *Transactions of the Royal Society of Edinburgh,* **23**, 157–170.

Thorp, J. H. and Covich, A. P. 1991. *Ecology and Classification of North American Freshwater Invertebrates.* Academic Press, San Diego, CA.

Titchmarsh, E. C. 1946. *Eigenfunction Expansions Associated with Second-Order Differential Equations,* vol. 1. Clarendon Press, Oxford.

~~Tu, P. N. V. 1991. *Introductory Optimization Dynamics.* Springer-Verlag, New York,~~ USA.

Tuckwell, H. C. 1988. *Elementary Applications of Probability Theory.* Chapman & Hall, London.

Tuljapurkar, S. and Caswell, H. 1997. *Structured-Population Models in Marine, Terrestrial, and Freshwater Systems.* Chapman & Hall, New York.

Turchin, P. 1990. Rarity of density dependence or population regulation with lags? *Nature,* **344**, 660–663.

Turchin, P. 1998. *Quantitative Analysis of Movement: Measuring and Modeling Population Redistribution in Animals and Plants.* Sinauer Associates, Inc., Sunderland, MA.

Turchin, P. and Taylor, A. D. 1992. Complex dynamics in ecological time series. *Ecology,* **73**, 289–305.

Twisleton-Wykeham-Fiennes, N. 1978. *Zoonoses and the Origins and Ecology of Human Disease.* Academic Press, London.

Vander Wall, S. B. and Balda, R. P. 1977. Coadaptations of the Clark's nutcracker and the pinyon pine for efficient seed harvest and dispersal. *Ecological Monographs,* **47**, 89–111.

van den Bosch, F., Hangeveld, R., and Metz, J. A. J. 1992. Analyzing the velocity of animal range expansion. *Journal of Biogeography*, **19**, 135–150.

Van Driesche, R. G. and Bellows, Jr, T. S. 1996. *Biological Control*. Chapman & Hall, New York.

Van Gemerden, H. 1974. Coexistence of organisms competing for the same substrate: an example among the purple sulfur bacteria. *Microbial Ecology*, **1**, 104–119.

Vatutin, V. A. and Zubkov, A. M. 1987. Branching processes. I. *Journal of Soviet Mathematics*, **39**, 2431–2475.

Vatutin, V. A. and Zubkov, A. M. 1993. Branching processes. II. *Journal of Soviet Mathematics*, **67**, 3407–3485.

Veldkamp, H. and Jannasch, H. W. 1972. Mixed culture studies with the chemostat. *Journal of Applied Chemistry and Biotechnology*, **22**, 105–123.

Verhulst, F. 1996. *Nonlinear Differential Equations and Dynamical Systems*. Springer-Verlag, Berlin.

Verhulst, P.-F. 1845. Recherches mathématiques sur la loi d'accroissement de la population. *Nouveaux Memoires de l'Académie Royale des Sciences et Belles Lettres de Bruxelles*, **18**, 3–38.

Volterra, V. 1926. Fluctuation in the abundance of a species considered mathematically. *Nature*, **118**, 558–560.

von Foerster, H. 1959. Some remarks on changing populations. In *The Kinetics of Cellular Proliferation*, F. Stohlman, Jr, editor. Grune & Stratton, New York, pp. 382–407.

Wallin, I. E. 1923. The mitochondria problem. *American Naturalist*, **57**, 255–261.

Wallin, I. E. 1927. *Symbionticism and the Origin of Species*. Williams & Wilkins, Baltimore, MD.

Waltman, P. 1983. *Competition Models in Population Biology*. SIAM, Philadelphia, PA.

Watson, H. W. and Galton, F. 1874. On the probability of extinction of families. *Journal of the Royal Anthropological Institute*, **4**, 138–144.

Wheldon, T. E. 1988. *Mathematical Models in Cancer Research*. Adam Hilger, Bristol.

Whelpton, P. K. 1936. An empirical method of calculating future population. *Journal of the American Statistical Association*, **31**, 457–473.

White, J. and Lloyd, M. 1975. Growth rates of 17- and 13-year periodical cicadas. *American Midland Naturalist*, **94**, 127–143.

White, J. and Strehl, C. 1978. Xylem feeding by periodical cicada nymphs on tree roots. *Ecological Entomology*, **3**, 323–327.

Wiggins, S. 1990. *Introduction to Applied Nonlinear Dynamical Systems and Chaos*. Springer-Verlag, New York.

Wilf, H. S. 1994. *Generatingfunctionology*. Academic Press, Boston, MA.

Williams, W. E. 1980. *Partial Differential Equations*. Oxford University Press, Oxford.

Winsor, C. P. 1932. The Gompertz curve as a growth curve. *Proceedings of the National Academy of Sciences of the United States of America*, **18**, 1–8.

Wolin, C. L. 1985. The population dynamics of mutualistic systems. In *The Biology of Mutualism: Ecology and Evolution*, D. H. Boucher, editor. Oxford University Press, New York, pp. 248–269.

Wolin, C. L. and Lawlor, L. R. 1984. Models of facultative mutualism: density effects. *American Naturalist*, **124**, 843–862.

Woodward, D. E. and Murray, J. D. 1993. On the effect of temperature-dependent sex determination on sex ratio and survivorship in crocodilians. *Proceedings of the Royal Society of London B*, **252**, 149–155.

Worster, D. 1994. *Nature's Economy: A History of Ecological Ideas.* Cambridge University Press, New York.

Wright, E. M. 1946. The non-linear difference-differential equation. *Quarterly Journal of Mathematics*, **17**, 245–252.

Yano, T. 1969. Dynamic behavior of the chemostat subject to substrate inhibition. *Biotechnology and Bioengineering*, **11**, 139–153.

Yule, U. 1924. A mathematical theory of evolution based on the conclusions of Dr. J. C. Willis, FRS. *Philosophical Transactions of the Royal Society of London*, **213**, 21–87.

Zwillinger, D. 1992. *Handbook of Differential Equations.* Academic Press, Boston, MA.

Author index

Subject index